Lecture Notes in Computer Science   6982

Commenced Publication in 1973
Founding and Former Series Editors:
Gerhard Goos, Juris Hartmanis, and Jan van Leeuwen

W0227407

Giuseppe Persiano (Ed.)

Algorithmic
Game Theory

4th International Symposium, SAGT 2011
Amalfi, Italy, October 17-19, 2011
Proceedings

 Springer

Volume Editor

Giuseppe Persiano
Università degli Studi di Salerno
Dipartimento di Informatica
84084 Fisciano (SA), Italy
E-mail: giuper@dia.unisa.it

ISSN 0302-9743   e-ISSN 1611-3349
ISBN 978-3-642-24828-3   e-ISBN 978-3-642-24829-0
DOI 10.1007/978-3-642-24829-0
Springer Heidelberg Dordrecht London New York

Library of Congress Control Number: 2011938671

CR Subject Classification (1998): I.6, H.5.3, J.1, K.6.0, H.3.5, J.4, K.4.4, G.1.2, F.2.2

LNCS Sublibrary: SL 3 – Information Systems and Application, incl. Internet/Web
and HCI

Typesetting: Camera-ready by author, data conversion by Scientific Publishing Services, Chennai, India

Printed on acid-free paper

Springer is part of Springer Science+Business Media (www.springer.com)

Preface

The present volume is devoted to the Fourth International Symposium on Algorithmic Game Theory (SAGT), an interdisciplinary event intended to provide a forum for researchers and practitioners to exchange innovative ideas and to be aware of each other's approaches and findings. The main focus of SAGT is on the study of the algorithmic aspects of game theory; typical questions include how scarce computational resources affect the way games between selfish agents are played and the impact of selfishness on the quality of the outcome of a multiplayer system. This is a departure from traditional algorithmic theory in which players are supposed to be cooperative.

The algorithmic approach to game theory has been applied primarily to problems from economics and computer science (e.g., auctions, network and routing problems). I believe though that this approach can be used to pose new questions and to give answers to problems in other fields like physics and biology and hope SAGT will be one of the fora that make this convergence happen.

SAGT 2011 took place in Amalfi (Italy) from October 17th to October 19th, 2011. The present volume contains all contributed papers that were presented at SAGT 2011 together with an abstract of the distinguished invited lectures of Bruno Codenotti (Consiglio Nazionale delle Ricerche, Pisa, Italy) and Xiaotie Deng (University of Liverpool, Liverpool, UK). The two invited lectures are found at the beginning of the volume and the regular papers appear in the order of presentation at the symposium.

In response to the call for papers, the Program Committee received 65 submissions of which 10 were co-authored by a Program Committee member. These submissions were handled by a special sub-committee that proposed to accept six. For the remaining submissions, 20 were selected for inclusion in the scientific program of the symposium after a detailed evaluation (each submission was read by at least three Program Committee members) and electronic discussion.

We wish to thank the creator of the EasyChair System, a free conference management system, which was very helpful in the selection of the scientific program.

July 2011   Giuseppe Persiano

Organization

SAGT 2011 was organized by the Dipartimento di Informatica "Renato M. Capocelli" of the Università degli Studi di Salerno, in cooperation with the ACM Special Interest Group on Electronic Commerce (ACM-SIGECOM) and the European Association for Theoretical Computer Science (EATCS).

Program Committee

Yonatan Aumann	Bar-Ilan University, Israel
Moshe Babaioff	Microsoft Research, USA
George Christodolou	University of Liverpool, UK
Xiatioe Deng	University of Liverpool, UK
Edith Elkind	Nanyang Technological University, Singapore
Amos Fiat	Tel Aviv University, Israel
Christos Kaklamanis	RACTI and University of Patras, Greece
Elias Koutsoupias	National and Kapodistrian University of Athens, Greece
Ron Lavi	The Technion – Israel Institute of Technology, Israel
Yishay Mansour	Tel Aviv University, Israel
Marios Mavronicolas	University of Cyprus, Cyprus
Peter Bro Miltersen	Aarhus University, Denmark
Muthu Muthukrishnan	Rutgers University, USA
Paolo Penna	Università di Salerno, Italy
Giuseppe Persiano	Università di Salerno, Italy (Chair)
Ariel Procaccia	Harvard University, USA
Amin Saberi	Stanford University, USA
Guido Schäfer	CWI and VU University Amsterdam, The Netherlands
James Schummer	Northwestern University, USA
Paul Spirakis	RACTI and University of Patras, Greece
Berthold Vocking	RWTH Aachen University, Germany
Yinyu Ye	Stanford University, USA

Steering Committee

Elias Koutsoupias	National and Kapodistrian University of Athens, Greece
Marios Mavronicolas	University of Cyprus, Cyprus
Dov Monderer	The Technion – Israel Institute of Technology, Israel

Christos Papadimitriou   University of California at Berkeley, USA
Giuseppe Persiano   Università di Salerno, Italy
Paul Spirakis   RACTI and University of Patras, Greece
                                 (Chair)
Berthold Vocking   RWTH Aachen University, Germany

Organizing Committee

Vincenzo Auletta   Università di Salerno, Italy (Co-chair)
Carlo Blundo   Università di Salerno, Italy
Diodato Ferraioli   Università di Salerno, Italy
Luigi Catuogno   Università di Salerno, Italy
Francesco Pasquale   Università di Salerno, Italy
Giuseppe Persiano   Università di Salerno, Italy (Co-chair)

External Reviewers

Agrawal, Shipra   Harks, Tobias   Papadopoulou, Vicky
Auletta, Vincenzo   Hoefer, Martin   Papakonstantinopoulou
Balcan, Maria-Florina   Jain, Shaili   Katia
Ben-Zwi, Oren   Kanellopoulos,   Pasquale, Francesco
Busch, Costas   Panagiotis   Pierrakos, George
Caragiannis, Ioannis   Kaporis, Alexis   Qi, Qi
Chen, Ning   Karanikolas, Nikos   Sun, Wei
Cole, Richard   Kash, Ian   Sun, Xiaorui
De Keijzer, Bart   Kontogiannis, Spyros   Trehan, Amitabh
Dobzinski, Shahar   Kovacs, Annamaria   Wang, Zizhuo
Dombb, Yair   Krysta, Piotr   Weinreb, Enav
Fanelli, Angelo   Kyropoulou, Maria   Wen, Tailai
Feldman, Michal   Lai, John   Xia, Lirong
Ferraioli, Diodato   Michaelis, Diethard   Zeinalipour-Yazti
Gairing, Martin   Mirrokni, Vahab   Demetrios
Gal, Shmuel   Panagopoulou   Zhang, Jinshan
Georgiou, Chryssis   Panagiota   Zohar, Aviv

Sponsoring Institutions

Dipartimento di Informatica "Renato M. Capocelli", Università degli Studi di
Salerno, Italy

Table of Contents

Session 3: Externalities

Session 4: Mechanism Design

Session 5: Complexity

Session 6: Network Games

Session 7: Pricing

Session 8: Routing Games

Computational Game Theory

Bruno Codenotti

Istituto di Informatica e Telematica,
Consiglio Nazionale delle Ricerche,
Pisa, Italy
bruno.codenotti@iit.cnr.it

Abstract. We first provide a quick background in Game and Economic Theory, and then discuss some fundamental computational questions arising in these areas. We will focus on the interplay between Game Theory and Computer Science, with an emphasis on some of the most challenging open questions.

G. Persiano (Ed.): SAGT 2011, LNCS 6982, p. 1, 2011.

Computation and Incentives of Competitive Equilibria in a Matching Market

Ning Chen[1] and Xiaotie Deng[2]

[1] Division of Mathematical Sciences, School of Physical and Mathematical Sciences,
Nanyang Technological University, Singapore
ningc@ntu.edu.sg
[2] Department of Computer Science, University of Liverpool, UK
xiaotie@liv.ac.uk

Abstract. Matching market and its many variants have been an intensively studied problem in Economics and Computer Science. In many applications centralized prices are used to determine allocations of indivisible items under the principles of individual optimization and market clearance, based on public knowledge of individual preferences. Alternatively, auction mechanisms have been used with a different set of principles for the determination of prices, based on individuals' incentives to report their preferences.

This talk considers matching markets run by a single seller with an objective of maximizing revenue of the seller, who employs a market equilibrium pricing for allocation. We will give a polynomial time algorithm to compute such an equilibrium given budget constraints, and show that the maximum revenue market equilibrium mechanism converges, under an optimal dynamic re-bidding sequence of the buyers, to a solution equivalent to the minimum revenue equilibrium under the true preferences of buyers, which in turn is revenue equivalent to a VCG solution.

We will also discuss other related issues as well as open problems.

1 Introduction

The commercial success of Google has been relied on a model for charging advertisers for placing their product information to respond to the clicks by curious surfers of the Internet. It has created a commercialization model characterized by the classical market extensively studied in the economics literatures with several important twists. Such scenario has been followed and emulated as well as modified to fit into different settings of web service systems. At the center of those commercialization models, there is a simple goal the designers trying to achieve: match advertisers' products and services to users, based on the signals, explicitly or implicitly expressed by the users, prioritized with respect to commercial value or social benefit to the system.

Search engines have relied on much improved methods to create such matchings. Prior to the sponsored search market for online advertising, emails sending

G. Persiano (Ed.): SAGT 2011, LNCS 6982, pp. 2–6, 2011.

advertising has been exploited as an effectively direct marketing replacement of the traditional bulk mail campaigns. The email marketing, however, suffered from its own success by creating a huge amount of junk mails to users with the often indiscrimative targeted users of such systems. The sponsored search approach of advertising is magnitude better in matching users' needs with merchants' products. While the techniques that match an ad with a signal from a searching user have been among the most important expertises of search engine companies, our focus will be on the pricing models designed to prioritize the order of advertisers presented to the users.

The advertising market can be characterized by a matching market, usually consisting of two parties of participants, buyers and products, where each buyer wants to purchase an item. When considering a single seller's market, the market maker owns all the products and sell them to the buyers by a mechanism to determine a matching of its products to the buyers together with prices. The seller holds all the power of choices on the selling mechanism, based on its own principle of maximizing revenue, social welfare, or serving any other purpose. On the other side of the game, the buyers hold the key information on their own private values for the products, and will respond in their best interests to the seller's strategies.

There have been two major classes of pricing principles in theoretical economics that could be applied for guiding the practice: auction mechanism and competitive market equilibrium. The former is concerned with providing a simple individual mechanism based on the revelation principle that reduces any dominant strategy rendering protocol to one where every player's optimal strategy is to speak the truth [20,15]. The latter, on the other hand, reveres the invisible hand [18] that balances up the supply-demand to achieve a market clearance condition [21]. Both paradigms have had profound influence in nurturing economic thoughts. However, they are rooted in different sets of rationalities that may reveal different properties.

There are deviations of the reality at a single seller's matching market from the traditional thought, especially with those related to and evolved from the sponsored search market. The market is repeated, dynamic, and moreover, changing in terms of the number of buyers and their compositions. On one hand, the seller could learn existing buyers and create new products and their combinations. On the other hand, buyers may also learn and coordinate based on information gathered through interplays at the marketplaces. Today's information and communications technology tools will make such analysis possible and in turn pose great challenges to algorithmic issues in market designs and market games.

2   First Approaches

The now well known GSP, for generalized second price auction, first deployed by Google at sponsored search market, makes the i-th highest bidder the winner of the i-th most profitable item (an ad displaying slot), and charges the bidder the $(i + 1)$-st highest biding price. Despite of the fact that a single bidder may manipulate by a false bid to improve its utility, it has been shown that the desired Nash equilibrium of the bidders will always out-perform the celebrated strategy-proof VCG protocol in terms of the seller's revenue and social welfare [12,19,14]. On the other hand, self-interest seeking advertisers, by the forward-looking best response [4], can limit the power of the seller to be equivalent in revenue to the long established VCG solution [7,6,13].

Auction based protocols have been studied and understood pretty well; we will therefore focus on another solution principle: centralized market equilibrium. There has been a practical concern in the market equilibrium solutions: budget, a major factor that has been neglected. Notwithstanding a fine and classical concept, it involves in demanding computation that may not be suitable for situations that require instant decisions (e.g., a query of sponsored keywords). Here an efficient computation method has become very important, which could create a scenario for algorithmic design and analysis to make a difference. In recent work [9], we develop a polynomial time algorithm to compute an equilibrium with budget constraints; our result, as well as [1,2], places market equilibrium a plausible solution for matching marketplaces.

A mechanism designer may still be worried about the market equilibrium solution concept in the matching market, even if a polynomial time algorithm is available. The matching market setting may allow several market equilibrium solutions: some may bring in more revenue to the market maker than others. It can be easily proven that, the minimum revenue equilibrium as a selling mechanism will rendre bidders truthful: everyone gets its maximum utility by bidding the truth. On the other hand, the maximum revenue equilibrium will be what the seller wants but the buyers could lie about their utility functions to manipulate the outcome of the game. In recent work [8], we show that the maximum equilibrium mechanism, in repeated interplays with the buyers' best responses to a solution when every participant gets a revenue equivalent to that of the minimum equilibrium mechanism. The maximum revenue market equilibrium is a more sophisticated version of the first price auction, while the minimum equilibrium is the corresponding counterpart of the second price auction. The convergence from the maximum revenue market equilibrium toward the minimum revenue equilibrium implies revenue equivalence between the two protocols in a deterministic and dynamic setting.

3   Future Challenges

An immediate effort would be to iron out some of the imperfection left out in this theoretical endeavor, and to answer the challenges to develop methodological guidance for the new problems and models emerged from the Internet. A largely ignored aspect in our discussion is competition, which always exists in the online advertising market. While some game theoretical work regarding the GSP auction was known [12,19,5,14], it is interesting to study the question for other mechanisms [17]; the matching market with perishable items such as clicks of Internet surfers makes it a unique market model to explore.

Despite of the fact that theoretical progress in dynamics of the matching market points to a convergence toward the VCG protocol, it still leave a lot to study further. Clearly being the minimum in revenue among all market equilibrium solutions, VCG is not what a seller would like to end up with [3], such as in the FCC spectrum auction [11]. Nisan et al. [16] independently showed a convergence result from first price auctions to VCG in the matching market. An intriguing question is under what circumstances we will have such convergence. Further buyers could consider every opportunity where they could make a marginal improvement by manipulation. In a recent work [10], we develop a concept of incentive ratio, measuring buyers' willingness to manipulate. It would be interesting to further explore the applicability of this solution concept in the matching market setting to look into possibilities overcoming the curse of thin revenue of the VCG protocol under certain circumstances.

References

1. Aggarwal, G., Muthukrishnan, S., Pal, D., Pal, M.: General Auction Mechanism for Search Advertising. In: WWW 2009, pp. 241–250 (2009)
2. Ashlagi, I., Braverman, M., Hassidim, A., Lavi, R., Tennenholtz, M.: Position Auctions with Budgets: Existence and Uniqueness. The B.E. Journal of Theoretical Economics 10(1), 20 (2010)
3. Ausubel, L., Milgrom, P.: The Lovely but Lonely Vickrey Auction. In: Cramton, P., Steinberg, R., Shoham, Y. (eds.) Combinatorial Auctions
4. Bu, T., Deng, X., Qi, Q.: Forward Looking Nash Equilibrium in Keyword Auction. IPL 105(2), 41–46 (2008)
5. Bu, T., Deng, X., Qi, Q.: Arbitrage opportunities across sponsored search markets. Theor. Comput. Sci. 407(1-3), 182–191 (2008)
6. Bu, T., Liang, L., Qi, Q.: On Robustness of Forward-looking in Sponsored Search Auction. Algorithmica 58(4), 970–989 (2010)
7. Cary, M., Das, A., Edelman, B., Giotis, I., Heimerl, K., Karlin, A., Mathieu, C., Schwarz, M.: On Best-Response Bidding in GSP Auctions. In: EC 2007, pp. 262–271 (2007)
8. Chen, N., Deng, X.: On Nash Dynamics of Matching Market Equilibria, CoRR abs/1103.4196 (2011)
9. Chen, N., Deng, X., Ghosh, A.: Competitive equilibria in matching markets with budgets. SIGecom Exchanges 9(1) (2010)
10. Chen, N., Deng, X., Zhang, J.: How Profitable are Strategic Behaviors in a Market? In: Demetrescu, C., Halldórsson, M.M. (eds.) ESA 2011. LNCS, vol. 6942, pp. 106–118. Springer, Heidelberg (2011)
11. Cramton, P.: The FCC Spectrum Auctions: An Early Assessment. Journal of Economics and Management Strategy 6(3), 431–495 (1997)
12. Edelman, B., Ostrovsky, M., Schwarz, M.: Internet Advertising and the Generalized Second-Price Auction. American Economic Review 97(1), 242–259 (2007)
13. Langford, J., Li, L., Vorobeychik, Y., Wortman, J.: Maintaining Equilibria During Exploration in Sponsored Search Auctions. Algorithmica 58, 990–1021 (2010)
14. Leme, R., Tardos, E.: Pure and Bayes-Nash Price of Anarchy for Generalized Second Price Auction. In: FOCS 2010 (2010)

15. Myerson, R.: Optimal Auction Design. Mathematics of Operations Research 6, 58–73 (1981)
16. Nisan, N., Schapira, M., Valiant, G., Zoha, A.: Best Response Auctions. In: EC 2011 (2011)
17. Pai, M.: Competition in Mechanism. ACM Sigecom Exchange 9(1) (2010)
18. Smith, A.: An Inquiry into the Nature and Causes of the Wealth of Nations. University of Chicago Press, Chicago (1977)
19. Varian, H.: Position Auctions. International Journal of Industrial Organization 6, 1163–1178 (2007)
20. Vickrey, W.: Counterspeculation, Auctions, and Competitive Sealed Tenders. Journal of Finance 16, 8–37 (1961)
21. Walras, L.: Éléments d'économie politique pure, ou théorie de la richesse sociale (Elements of Pure Economics, or the theory of social wealth) (1874)

Repeated Budgeted Second Price Ad Auction[*]

Asaph Arnon[1] and Yishay Mansour[2]

[1] School of Computer science, Tel Aviv University and Google Tel Aviv.
asapha@google.com
[2] School of Computer science, Tel Aviv University
mansour@cs.tau.ac.il

Abstract. Our main goal is to abstract existing repeated sponsored search ad auction mechanisms which includes budgets, and study their equilibrium and dynamics. Our abstraction has multiple agents biding repeatedly for multiple identical items (such as impressions in an ad auction). The agents are budget limited and have a value for per item. We abstract the repeated interaction as a one-shot game, which we call *budget auction*, where agents submit a bid and a budget, and then items are sold by a sequential second price auction. Once an agent exhausts its budget it does not participate in the proceeding auctions.

Our main result is that if agents bid conservatively (never bid above their value) then there always exists a pure Nash equilibrium. We also study simple dynamics of repeated budget auctions, showing their convergence to a Nash equilibrium for two agents and for multiple agents with identical budgets.

1   Introduction

Auctions have become the main venue for selling online advertisements. This trend started in the sponsored search advertisements (such as, Google's AdWords, Yahoo!'s Search Marketing and Microsoft's AdCenter), and expended to the display advertisement (such as, Double click Ad Exchange [19]). This trend has even propagated to classical advertisement media, such as TV [20].

There are a few features that are shared by many of those auctions mechanisms. First, the price is set using a second price (or a generalized second price (GSP)) with the motivation that users should try to bid their utility rather than search for an optimal bid value. Second, there are daily budgets that cap the advertiser's total payment in a single period (e.g., day). Our main goal is to abstract a model for such existing auctions, and study its equilibria and dynamics.

It is worthwhile to expand on the role of budgets in auctions. The budget allows the advertiser to cap its spending in a given period (day). This is an important feature to the advertiser, for a few reasons. First, many advertisers

---

[*] This research was supported in part by the Google Inter-university center for Electronic Markets and Auctions, by a grant from the Israel Science Foundation, by a grant from United States-Israel Binational Science Foundation (BSF), and by a grant from the Israeli Ministry of Science (MoS).

G. Persiano (Ed.): SAGT 2011, LNCS 6982, pp. 7–18, 2011.

are *brand advertisers*, whose goal is to promote their brand and they do not see an immediate return on their advertisement investments. For such an advertiser, a budget is the main tool to control the expense of an ad campaign. Second, even merchants that can test their return on investment (ROI) from ads, many times are budget limited, due to the corporate financial structure and the way budgets are allocated throughout the corporation.

There has been an increasing research interest in the role of budgets in auctions, since budgets significantly influences the strategic behavior of agents. A very interesting line of research is constructing incentive compatible mechanism for an auction with budgets [7,11,14]. Another line of research has been maximizing the auctioneer's revenue [1,18,4,21,3]. In this work we have a very different agenda. We would like to abstract the existing mechanisms and study their equilibria and dynamics.

In our abstraction, as with any abstraction, we made a few compromises. The nature of the online advertisement auctions, is that there is a huge number of daily impressions, and agents compete repeatedly for those impressions. In some systems the advertisers input their budget limit explicitly (for example, in the Double Click Ad Exchange [19], Google's AdWords, TV [20], etc.), and the system bids on their behalf until either their budget is exhausted or the day ends. We abstract each day as a single-shot game, where the advertiser sets at the start of the day its bid and budget. (A similar conceptual abstraction to a one-shot game was done for studying the truthfulness of click through rates [5,10].)

More concretely, each agent has a private value (for each item) and a private budget (which caps its total spend in a day). Agents bid for multiple identical divisible items.[1] Each day, the agents submit a bid and a budget to the auctioneer, which conceptually runs a sequence of second price auctions with some fixed minimum price. The auction terminates when all items have been sold or all agents have exhausted their budget. This sequential auction is a one-shot *budget auction*. The one-shot budget auction abstracts a repeated sponsored search auction for a single slot or a single display advertisement [19].

Our main focus is studying the properties and the existence of pure Nash equilibrium in a budget auction. We first observe that in a budget auctions, submitting the true budget is a dominant strategy, while bidding the true value is not a dominant strategy. The existence of pure Nash equilibrium depends on the assumptions regarding the bids of losing agents. For the case of two agents or multiple agents with identical budgets we show that there exists a pure Nash equilibria, even when the losing agents are restricted to bid their true value. For the case of multiple agents with different budgets, if losing agents are restricted

---

[1] While technically, advertisement impressions are clearly not divisible, due to the large volume of impressions, this is a very reasonable abstraction. Although using a single devisable item is equivalent, we found it less 'natural' for modeling ad auctions. In a discrete model our main results (PNE existence) will require an additional assumption which is that Critical Bids (see definition in the next chapter) are epsilon-separated. Otherwise we can only prove the existence of ϵ-Nash equilibrium.

to bid their true value then there are cases where no pure Nash equilibrium exists. Our main result is that if we relax this restriction, and assume that the losing agents bid conservatively, i.e., any value between the minimum price and their value, then there always exists a pure Nash equilibrium.

We also study simple dynamics of repeated budget auctions with myopic agents. For the dynamics we use the *Elementary Stepwise System* [22], where in each day one non-best-responding agent modifies his bid to a best response, and bid values are discrete. We prove that these repeated budget auction converge to a Nash equilibrium for the case of two agents and for the case of multiple agents with identical budgets (under some restrictions).

To illustrate our results for the repeated budget auction we ran simulations. We observed two distinct bidding patterns: either smooth convergence to an equilibrium or a bidding war cycle. The smoothed convergence is observed for a very wide range of parameters and suggests that the convergence is much wider than we are able to prove.

Related Work: The existence of a pure Nash equilibrium in GSP sponsored search auction was shown in the seminal works [13,23]. (For equilibrium in other related models see [21].) There are many works on dynamics of bidding strategies in Ad Acutions, including theoretical and empirical works [15,12,2,8,9,17].

Paper Outline: Section 2 presents the budget auction model and derives some basic properties of budget auction. Section 3 studies the existence of pure Nash equilibria in budget auctions. Section 4 analyzes the dynamics of budget auctions when played repeatedly, and presents simulations. (The proofs are omitted due to lack of space.)

2   The Model

The budget auction model has a set of k agents, $K = \{1, ..., k\}$, bidding to buy N identical divisible items. Each agent $i \in K$ has two private values: his daily budget $\hat{B}_i > 0$, and his value for a single item $v_i > 0$. His utility u_i depends on the amount of items he received x_i and the price he paid p_i, and $u_i(x_i, p_i) = x_i(v_i - p_i)$ as long as he did not exceed his budget (i.e., $x_i p_i \leq \hat{B}_i$), and $u_i = -\infty$ if he exceeded his budget[2] (i.e., $x_i p_i > \hat{B}_i$).

The auction proceeds as follows. The auctioneer sets a minimum price p_{min}, which is known to the agents (she ignores bids below the minimum price). Each agent $i \in K$ submits two values, his bid $b_i > 0$ and his budget $B_i > 0$. Therefore, the auction's input is a vector of bids $\boldsymbol{b} = (b_1, b_2, ..., b_k)$, and a vector of budgets $\boldsymbol{B} = (B_1, B_2, ..., B_k)$. The output of the auction is an allocation $\boldsymbol{x} = (x_1, x_2, ..., x_k)$, such that $\sum_{i \in K} x_i \leq N$ and prices $\boldsymbol{p} = (p_1, p_2, ..., p_k)$, such that $p_i \in [p_{min}, b_i]$. Agent i is charged $x_i p_i$ for the x_i items he receives.

The allocation and prices are calculated in the following way. Initially, the auctioneer renames the agents such that $b_1 \geq b_2 \geq ... \geq b_k \geq p_{min}$ (when equal

---

[2] This hard budget approach is used also in [11]. Another way to ensure agents don't exceed their budget is if agents deposit their budget to the auctioneer (instead of just reporting it), and the auctioneer returns unused budget at the end of the auction.

Table 1. An example of a budget auction with four agents, $N = 100$ and $p_{min} = 0$

Agent	Private		Submitted		Outcome			
	$\hat{B}$	v	B	b	x	p	u	Type
A	20	2.0	20	1.0	20	1.0	20	$Winner$
B	25	1.5	25	1.0	50	0.5	50	$Winner$
C	30	1.5	20	0.5	30	0.3	36	$Border$
D	20	0.5	20	0.3	0	0.0	0	$Loser$

bids are sorted lexicographically), we later refer to this index also as *ranking*. First, agent 1 receives items at price $p_1 = b_2$ until he runs out of budget or items, i.e., $x_1 = \min(N, B_1/p_1)$. Then, if there are still items left for sale, agent 2 pays a price $p_2 = b_3$, for $x_2 = \max\{0, \min(N - x_1, B_2/p_2)\}$ items. In general, agent i has allocation $x_i = \max\{0, \min(N - \sum_{j=1}^{i-1} x_j, B_i/p_i)\}$ and price $p_i = b_{i+1}$, for $i \in [1..k-1]$ and $p_k = p_{min}$. The auction is completed either when all items are sold, or when all agents exhaust their budgets. Obviously, if all items are sold to agents with higher rank than agent i, then $x_i = 0$ and $u_i = 0$. Note that by this definition, the allocation of items to agents will never exceed the supply N, i.e., $\sum_{i \in K} x_i \leq N$.

Given the outcome of the budget auction, we can split the agents into three different categories: *Winner Agents*, *Loser Agents* and a *Border Agent*. A *Border Agent* is the lowest ranked agent that gets a positive allocation, i.e., h is a Border Agent if $h = \max\{i : x_i > 0\}$. Any agent $j > h$ has $x_j = 0$ and is called a *Loser Agent*. Any agent $i < h$ is called a *Winner Agent* and has $x_i = B_i/p_i > 0$, i.e., winner agents exhaust their budgets. (See Table 1 for an example.)

It is worth making a few remarks regarding our model. Our main goal is to abstract a repeated GSP auction with budgets. We make few important simplifying assumptions: First, that budgets and bids are set once, which simplifies the game to be a one-shot game. This assumption is equivalent to assuming that the agents do not modify their bids. Second, we consider only a single item with a know quantity. Third, we assume that the items are divisible and prices are continuous, which are both a very accurate approximation in a sponsored search setting (where the number of impressions is usually huge, and prices are discritized at a very fine level). Given those assumptions, our model gives the same outcome as a repeated GSP auction.

For the most part we assume that agent i submits bids in the range $[p_{min}, v_i]$. The assumption that $b_i \geq p_{min}$ is with out loss of generality, since the auctioneer ignores agents that bid below p_{min}. The assumption that agents do not bid above their true value, i.e., $b_i \leq v_i$, is a very reasonable and realistic assumption, and was termed *conservative bidding* in [16,6].[3]

---

[3] Theoretically, agent i might profit by over bidding his value, since it increases the price of the agent $i-1$ who ranked above him and therefore, decreases his allocation x_{i-1}. This will leave more items for agent i and might increase his own allocation x_i. Nevertheless, such bidding might expose the agent to negative utility, since he might pay more than his value.

In a *Pure Nash Equilibrium* (PNE) no agent $i \in K$ can gain by unilaterally changing his submitted bid b_i and budget B_i. Formally, for a bid vector $\boldsymbol{b}$ and a budget vector $\boldsymbol{B}$, let $\boldsymbol{b}_{-i}$ and $\boldsymbol{B}_{-i}$ be the submitted bids and budgets, respectively, of all agents except agent i. The pair $(\boldsymbol{b}, \boldsymbol{B})$ is a PNE, if for each agent i, and any deviation b_i' and B_i', we have $u_i((b_i', \boldsymbol{b}_{-i}), (B_i', \boldsymbol{B}_{-i})) \leq u_i(\boldsymbol{b}, \boldsymbol{B})$.

Not surprisingly, bidding the true value is not a dominant strategy, and agents can bid lower than their true value in order to maximize their utility. We observe that submitting the true budget is a dominant strategy.

Lemma 2.1. *In a budget auction, bidding the true value is not a dominant strategy, while submitting the true budget is a dominant strategy.*

It is instructive to compare the allocation and prices of the budget auction to the *Market Equilibrium Price*, which is the price that equalizes the supply and demand. Figure 1 shows the computation of the market equilibrium price, for the example in Table 1.

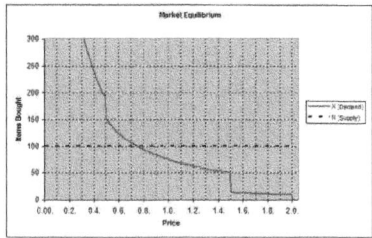

Fig. 1. The Market equilibrium price of the example in Table 1 is $p_{eq} = 0.75$, as at that point the aggregated demand equals the supply. Note that at the vertical drops, the aggregated demand is an interval and not a point. This happens when the price equals the values (v_i) of the different agents (in the example this occurs at 0.5, 1.5 and 2.0)

One can impose the market equilibrium price to be the outcome of the budget auction in two simple alternative ways. The first is that the auctioneer can set the minimum price to be the market equilibrium price, i.e., $p_{min} = p_{eq}$. The second is by adding a dummy agent, that bids p_{eq} with an infinite budget. In both cases we will have a PNE when each 'original' agent i bids $b_i = \min\{v_i, p_{eq}\}$. Clearly, any agent that gets an allocation would pay p_{eq}, and the property of market equilibrium price guarantee that all items are sold. Both alternatives are not satisfactory. First since the clearing price is unknown (intuitively, this why the auctioneer is holding an auction in the first place). In addition, in the second alternative the dummy agent risks negative utility due to his non-conservative bidding (in both his bid and budget) just to enforce the market equilibrium price. For this reason we would concentrate on conservative biding, which guarantee a non-negative utility.

Table 2. Two possible outcomes of the budget auction with 4 agents, $N = 100$ and $p_{min} = 0$. Option I is the only 'candiadte' for a PNE. Nevertheless, agent C can deviate and improve his utility by underbidding the loser agent (option II). This example shows that when loser agent bid their values, the are cases without a PNE.

Agent	Private Values			I - bidding critical value					II - under bidding				
	$\hat{B}_i$	v_i	c_i	b_i	p_i	x_i	u_i	type	b_i	p_i	x_i	u_i	type
A	40	2.0	1.143	1.143	1.143	35	30	winner	1.143	1.143	35	30	winner
B	40	2.0	1.143	1.143	1.143	35	30	winner	1.143	1.0	40	40	winner
C	40	2.0	1.143	1.143	1.0	30	30	border	$1.0 - \epsilon$	0.0	17	34	border
D	8	1.0	1.0	1.0	0	0	0	loser	1.0	$1.0 - \epsilon$	8	0	winner

3   Pure Nash Equilibrium

In this section we study the existence of a Pure Nash Equilibrium (PNE) in budget auctions. Our main result is that under mild conditions a PNE does exists for every budget auction.

We start by showing properties that any PNE in a budget auction must have, we then define the notion of *critical bid*, which intuitively is the bid which make the agent indifferent between being a winner or a border agent, and we complete by proving that PNE exist in budget auctions.

3.1   Properties of a Pure Nash Equilibrium

We show that in any PNE all winner agents pay the same price, which implies that all winner agents and the border agent bid the same. In addition, we show that this price is at most the Market Equilibrium price.

Lemma 3.1. *In any PNE, all winner agents pay the same price p, the border agent pays a price $p' \leq p$, and any loser agent j (if exists) has value $v_j \leq p$. In addition, p is at most the market equilibrium price, i.e., $p \leq p_{eq}$.*

It seems that one of the critical assumptions in our model is regarding the bids of the loser agents. As shown before, if we do not assume that agents are conservative, we can force a PNE at the market equilibrium price. (See earlier discussion regarding the drawbacks of such an equilibrium). A rather natural assumption is that loser agents bid truthfully, i.e., $b_j = v_j$ and $B_j = \hat{B}_j$ (at equilibrium, they will get a utility of at most zero with any bid). The example in Table 2 shows a case where there is no PNE when loser agents are restricted to report their true value and true budget, and is summarized in the following claim.

Lemma 3.2. *If the loser agents are restricted to bid their true value and budget, then there exists a budget auction with no pure Nash equilibrium.*

3.2   Critical Bid

The *critical bid* plays an important role in our proof of the existence of a PNE. Intuitively, when other agents bid low, an agent could prefer the top rank (as it is cheap). Similarly, when other agents bid high, an agent could prefer the bottom rank, and get the 'leftover' items at the minimum price. The critical bid models the transition point between these two strategies. Specifically, consider the case when all the agents bid the same value, then critical bid is the value, for which an agent is indifferent between being a winner agent and a border agent (receiving the remaining items at minimum price).

Definition 3.1. *Consider an auction with the set of agents K and a minimum price of p_{min}. The* critical bid *for agent i is the bid value $x = c_i(K, p_{min})$, such that when all agents participating in the auction bid x, i.e., $\boldsymbol{b} = (x, ..., x)$, agent i is indifferent between the top rank (being a winner agent) and the bottom rank (being a border agent).*

Obviously each agent has potentially a different critical bid. When clear from the context we denote the critical bid of agent j by c_j. A function that would be of interest is $\varphi_k(p_{min}) = \min_{1 \leq i \leq k} \{c_i(K, p_{min})\}$ which is the lowest critical bid among agents in the set K with a minimum price p_{min}. The following 3 lemmas describe the critical bid properties with respect to the Market Equilibrium Price, and how the critical bid reflects agents incentives.

Lemma 3.3. *Let $c_j(K, p_{min})$ be the critical bid of agent j, then: (a) $c_j(K, p_{min}) \in [p_{min}, v_j]$, and (b) if $\boldsymbol{b} = (x, ..., x)$ then for $x < c_j(K, p_{min})$ agent j prefers the top rank and for $x > c_j(K, p_{min})$ agent j prefers the bottom rank.*

Lemma 3.4. *Any agent's critical value is at most the Market Equilibrium Price.*

Lemma 3.5. *Consider a bid vector $\boldsymbol{b} = (b_1, \ldots, b_k)$. Then: (a) The top ranked agent, or any winner agent $j \in K$, cannot improve his utility by bidding higher, i.e., $b'_j > b_j$, (b) The bottom ranked agent, or any loser agent $j \in K$, cannot improve his utility by bidding lower, i.e., $b'_j < b_j$, and (c) If every agent $i \in K$ bids $b_i = c_j$ (agent j critical bid), then agent j cannot improve his utility by changing his bid.*

3.3   PNE Existence: Special Cases

In this section we prove the existence of a PNE in a budget auction in two interesting special cases: only two agents and multiple agents with identical budgets. The proofs of these cases would be latter extended to establish the general theorem, that a PNE exists for budget auction with any number of agents.

Two Agents: We start by characterizing the PNE for only two agents, which will later be the base of our induction for proving PNE existence for multiple agents.

Theorem 3.1. *Assume that we have two agents with $c_2 \leq c_1$. In the case that $B_1/v_2 < N$ then any bids $b_1 = b_2 \in [c_2, \min\{v_2, c_1\}]$ are a PNE. In the case that $B_1/v_2 \geq N$ then any bids $b_1 \in [v_2, v_1]$ and $b_2 \in [p_{min}, v_2]$ are a PNE. Those are the only PNEs where agents submit their true budget and bid at most their value.*

Agents with identical budgets: The fact that agents have identical budgets prevents winner agents from underbidding loser agents. This allowed us to assume that loser agents bid truthfully both their budget and value.

Theorem 3.2. *There exists a PNE for any number of agents with identical budgets, where agents submit their true budget ($B_i = \hat{B}_i$), bid $b_i \in [p_{min}, v_i]$, and loser agents bid $b_i = v_i$.*

For the general case of different budgets, if we restrict loser agents to bid their true value and budget, then there are examples where a PNE does not exists (Table 2). The main idea for the general case would be to relax this restriction and let the loser agents submit their true budget, and to bid conservatively, i.e., any bid at most their value.

3.4   PNE Existence: General Case

In this subsection we prove that every budget auction with any number of agents has a Pure Nash Equilibrium where the agents submit their true budget and bid at most their value.

Assume that, $v_1 \geq v_2 \geq ... \geq v_k$. For $h \leq k$, let $S_h = \{1, 2, ..., h\}$, such that $S_k = K$. The following lemma shows that if there is a critical bid which is lower than the value of all h agents, then there is a PNE.

Lemma 3.6. *If the lowest critical bid is lower than the value of any agent, i.e., $c_j = \varphi_h(p_{min}) < v_h$, then $\mathbf{b} = (c_j, ..., c_j)$ is a PNE for S_h, where agent j is the border agent and other agents are winner agents.[4]*

The following lemma shows that by modifying the minimum price we can modify the price p that winner agents pay in a PNE.

Lemma 3.7. *Let $\mathbf{b}^1$ be a PNE for the set S_h of h agents and minimum price p_{min}, such that all winner agents pay price $p < v_h$, the border agent pays p_{min}, and there are no loser agents. Then for every $p^* \in [p, v_h]$ there exists a minimum price $p^*_{min} \in [p_{min}, v_h]$ and an agent $j \in S_h$ such that there is a PNE $\mathbf{b}^2$ in which every agent $i \neq j$ is a winner agent and pays p^*, agent j is the border agent and pays $p_j = p^*_{min}$, and there are no loser agents.*

The following lemma is essentially our inductive step in the proof of the PNE. It shows that we can increase the number of agents in a PNE by introducing a new agent with a value lower than the value of any of the previous agents.

---

[4] We assume that agent j slightly underbids c_j, and we ignore this small perturbation.

Lemma 3.8. *Let b^1 be a PNE with h agents and a minimum price p_{min}, such that all winner agents pay price p. If there is a new agent $h+1$ such that (a) $v_h \geq v_{h+1}$, and (b) For every $i \in S_h$ the new critical bid $c_i(S_{h+1}, p_{min}) \geq v_{h+1}$, then we can define a b^2 which is a PNE for S_{h+1} with the same minimum price p_{min}, where agent $h+1$ is a loser agent.*

The following is our main theorem, regarding the existence of a PNE in a budget auction. It shows that when all the agents bid conservatively, there is always a PNE.

Theorem 3.3. *There exists a PNE for any number of agents, where agents submit their true budget ($B_i = \hat{B}_i$) and bid at most their value ($b_i \leq v_i$).*

4   Repeated Budget Auction

In this section we analyze the dynamics of the budget auction when it is played multiple times with myopic agents. Our goal is to exhibit simple dynamics that converge to an equilibrium in this repeated setting. We are able to show the convergence of two agents, under rather general assumptions, and the convergence of multiple agents with identical budget, under more restrictive assumption (the most important is that the agents start with low bids). Our dynamics assumes that loser agent bid their value, and therefore we know that there are values for which there is no PNE (see Table 2). In such cases, clearly, our dynamics will not converge. assumption important is that the agents start with low bids). Our dynamics assumes that loser agent bid their value, and therefore we know that there are values for which there is no PNE (see Table 2). In such cases, clearly, our dynamics will not converge.

4.1   Bidding Strategies and Dynamics

We first outline our assumption regarding the way the agents select their budget and bids. For a one-shot budget auction reporting the true budget is a dominant strategy (Claim 2.1), so we assume agents always report their true budget (although, technically, in the repeated auction setting it is not a dominant strategy anymore). Since even for a one-shot budget auction, bidding true value is not a dominant strategy (Claim 2.1), we should definitely observe agents bidding differently than their value. We assume that bids are from a discrete set, namely $b_i^t = \epsilon \cdot \ell$ for some integer ℓ.

Best Response: We assume that agents are myopic, and when modifying their bid, they are performing a best response to the other agents' bids. Since there could be many bids which are best response, we specify a unique bid that is selected as BR_i, as follows. Let the $BRS_i(\boldsymbol{b}_{-i})$ be the set of (discrete) bids that maximizes agent's i utility given the bids $\boldsymbol{b}_{-i}$ of other agents. Let $x = l \cdot \epsilon = \min\{BRS_i(\boldsymbol{b}_{-i})\}$. (This implies that for every $y = l' \cdot \epsilon < x$, we have $u_i(\boldsymbol{b}_{-i}, y) < u_i(\boldsymbol{b}_{-i}, x)$, and for every $y = l' \cdot \epsilon > x$, we have $u_i(\boldsymbol{b}_{-i}, y) \leq u_i(\boldsymbol{b}_{-i}, x)$.) Let

$$BR_i(\boldsymbol{b}_{-i}) = \begin{cases} x & \text{if } u_i(\boldsymbol{b}_{-i}, x) > 0 \\ v_i & \text{if } u_i(\boldsymbol{b}_{-i}, x) = 0 \end{cases}$$

Note that an agent for which any best response yields zero utility, bids his true value.

Dynamics: After each daily auction we compute for each agent its best response. If all the agents are performing a best response, the dynamics terminates (in a PNE). Otherwise, a single agent, which is not playing best response, is selected by a centralized *Scheduler*, and changes his bid using the specific *Best Response* we described.

We use the following notation: b_i^t is the bid of agent i at day t. It is important to note that: (i) Budget restriction is daily - meaning that agent i can spend up to $\hat{B}_i$ each day, (ii) Agents have full information: they know the number of items (N) the minimum price (p_{min}), and after each day they observe the bids ($\boldsymbol{b}$) budgets ($\boldsymbol{B}$) prices ($\boldsymbol{p}$) and allocations ($\boldsymbol{x}$) of the previous days. Nevertheless, each agent i true value v_i and real budget $\hat{B}_i$ are private information.

Scheduler: We model the dynamics as an *Elementary Stepwise System* (ESS) [22] with a scheduler. The scheduler, after each daily auction, selects a single agent that changes his bid to his best response. We considered the following schedulers: (i) *Lowest First* - From the set of agents that are not doing best response, the lowest ranked agent is selected. (ii) *Round Robin* - Selects agents by order of index in a cyclic fashion. (iii) *Arbitrary scheduler* - Selects arbitrarily from the set of agents that are not doing best response.

4.2   Convergence

In this section we study the converges of the repeated budget auction to a PNE. We start with two agents, and generalize it to any number of agents with identical budgets.

Theorem 4.1. *For a repeated budget auction with two agents and discrete bids $b_i^t = l \cdot \epsilon \in [p_{min}, v_i]$, the ESS dynamics with any scheduler and any starting bids converges to a PNE.*

Next we prove that a repeated budget auction with any number of agents, with identical budgets and different values, converge to a PNE. However, for our proof we need to make sure that no two critical bids are equal (which is guaranteed if no two agents have the same value). In addition, we assume that the aggregated demand at minimum price exceeds the supply N. (Otherwise, it is an uninteresting case where all critical bids equal p_{min} and this is a PNE.) For agents with identical budgets, B, this implies that $Bk > Np_{min}$. We can now state the convergence theorem for agents with identical budgets.

Theorem 4.2. *A repeated budget auction with any number of agents with identical budgets and different values, with starting bids of p_{min}, and ESS dynamics with Lowest First scheduler, converges to a PNE.*

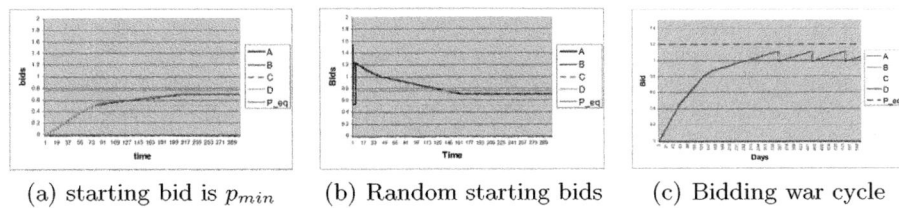

| (a) starting bid is p_{min} | (b) Random starting bids | (c) Bidding war cycle |

Fig. 2. Simulations of repeated budget auction using a 'Round Robin' scheduler. The agents in (a) and (b) are described at Table 1. In (a) their starting bid is the minimum price while in (b) they start at a random bid. In both cases the auction converged to a PNE, in which agents A and C are winner agents, agent B is the border agent, and agent D is a loser agent. In (c) the agent participating are described at Table 2, and the simulation shows a Bidding War cycle pattern.

4.3 Simulations

This section shows simulations of dynamics in budget auctions, which can give some intuition about typical bidding patterns of myopic agents. We simulated an ESS dynamics with a Round Robin scheduler and $\epsilon = 0.01$.

Our simulations show two bidding patterns: smoothed convergence to an equilibrium and a bidding war cycle. Similar patterns where observed by Asdemir [2] for the case of two symmetrical agents (identical budget and value).

Convergence: In our theoretical results, we only managed to prove convergence to PNE under strict restriction. Our simulations, however, shows that in some cases even when we relax these restriction the auction converge (See Figure 2 (a) and (b) for examples of convergence to a PNE with different budgets, different starting bids and a Round Robin scheduler).

Bidding War: Auctions that do not converge to an equilibrium follow a 'Bidding War Cycle' pattern as shown in Figure 2 (c). In this pattern some agents out bid each other, up to a certien price at which one agent drops his bid, and the other agents follow. Then, the agents continue to out bid each other and start a new cycle. This pattern was also spotted in real data collected by Edelman and Ostrovsky [12] from Overture search engine which they referred to as 'Sawtooth' pattern. It is worth mentioning that Overture used a first price auction mechanism, in which the existence of this pattern is less surprising. This bidding war pattern matches our theoretical results that show cases where there is no PNE if the loser agents is restricted to bid truthfully (Table 2).

References

1. Andelman, N., Mansour, Y.: Auctions with budget constraints. In: Hagerup, T., Katajainen, J. (eds.) SWAT 2004. LNCS, vol. 3111, pp. 26–38. Springer, Heidelberg (2004)
2. Asdemir, K.: Bidding patterns in search engine auctions. In: In Second Workshop on Sponsored Search Auctions, ACM Electronic Commerce, Press (2006)

3. Azar, Y., Birnbaum, B.E., Karlin, A.R., Mathieu, C., Nguyen, C.T.: Improved approximation algorithms for budgeted allocations. In: Aceto, L., Damgård, I., Goldberg, L.A., Halldórsson, M.M., Ingólfsdóttir, A., Walukiewicz, I. (eds.) ICALP 2008, Part I. LNCS, vol. 5125, pp. 186–197. Springer, Heidelberg (2008)
4. Azar, Y., Birnbaum, B.E., Karlin, A.R., Nguyen, C.T.: On revenue maximization in second-price ad auctions. In: Fiat, A., Sanders, P. (eds.) ESA 2009. LNCS, vol. 5757, pp. 155–166. Springer, Heidelberg (2009)
5. Babaioff, M., Sharma, Y., Slivkins, A.: Characterizing truthful multi-armed bandit mechanisms: extended abstract. In: ACM Conference on Electronic Commerce, pp. 79–88 (2009)
6. Bhawalkar, K., Roughgarden, T.: Welfare guarantees for combinatorial auctions with item bidding. In: SODA (2011)
7. Borgs, C., Chayes, J., Immorlica, N., Mahdian, M., Saberi, A.: Multi-unit auctions with budget-constrained bidders. In: EC 2005: Proceedings of the 6th ACM Conference on Electronic Commerce, pp. 44–51 (2005)
8. Borgs, C., Immorlica, N., Chayes, J., Jain, K.: Dynamics of bid optimization in online advertisement auctions. In: Proceedings of the 16th International World Wide Web Conference, pp. 13–723 (2007)
9. Cary, M., Das, A., Edelman, B., Giotis, I., Heimerl, K., Karlin, A.R., Mathieu, C., Schwarz, M.: Greedy bidding strategies for keyword auctions. In: ACM Conference on Electronic Commerce, pp. 262–271 (2007)
10. Devanur, N.R., Kakade, S.M.: The price of truthfulness for pay-per-click auctions. In: ACM Conference on Electronic Commerce, pp. 99–106 (2009)
11. Dobzinski, S., Nisan, N., Lavi, R.: Multi-unit auctions with budget limits. In: Proc. of the 49th Annual Symposium on Foundations of Computer Science, FOCS (2008)
12. Edelman, B., Ostrovsky, M.: Strategic bidder behavior in sponsored search auctions. In: Workshop on Sponsored Search Auctions, ACM Electronic Commerce, pp. 192–198 (2005)
13. Edelman, B., Ostrovsky, M., Schwarz, M.: Internet advertising and the generalized second price auction: Selling billions of dollars worth of keywords. American Economic Review 97 (2005)
14. Fiat, A., Leonardi, S., Saia, J., Sankowski, P.: Single valued combinatorial auctions with budgets. In: ACM Conference on Electronic Commerce 2011, pp. 223–232 (2011)
15. Kitts, B., LeBlanc, B.J.: Optimal bidding on keyword auctions. Electronic Markets 14(3), 186–201 (2004)
16. Leme, R.P., Tardos, É.: Pure and bayes-nash price of anarchy for generalized second price auction. In: FOCS, pp. 735–744 (2010)
17. Markakis, E., Telelis, O.: Discrete strategies in keyword auctions and their inefficiency for locally aware bidders. In: Saberi, A. (ed.) WINE 2010. LNCS, vol. 6484, pp. 523–530. Springer, Heidelberg (2010)
18. Mehta, A., Saberi, A., Vazirani, U., Vazirani, V.: Adwords and generalized online matching. J. ACM 54(5), 22 (2007)
19. Muthukrishnan, S.: Ad exchanges: Research issues. In: Leonardi, S. (ed.) WINE 2009. LNCS, vol. 5929, pp. 1–12. Springer, Heidelberg (2009)
20. Nisan, N.: Google's auction for TV ads. In: Fiat, A., Sanders, P. (eds.) ESA 2009. LNCS, vol. 5757, pp. 553–553. Springer, Heidelberg (2009)
21. Nisan, N., Roughgarden, T., Tardos, E., Vazirani, V.: Algorithmic Game Theory, Cambridge (2007)
22. Orda, A., Rom, R., Shimkin, N.: Competitive routing in multiuser communication networks. IEEE/ACM Trans. Netw. 1(5), 510–521 (1993)
23. Varian, H.R.: Position auctions. International Journal of Industrial Organization (2006)

Prompt Mechanism for Ad Placement over Time

Yossi Azar* and Ety Khaitsin

Blavatnik School of Computer Science,
Tel-Aviv University,
Tel-Aviv 69978, Israel
azar@tau.ac.il, avienda1@gmail.com

Abstract. Consider video ad placement into commercial breaks in a
television channel. The ads arrive online over time and each has an
expiration date. The commercial breaks are typically of some uniform
duration; however, the video ads may have an arbitrary size. Each ad
has a private value and should be posted into some break at most once
by its expiration date. The player who own the ad gets her value if
her ad had been broadcasted by the ad's expiration date (obviously, after ad's arrival date), and zero value otherwise. Arranging the ads into
the commercial breaks while maximizing the players' profit is a classical
problem of ad placement subject to the capacity constraint that should
be solved truthfully. However, we are interested not only in truthfulness
but also in a prompt mechanism where the payment is determined for
an agent at the very moment of the broadcast. The promptness of the
mechanism is a crucial requirement for our algorithm, since it allows a
payment process without any redundant relation between an auctioneer
and players. An inability to resolve this problem could even prevent the
application of such mechanisms in a real marketing process. We design
a 6-approximation prompt mechanism for the problem. Previously Cole
et al considered a special case where all ads have the same size which
is equal to the break duration. For this particular case they achieved a
2-approximation prompt mechanism. The general case of ads with arbitrary size is considerably more involved and requires designing a new
algorithm, which we call the Gravity Algorithm.

1 Introduction

Advertising has long since been ubiquitous in the world of trade activities. Recently, with online technologies participating in and transforming the economics
realm, the WEB space naturally becomes an ads space. Consequently, mechanisms that help to arrange ads in physical and online space draw our attention.
Consider a display ad space with fixed capacity, which contains a new set of ads
every day. The ads arrive online over time, each one has parameters of size, value,
arrival and expiration date before which it should be posted. Each day ads of a

---

* Partially supported by the Israeli Science Foundation (grant No. 1404/10) and by
the Google Inter-university center for Electronic Markets and Auctions.

G. Persiano (Ed.): SAGT 2011, LNCS 6982, pp. 19–30, 2011.
© Springer-Verlag Berlin Heidelberg 2011

fixed total size can be published. It is natural to assume that the size, arrival and expiration date are fixed parameters of the ad and cannot be reported falsely. Therefore we regard these parameters as public information. One may think for example of a poster advertising a concert. It has a fixed physical size and the date of the concert event determines the ad's expiration date. The only private information of an ad is its value. The ads are published for the entire day. The player gets her value if her ad had been published by the ad's expiration date (obviously, after the ad's arrival date), and zero value otherwise.

We are interested in a truthful (incentive compatible) mechanism. This is an algorithm which gets an input from selfish players who try to maximize their profit, and motivates them to report their private information truthfully. This goal is usually achieved by means of manipulating the payments collected from players depending on the algorithm's outcome. Our mechanism belongs to the single parameter problems domain, which is a well studied and understood class of problems. The single parameter truthful mechanisms are equivalent to monotone algorithms, which means that the winner would still win if she reports a higher value (keeping other players' values fixed). The mechanism charges every winner a critical price that is determined by a threshold, below which the player loses and above which wins. This critical price can be computed in polynomial time.

A classical technique to achieve truthfulness involves using VCG. Unfortunately, VCG often cannot be applied to online models, since it requires the exact optimality which rarely can be achieved in an online fashion, even with unbounded computational power. Moreover, even when exact optimality can be achieved, the payment cannot be determined at the time of service, since it depends on future events. This suggests the design of *prompt mechanisms*, meaning that a player gets to know the price at the time of publication - unlike the standard online truthful algorithm where the critical price may be determined only in the future.

This model is general and may have numerous and various applications. The ad space may be a physical newspaper sheet with new ads being published on it daily. Another example is a billboard that displays a set of ads on a fixed space with changes every specific time period.

Also, the given space may be a virtual one, and applied to online marketing. For example, Google TV deals with TV companies that offer ad slots, and advertisers that have ads to be published. Google puts together a schedule of ads that fills a commercial break (ads may have different length, but the breaks are typically of the same duration), and sends this schedule to a TV company, which broadcasts it as it is.

It is also worthwhile to mention that our problem can model a buffer management or a broadcast problem. Specifically, we consider the classical model of packets that need to be transmitted through an output port. Each player has a packet with value, length, arrival and expiration date. Every time step packets with bounded total size can be transmitted. The switch has to decide which packets would be transmitted at each time step while maintaining the size constraint. The goal is to design a truthful algorithm which transmits packets with (approximate) maximum total value.

Prompt mechanisms were introduced in [11]. As discussed there, several reasons for determining the payment at the time of the service exist: first, a player does not know how much money she has after winning until she leaves, and hence cannot participate in another auction. Second, prompt mechanisms can help to resolve some problems of the payment verification. There are two possible ways to ensure a payment and each of them is problematic in a distinct way: for example, if a player pays long after she had won, she may try to get out of the payment. Otherwise, a winner can be demanded to submit an empty check, and the real amount will be deducted later, which requires a considerable trust from the player. In prompt algorithms these situations are avoided, allowing a player to know the price at the moment of her win. Although the concept of prompt models is quite new in the mechanism design field, it has already drawn an attention as can be seen in [11,4].

The model studied in [11] is close to the particular case of ours specifically, they also consider ads arriving over time. In contrast to our model they considered the special case where the size of each ad is 1, and each day a single ad is published. They have presented a prompt mechanism which yields the approximation ratio of 2 and shown that this is the best possible result for prompt mechanisms. It is natural to ask how to design an algorithm that deals with ads of general sizes. We note that the algorithm of [11] is tailored to the unit size case, hence extending the algorithm to deal with ads of arbitrary sizes is by no means obvious. In addition, the problem of ads with arbitrary size is considerably more complicated than the unit size case, as we would need to deal with integrality issues. Note that the 2 lower bound for the unit size case naturally holds for ads of arbitrary size.

Our Results. We solve the general problem with ads of different sizes.

- Our main result is a truthful, prompt algorithm for online ad placement over time problem, which attains a 6-approximation ratio to the social welfare.
- For relatively small ads ($size \leq \epsilon$) the approximation ratio is $2 + O(\epsilon)$.

We also extend the scope of our study to a more general model, the restricted assignment model, where each ad has a finite set of time periods at which it could be posted. Taking TV ads for example, there are some ads that should be published in mornings or at peak hours only (each time unit is a commercial break). Every player wants her ad to be published during one of those time periods, otherwise she gets zero value. Our algorithm can be easily generalized and modified to answer this case, and attain a 6-approximation ratio (and $2 + O(\epsilon)$ ratio for small ads of size at most ϵ).

Typically, the approximation ratio is calculated for an integral mechanism versus an integral optimum solution. In our case, the achieved approximation ratio also holds for an integral mechanism versus fractional optimum (meaning optimal algorithm that can accept parts of the ads and gain partial profit respectively).

The algorithm treats ads with size $> \frac{1}{2}$ and size $\leq \frac{1}{2}$ separately. The big size ads are treated similarly to the uniform unit size case ([11]). Our main contribution is designing the Gravity Algorithm which focuses on the small ads. It has a tentative schedule of small ads for each day, and always prefers ads with higher density (i.e., the ratio of value to size) even if ad's value is small. The crucial detail of the algorithm is a choice of the time step which a newly arriving ad would be assigned to. It is determined by comparing the densities in the available time steps at a "depth" which depends on the size of the new ad.

Related Works. In a generalized full information setting, the problem is similar to the Multiple Knapsack problem. Multiple Knapsack is a well known problem with a lot of peculiar variations and useful applications ([13]). Its basis, the Knapsack problem, is NPC hard, therefore only approximation algorithms exist. For the Knapsack problem a FPTAS algorithm is available [16] while for the Multiple Knapsack problem there is a PTAS algorithm [8,14]. Our model is intimately related to online MKP with preemption. Designing a truthful model for MKP problem has been considered: the authors of [7] have obtained an approximation ratio of $\frac{e}{e-1} \approx 1.582$ in an offline fashion with bins of the same capacity (this case was also studied by [3]). In an online fashion a $2 + \epsilon$ approximation ratio is achieved in [7]. Note that the latter model is very similar to ours; however, our algorithm is prompt while their algorithm does not seem to be extendable to satisfy the promptness requirement.

The special case of our model with uniform unit size ads received a lot of attention in scheduling, packet management and multiple unit auctions frameworks. Online, truthful auction with expiring items was studied by [7], who presented a truthful 2-competitive mechanism. Non-truthful online model of unit jobs scheduling is more widely studied with more thorough results being obtained: the best known deterministic online algorithm has 1.828 competitive ratio [12,17], the lower bound for competitive ratio of deterministic online algorithm is $\frac{2}{\sqrt{5}+1} \approx 1.618$ [2,10]. Giving the randomized setting, the best known online algorithm yields a competitive ratio of $\frac{e}{e-1} \approx 1.582$ [9,5], and it is known that no online algorithm can obtain a ratio better than 1.25 [10]. Note that in offline settings this problem can be solved optimally. If the ads are not of size 1 but of any other uniform size $\frac{1}{k}$ then [9,5] obtain an approximation ratio of $(1 - (\frac{k}{k+1})^k)^{-1}$; with $k \to \infty$ the ratio tends to $\frac{e}{e-1} \approx 1.582$. The lower bound for this case is 1.17 [15].

The ad placement problem has already been widely studied in various fashions, including online algorithms and truthful mechanisms, for example [6,1,18]. One of the closest works is [1], studying an online model of ad auctions as a single knapsack problem, and designing a truthful mechanism that maximizes the revenue for this problem. However, most of those works have some differences from ours model. In most of them, for example, there is a factor of frequency (the number of times a single ad should appear), or some other distinctions.

2 The Model

Consider an ads display space of size 1 for a period of n days. Different ads of total size of at most 1 can be published each day. There are players arriving online, each has an ad to be published. An ad is represent by a tuple (s, v, a, e) where $s \in (0, 1]$ refers to its size, $v \in \mathbb{R}^+$ - the players value, and a, e - are the arrival and expiration time. A player wants her ad to be published once before the expiration time. If so, the benefit v is gained; otherwise, the player gets no benefit. The ad's value is a private information of a player, while the rest of the information is public. The algorithm should be incentive compatible. In a single parameter environment it is well that it is equivalent to being a monotone algorithm. Specifically, there is a threshold value above which the player wins and below which she loses. Also, the algorithm has to be prompt, which means the threshold value can be calculated at the very publishing moment. The goal of the algorithm is to maximize the social welfare which is the total value of all published ads.

The further structure of the paper: in the section 2.2 we describe the Gravity algorithm. In the section 3 we prove its truthfulness and promptness. In the section 4 a proof of the approximation ratio is supplied.

2.1 The Gravity Algorithm

Definition 1. Publishing window *of an ad is the time period between its arrival and expiration time.*

Definition 2. *Let density of an ad be* $d = v/s$. *When comparing two distinct ads, we call an* heavier *ad to the one with the higher density.*

The Gravity algorithm maintains a tentative ads schedule for each day. Whenever it handles a new ad's arrival event, it *assigns* it to one of those days. After the assignment there is a test for the ad to enters to the day schedule. If it passes the test it is incorporated into this schedule of that day, otherwise it is rejected. The assignment of the ad to a day is final. Once an ad is assigned to some day either it will be published at that day or it will be rejected. Our algorithm treats *big* ads (ads of size $s > 1/2$) and *small* (ads of size $s \leq 1/2$) separately. Hence, for every day it maintains two alternative schedules: one with a single *big* ad, and the other one of some *small* ads. At the daily publishing event only one of those daily schedules is actually published. The algorithm deals with the tentative daily schedule as if it is a physical bins (with fixed capacity 1). Whenever an ad is assigned to some day t, we say that it is assigned to bin b, where b is bin_{small} or bin_{big} depending on the ad's size. When a new *big* ad enters a bin, the previous ad in that bin is rejected. For the *small bins*, the algorithm maintains the ad of a daily schedule sorted by density. An heavier ad sinks deeper inside the bin and the possibly empty space in the bin is at the top. Every ad in the bin defines an interval in $[0, 1]$ where it is located. When a new small ad enters the bin, the algorithm inserts the ad to the sorted by density list. Note that the interval of the lighter ads in the bin will shift upwards. Moreover, that at this moment the total size of the ads may exceed 1. Hence all ads above 1 are rejected. There can be at most one ad whose open interval contains 1 and hence it fits the bin's capacity only partially. We call it a *partial* ad and it is also rejected. Yet, the algorithm treats it as if it remained inside, while being "cut" at 1. Its value is reduced so that its density remains the same. Note that although the algorithm treats a partial ad as if it is inside the bin, it does not participate in the publishing event.

Definition 3. *Define D_b^x to be the density at the depth x in the bin b which is the density of the ad whose interval in $[0,1]$ contains the point $1-x$. If the depth x is between two ads, then the it is the lighter density between the two.*

Definition 4. *For any bin b define V_b to be the sum of values of all ads in b.*

Algorithm 1. Gravity Algorithm - contains two parts: Publishing event and Ad arrival event.

Publishing Event: Day t publishing - let b be $bin(t)$
if $\sum_{p \in b_{small}} v(p) \geq v(b_{big})$ **then**
    ads in b_{small} are published and ad in b_{big} is rejected
else
    ad in b_{big} is published and ads in b_{small} are rejected
end if

Ad Arrival Event: An ad $p = (s, v, a, e)$ just arrived
if $s \leq 1/2$ **then**
    Let b_{small} be the bin for which D_b^s is minimal in *publishing window* of p
    if $D_{b_{small}}^s < d_p$ **then** p enters the bin b_{small}
    (ads in this bin are reordered, and some possibly rejected or become *partial*)
    else p is rejected
    end if
else
    Let b_{big} be the bin for which V_b is minimal in *publishing window* of p
    if $V_{b_{big}} < v$ **then** p enters the bin b_{big} and the current ad in b_{big} is rejected
    else p is rejected
    end if
end if

The following observation follows from the definition of the algorithm and the usage of partial ad.

Observation 1. *The density D_b^x for any bin b and $0 \leq x \leq 1$ is non decreasing function throughout the execution of the algorithm.*

Our main result is the following theorem:

Theorem 2. *The Gravity Algorithm is truthful, prompt and 6-competitive.*

The theorem is proved in sections 3 and 4. In section 3 we show truthfulness and promptness. In section 4 we analyze the approximation ratio.

3   Truthfulness

In this section we prove the truthfulness and the promptness of the algorithm.

Lemma 1. *The Gravity Algorithm is truthful (i.e. monotone).*

Proof. We will prove that the algorithm is truthful by showing it is monotone. Suppose that a player i with an ad $p = (s, v, a, e)$ is published as a part of the bin b. We have to show that if she reports $v' \geq v$ then her ad is still published. If $s > 1/2$, then the proof is identical to the one provided in [11] and it is omitted.

Next we focus on small ads, i.e. $s \leq 1/2$. We prove by induction on all ads arriving after p, that the state of all the bins remains exactly the same (albeit the false value report) up to the ads' order inside b. That means that the bin b contains the same ads at the publishing event, and p is published again. Before the arrival of p the flow of the algorithm is exactly as before. At its arrival p is assigned to a certain bin independently of its value. As p would be accepted with the original value, it will also be accepted with a new higher value and density. Now we will consider one by one all further arriving ads and prove that the state after each arrival stays the same as in the original case.

Let $c = (s_c, v_c, a_c, e_c)$ be an arbitrary ad arriving after p. Note that the *density at depth* s_c in the bin b became larger after p's value increased. One of the following three cases holds:

- If originally c is assigned to other bin, it will be assigned there again, and the algorithm takes care of c in the same way.
- If originally c is assigned to the bin b, and ad p is deeper inside the bin than s_c (i.e. the interval of p ends below $1 - s_c$), then we know that c does not compete with p. In the untruthful case, since p is in the same place or perhaps deeper, the algorithm acts in the same way as before.
- If originally c is assigned to the bin b and competes with p then there can be 2 possibilities (notice that in original case p wins, which means that c is rejected when arrived):
  - Ad c is assigned to bin b again; then it will be rejected again.
  - Ad c is assigned now to some other bin h: that means that $D_b^{s_c} \leq D_h^{s_c} \leq D'^{s_c}_b$. We know that originally c was rejected, which means $d_c \leq D_b^{s_c}$, then $d_c \leq D_h^{s_c}$ thus c is rejected from bin h as in the original case.

This flow works for all ads that arrive after p, and then p is still published with the increased value. The algorithm is monotone, and consequently, truthful. ∎

3.1   Promptness

Recall that the algorithm is monotone; hence, the critical price is well defined. It is easy to see that every ad can be published only within the bin it was assigned to. Moreover, the publishing of this ad does not depend on the ads that would arrive after the publishing moment. This means that the critical price can be calculated at the very publishing moment - which makes the algorithm prompt.

4 Approximation Ratio

In this section we compute the approximation ratio. We do it separately for *big* and *small* ads, and show that for *small* ads the ratio is 4, and for *big* ads it is 2. Then we show how to combine the analysis and achieve the total approximation ratio of 6 for the entire algorithm.

4.1 Big Ads Approximation Ratio

In this subsection we assume that all ads are *big* (i.e. of sizes larger than $\frac{1}{2}$). For the *big* ads the proof is identical to proof in [11], in which they prove that $\text{Opt}_{\text{big}} \leq 2 \cdot \text{Alg}_{\text{big}}$. Here the proof is omitted.

4.2 Small Ads Approximation Ratio

Within this subsection we assume that all ads are *small* (i.e. of sizes smaller than or equal to $\frac{1}{2}$).

Theorem 3. $\text{Opt}_{\text{small}} \leq 4 \cdot \text{Alg}_{\text{small}}$.

Proof. We fractionally match every ad that was published in Opt to some bin. Then for every bin we prove that the total value of all ads that were matched to that bin is at most 4 times the total value of ads in the bin of Alg .By summing by all ads of Opt and all bins of Alg we get a ratio of at most 4 between the total social benefit and Alg.

Definition 5. *Let O be the set of all ads that were published in* Opt. *Specifically* $O = \cup_{i=1}^{i=n} O_i$ *where O_i is the set of ads that were assigned to bin i in* Alg.

Definition 6. *Given $o \in O$, define $Home(o)$ as the bin where o was published in* Opt.

Now we will describe the fractional matching. An ad o in O_i will be matched by one (or sometimes two) of those rules:

Rule 1. If an ad o was rejected at arrival event then it is matched to the bin $Home(o)$.

Rule 2. If an ad o had entered to bin i but was preempted later, then it is fractionally matched to the $Home$ bins (in Opt) of the ads in O that entered the bin after o's preemption. Let $O_i = (o_1, o_2, ... o_{last})$ be an ordered set, ordered by arriving timer and assume o is r'th in the order so $o = o_r$. For every $k > r$ we define the *paying* function. Let $o_k \in O_i$ be an ad that was in the bin i and arrived after the ad o_r had been preempted; o_k will *pay* for o_r a fractional part as described below. Note that some fraction of the ad o_r may remain "unpaid" for, and it is the rule 3 that will take care of it. *Pay* is a recursive function defined as follows (it is defined only for $k > r$):

$$pay(o_r, o_k) = Min(size(o_k) - \sum_{l=1}^{r-1} pay(o_l, o_k), size(o_r) - \sum_{l=r+1}^{k-1} pay(o_r, o_l)).$$

Whenever we say that the ad o_k pays for a part of the ad o_r, the meaning is that the part of o_i is viewed as a separate ad of size $pay(o_r, o_k)$ and matched to $Home(o_k)$. Several ads can *pay* for o_r, so that the sum of parts they *paid* for is up to the size of o_r. It may happen as well that one ad pays for several ads but the sum of these ads or ads' parts is always up to the paying ad's size. If no one pays for a part of some ad, this part would be matched by the rule 3.

Rule 3. If an ad o were not preempted, then it is matched to bin i. In addition, the parts of ads which were not matched in rule 2 will also be matched to the bin i.

This way, every ad in O is matched via one (or more) of those matching rules to a bin (or fractionally to several bins). For every bin and every matching rule we will calculate the ratio between the values of the ads that were matched to this bin and the values of the ads published in the bin in Alg. Then we calculate the general ratio.

Definition 7. *Given a set of ads (or parts of ads) A. Then let $S(A) = \sum_{p \in A} s_p$ be their total size and $V(A) = \sum_{p \in A} v_p$ be their total value. Let Z_j be the set of ads published at bin j by the algorithm not including the* partial ad *and let $V_j = V(Z_j)$.*

It is enough to show that the total value of all ads that were matched to bin j, does not exceed $4V_j$. We show it by bounding the total value of all ads that were matched to j by every one of the rules:

Rule 1

Let $G_1(j)$ be the set of all ads that were matched by the first rule: they were published in Opt at bin j (since we have fixed j we omit the index and denote it as G_1). These ads were assigned to some bins in Alg and rejected immediately at their arrival event. Let d_{max} be the maximal density of an ad in G_1. We can conclude that d_{max} is at least as the average density of the set G_1. Let an ad $p = (s, v, a, e) \in G_1$ be a one for which $v/s = d_{max}\}$. The ad p was assigned to some bin h. Note that p could have been assigned to bin j, and hence $D_h^s \leq D_j^s$ (when p arrives). By that and by the fact that p was rejected when arrived, we derive that $d_{max} \leq D_h^s \leq D_j^s$. Since $s \leq 1/2$, then at the moment of p's arrival the bin j was filled up to a half with ads of density at least d_{max}. The observation before implies that the density at any depth the bin is non-decreasing. Hence, at j's publishing event the bin was full up to half with ads of density at least d_{max} (this half does not contain the *partial ad*, so we can use V_j). Hence

$$V_j \geq 1/2 \cdot d_{max} \rightarrow d_{max} \leq 2V_j \rightarrow V(G_1) \leq 2 \cdot S(G_1) \cdot V_j.$$

Rule 2

Let $G_2(j)$ be the set of all ads or ads' parts that have been matched by the second rule: they were *paid* by ads or ads' parts that were published at bin j in Opt (since we have fixed j we omit the index and denote it as G_2). We call the set of those *paying* ads or ads' parts G_2'. For example, if an ad of size $1/2$ has *paid* for two other ads, each being of the size $1/6$, then the *paying* ad is divided into two different ads of size $1/6$ and the same density; the rest $1/6$ is ignored. For each $p \in G_2$, define $Payers(p) = \{q \in G_2' | q$ *paid on* $p\}$. Notice that $s_p = S(Payers(p))$ and $S(G_2) = S(G_2')$. By the monotone density observation the density of any ad in $Payers(p)$ is at least d_p. Let d_{max} be the maximal density in G_2 and $p = (s, v, a, e) \in G_2$ for which $v/s = d_{max}$. Ad p was assigned to a bin h. Let $q \in Payers(p)$ such that q was assigned to the bin h. Note that q could have been assigned to the bin j and hence at the moment q's arrival $D_h^{s_q} \leq D_j^{s_q}$. At this moment p has been preempted already; this means $d_{max} \leq D_h^{s_p}$ and $s_q \leq s_p$ which implies that $d_{max} \leq D_h^{s_q} \leq D_j^{s_q} \leq D_j^{1/2}$. We conclude that at the moment of q's arrival, the bin j is full up to half with ads of density at least d_{max}. The density inside the bin is monotone non-decreasing over time and at j's publishing event the bin is filled up to a half with ads of a density at least d_{max} (this half does not contain the *partial ad*, so we can use V_j). Hence

$$V_j \geq 1/2 \cdot d_{max} \rightarrow d_{max} \leq 2V_j \rightarrow V(G_2) \leq 2 \cdot S(G_2) \cdot V_j.$$

Rule 3

Let $G_3(j)$ be the set of all ads (or parts of ads) that are currently matched by third rule: they had entered bin j and were either published or preempted but were not paid for (since we have fixed j we omit the index and denote it as G_3). Unlike G_1 and G_2, we view G_3 as evolving over the time.

Lemma 2. *At any time $S(G_3) \leq 1$ where $S(G_3)$ is the sum of the sizes of the ads in G_3. The proof is omitted.*

Now, as we know that $S(G_3) \leq 1$, we will bound $V(G_3)$. We divide G_3 into 2 sets: $G_{in} \cup G_{out} = G_3$, G_{in} are the ads that have been finally published (may include the *partial ad*), and G_{out} are ads that have been preempted. Let P_j be the set of ads published at bin j including (incuding) the *partial ad*. Note that P_j is Z_j union with the the *partial ad* in j. We also divide P_j (ads that were published in Alg at bin j), into 2 groups $P_j = P_j^1 \cup P_j^2$ such that $G_{in} = P_j^1$ and $P_j^2 = P_j - P_j^1$. Clearly $V(G_{in}) = V(P_j^1)$ and we will show $V(G_{out}) \leq V(P_j^2)$. If G_{out} is empty we are done. Else we know that $S(P_j) = 1$, because if an ad was preempted from a bin, this bin will be always filled with ads and a partial ad. By Lemma 2, $S(G_{out}) \leq 1 - S(G_{in}) = 1 - S(P_j^1) = S(P_j^2)$. Let d_{max} be the maximal density of ads in G_{out}, $d_{max} \geq V(G_{out})/S(G_{out})$. Observe that every ad in P_j^2 has at least density d_{max}. We obtain that $V(G_{out}) \leq d_{max} \cdot S(G_{out}) \leq d_{max} \cdot S(P_j^2) \leq V(P_j^2)$. We put it together and conclude:

$$V(G_3) = V(G_{in}) + V(G_{out}) \leq V(P_j^1) + V(P_j^2) = V(P_j)$$
$$= V(Z_j) + v_{partial\ ad} = V_j + v_{partial\ ad}$$

It remains to handle is the *partial ad* that was viewed as a regular ad but eventually does not get published. The size of the *partial ad* is less than half and it has the least density in the bin. Hence $v_{partial\ ad} \le V_j$ which implies that $V(G_3) \le 2 \cdot V_j$.

Summary

Now we sum these 3 results together. For all j we have

1. $V(G_1(j)) \le 2S(G_1(j)) \cdot V_j$
2. $V(G_2(j)) \le 2S(G_2(j)) \cdot V_j$
3. $V(G_3(j)) \le 2V_j$

Notice that $G_1(j)$ and $G_2'(j)$ both were published in Opt in bin j, meaning that their size together is up to 1, and as we already know $S(G_2(j)) = S(G_2'(j))$. Then we get,

$$V(G_1(j)) + V(G_2(j)) + V(G_3(j)) \le 2(S(G_1(j)) + S(G_2(j)))V_j + 2V_j$$
$$\le 2V_j + 2V_j = 4V_j.$$

Now we sum over all j to obtain:

$$V_{OPT}^{small} = \sum_{j=1}^{n} V(G_1(j)) + V(G_2(j)) + V(G_3(j)) \le 4\sum_{j=1}^{n} V_j = 4V_{ALG}^{small}.$$

∎

Proof of Theorem 2. The Gravity Algorithm is truthful by the Lemma 1, and also prompt. We have calculated the approximation ratio separately for *big* and *small* ads, and shown that for *small* ads the ratio is 4 in Theorem 3, and for *big* ads it is 2 (the proof for *big* ads is the same as in [11]). This implies that the total approximation ratio of the algorithm is 6 - details are omitted.   ∎

5   Concluding Remarks

We designed a prompt mechanism for ads with arbitrary sizes that are placed over time. Our mechanism is truthful, prompt and achieves 6-approximation. It would be interesting to know the best approximation for prompt mechanism as the best lower bound is 2. Moreover, for ads of relatively small sizes the lower bound does not hold and it may be possible to get an approximation better than 2 for prompt mechanisms.

Acknowledgements. The authors would like to express sincere gratefulness to Iftah Gamzu for his valuable contribution to the present work.

References

1. Aggarwal, G., Hartline, J.D.: Knapsack auctions. In: Proceedings of the Seventeenth Annual ACM-SIAM Symposium on Discrete Algorithms, SODA (2006)

2. Andelman, N., Mansour, Y., Zhu, A.: Competitive queueing policies for qos switches. In: Proceedings of the Fourteenth Annual ACM-SIAM Symposium on Discrete Algorithms, SODA 2003, pp. 761–770. Society for Industrial and Applied Mathematics, Philadelphia (2003)
3. Azar, Y., Gamzu, I.: Truthful unification framework for packing integer programs with choices. In: Aceto, L., Damgård, I., Goldberg, L.A., Halldórsson, M.M., Ingólfsdóttir, A., Walukiewicz, I. (eds.) ICALP 2008, Part I. LNCS, vol. 5125, pp. 833–844. Springer, Heidelberg (2008)
4. Babaioff, M., Blumrosen, L., Roth, A.L.: Auctions with online supply. In: Fifth Workshop on Ad Auctions (2009)
5. Bartal, Y., Chin, F.Y.L., Chrobak, M., Fung, S.P.Y., Jawor, W., Lavi, R.: Online competitive algorithms for maximizing weighted throughput of unit jobs. In: Diekert, V., Habib, M. (eds.) STACS 2004. LNCS, vol. 2996, pp. 187–198. Springer, Heidelberg (2004)
6. Borgs, C., Immorlica, N., Chayes, J., Jain, K.: Dynamics of bid optimization in online advertisement auctions. In: Proceedings of the 16th International World Wide Web Conference, pp. 13–723 (2007)
7. Chekuri, C., Gamzu, I.: Truthful mechanisms via greedy iterative packing. In: Proceedings 12th International Workshop on Approximation Algorithms for Combinatorial Optimization Problems, pp. 56–69 (2009)
8. Chekuri, C., Khanna, S.: A polynomial time approximation scheme for the multiple knapsack problem, vol. 35, pp. 713–728. Society for Industrial and Applied Mathematics, Philadelphia (2005)
9. Chin, F.Y.L., Chrobak, M., Fung, S.P.Y., Jawor, W., Sgall, J., Tich, T.: Online competitive algorithms for maximizing weighted throughput of unit jobs. 4, 255–276 (2006)
10. Chin, F.Y.L., Fung, S.P.Y.: Online scheduling with partial job values: Does time-sharing or randomization help? 37, 149–164 (2003)
11. Cole, R., Dobzinski, S., Fleischer, L.: Prompt mechanisms for online auctions. In: Monien, B., Schroeder, U.-P. (eds.) SAGT 2008. LNCS, vol. 4997, pp. 170–181. Springer, Heidelberg (2008)
12. Englert, M., Westermann, M.: Considering suppressed packets improves buffer management in qos switches. In: Proceedings of the Eighteenth Annual ACM-SIAM Symposium on Discrete Algorithms, SODA 2007, pp. 209–218. Society for Industrial and Applied Mathematics, Philadelphia (2007)
13. Fidanova, S.: Heuristics for multiple knapsack problem. In: IADIS AC, pp. 255–260 (2005)
14. Kellerer, H.: A polynomial time approximation scheme for the multiple knapsack problem. In: Hochbaum, D.S., Jansen, K., Rolim, J.D.P., Sinclair, A. (eds.) RANDOM 1999 and APPROX 1999. LNCS, vol. 1671, pp. 51–62. Springer, Heidelberg (1999)
15. Kesselman, A., Lotker, Z., Mansour, Y., Patt-shamir, B., Schieber, B., Sviridenko, M.: Buffer overflow management in qos switches. SIAM Journal on Computing, 520–529 (2001)
16. Lawler, E.L.: Fast approximation algorithms for knapsack problems. In: Proceedings of the 18th Annual Symposium on Foundations of Computer Science, pp. 206–213. IEEE Computer Society, Washington, DC (1977)
17. Li, F., Sethuraman, J., Stein, C.: Better online buffer management. In: Proceedings of the Eighteenth Annual ACM-SIAM Symposium on Discrete Algorithms, SODA 2007, pp. 199–208. Society for Industrial and Applied Mathematics, Philadelphia (2007)
18. Zhou, Y., Chakrabarty, D., Lukose, R.: Budget constrained bidding in keyword auctions and online knapsack problems. In: Papadimitriou, C., Zhang, S. (eds.) WINE 2008. LNCS, vol. 5385, pp. 566–576. Springer, Heidelberg (2008)

The Multiple Attribution Problem
in Pay-Per-Conversion Advertising

Patrick Jordan, Mohammad Mahdian, Sergei Vassilvitskii, and Erik Vee

Yahoo! Research, Santa Clara, CA 95054
{prjordan,mahdian,sergei,erikvee}@yahoo-inc.com

Abstract. In recent years the online advertising industry has witnessed a shift from the more traditional pay-per-impression model to the pay-per-click and more recently to the pay-per-conversion model. Such models require the ad allocation engine to translate the advertiser's value per click/conversion to value per impression. This is often done through simple models that assume that each impression of the ad stochastically leads to a click/conversion independent of other impressions of the same ad, and therefore any click/conversion can be attributed to the last impression of the ad. However, this assumption is unrealistic, especially in the context of pay-per-conversion advertising, where it is well known in the marketing literature that the consumer often goes through a *purchasing funnel* before they make a purchase. Decisions to buy are rarely spontaneous, and therefore are not likely to be triggered by just the last ad impression. In this paper, we observe how the current method of attribution leads to inefficiency in the allocation mechanism. We develop a fairly general model to capture how a sequence of impressions can lead to a conversion, and solve the optimal ad allocation problem in this model. We will show that this allocation can be supplemented with a payment scheme to obtain a mechanism that is incentive compatible for the advertiser and fair for the publishers.

1 Introduction

In 2009 Internet ad revenues totaled \$22.7B, of which sponsored search and display advertising accounted for 47% and 22%, respectively [14]. Although still a relatively nascent industry, the mechanism for advertising on the internet has evolved considerably over the past two decades. Initially, ads were sold on a purely CPM (cost-per-mille) basis, and it was the number of impressions that determined the payment made by the advertiser. As the marketplace matured, publishers allowed advertisers to pay per click (CPC basis), and, more recently per action [12] or conversions (CPA basis).

Auction mechanisms play a critical role in both of these formats [8] and the celebrated Generalized Second Price (GSP) mechanism has been extensively studied and analyzed [7,1]. A crucial assumption behind the analyses is a simplistic model of user behavior, namely that the probability of the user clicking on the ad is *independent* of the number of times the user has previously viewed

G. Persiano (Ed.): SAGT 2011, LNCS 6982, pp. 31–43, 2011.

the ad. This is equivalent to assuming that showing the ad does not result in changes to future user behavior. The flaw in this reasoning is best illustrated by the fact that an ad loses its effectiveness over time, and the click probabilities are not going to be identical for the first and the thousandth view of the same ad by the same user [11]. If the user has not reacted (via a click or conversion) to an ad after the first 999 impressions, it is highly unlikely that the thousandth one is going to change her mind. Conversely, the first few impressions may result in a superlinear increase in conversion probability (much like having a second friend in a group increases greatly increases the probability of the user joining [2]).

This notion is known as a purchase funnel [3], and has been at the core of the marketing literature for almost a century [15]. In online advertising various, this is recognized as a major issue (see, for example, [6,10,5]), and analytics tools and ad hoc methods have been developed to reflect the consequences of this type of user behavior. For example, the practice of frequency capping [4,9], whereby an advertiser limits the number of exposures of his ad to any user, is a crude way to optimize for the fact that an ad loses its effectiveness after a certain number of views. On the other hand, to the best of our knowledge, current mechanisms do not reward the publisher for displaying an ad that may not result in a click until its second or third view.

As a concrete example, consider an ad that never results in a click on the first impression, but always results in a click on the second impression. (We go through a more elaborate example in the next Section.) In this case, even with perfect click probability estimation the ad will never be shown, since every publisher does a myopic optimization, and the ad in question is guaranteed to have a zero payoff on its first view. In order to create proper incentives to the publishers, the mechanism designer must recognize that a given click or conversion is not simply result of the actions of the last publisher (as it is attributed today), but rather a result of the *aggregate* actions of all of the previous publishers. Therefore, to ensure maximum efficiency, one must attribute the conversion (and the payoffs that go with it) to all of the publishers along the chain.

In this work we mathematically formulate the *multiple attribution problem* and explore the proper method for transforming a bid per conversion to an effective bid per impression to ensure maximum efficiency. We remark that the multiple attribution problem is not only relevant to web advertising scenarios. For example, consider the problem faced by a website designer facing an increase in user traffic. Is that increase due to the last change made on the site, or is it due to the continuous work and the multitude of changes done over the past year. Similarly, suppose a brick and mortar retailer is losing clients to an online merchant. How much of that loss should be attributed to the recent history, and how much to an effect accumulated over a longer time horizon.

In addition to the optimal allocation problem in a multiple attribution setting, we explore the associated pricing problem. This problem is complicated by two constraints: a pay-per-conversion advertiser must pay only when a conversion occurs; and different impressions might be served on different publishers, and therefore it also matters how the payment of the advertiser is split between these publishers. While the first constraint can be satisfied easily, we can only prove that we can simultaneously satisfy both constraints in a special case where the opportunity cost is a constant. Our proof uses the max-flow min-cut theorem.

The rest of this paper is organized as follows: in the next section, we show how attributing a conversion to the last impression can lead to inefficiencies in the market. This motivates a model (defined in Section 3) that assumes that the user follows a Markovian process. The optimal allocation problem for this model is formulated in Section 4 as a Markov Decision Process. We solve the Bellman equations for this process in Section 5, getting a closed-form solution in a special case and a method to compute the values and the effective bid-per-impression in general. In Section 6, we prove that the allocation mechanism admits a pay-per-conversion payment scheme that is incentive compatible for the advertiser and (in the case that the opportunity cost is a constant) fair for the publishers. We conclude in Section 7 with a discussion of how our results can be generalized and applied in practice.

2   Inefficiency of the Last- Impression Attribution Scheme

The model that attributes each conversion to the last impression of the ad is built on the assumption that upon each impression, the user stochastically decides whether or not to purchase the product, independent of the number of times she has previously seen the ad. However, this is not an accurate assumption in practice, and when this assumption is violated, the last-impression attribution scheme can be inefficient. Here we explain this with a simple scenario: focus on one pay-per-conversion advertiser that has a value of $1 per conversion. Assume a user sees the ad of this advertiser four times on average. The probability of converting after viewing the ad for the first time is 0.02, and after the second viewing this probability increases to 0.1. The third and the fourth viewing of the ad will not lead to any conversions. Also, assume that this ad always competes with a pay-per-impression ad with a bid of 4 cents per impression.

First, consider a system that simply computes the average conversion rate of the ad and allocates based on that. This method would estimate the conversion rate of the ad at $(0.02 + 0.1 + 0 + 0)/4 = 0.03$. Therefore, the ad's effective bid per impression is 3 cents and the ad will always lose to the competitor. This is inefficient, since showing the ad twice gives an average expected value of 6 cents per impression, which is more than the competitor.

If we employ frequency capping and restrict the ad to be shown at most twice to each user, the above problem would be resolved, but another problem arises. In this case, the average conversion rate will be $(0.02+0.1)/2 = 0.06$, and the ad will win both impressions. This is indeed the efficient outcome, but let us look at this outcome from the perspective of the publishers. If the two impressions are on different publishers, the first publisher only gets 2 cents per impression in expectation, less than what the competitor pays. This is an unfair outcome, and means that this publisher would have an incentive not to accept this ad, thereby creating inefficiency.

Finally, note that even if the conversion rate is estimated accurately for each impression, still the usual mechanism of allocating based on expected value per impression is inefficient, since it will estimate the expected value per impression at 2 cents for the first impression. This will lose to the 4 cent competitor, and never gives the ad a chance to secure the second, more valuable, impression.

3   The Model

In this section we formalize a model that captures the fact that the user goes through a purchase funnel before buying a product, and therefore the conversion probability of an ad depends on the number of times the ad is shown. We model the user's behavior from the perspective of one pay-per-conversion advertiser A. We have T opportunities to show an ad to the user, where T is a random variable. For simplicity, we assume that T is exponentially distributed. This means that there is a fixed drop-out probability $q \in (0, 1)$, and every time a user visits a page on which an ad can be shown, there is a probability q that she will drop out after that and will not come back to another such page. Every time we have an opportunity to show an ad to this user, we must decide whether to show A's ad or the competitor's ad. Assume the value per impression of the competitor's ad is R. In other words, R is the opportunity cost of showing A's ad. We assume that R is a random variable, and is independently and identically distributed each time. We will present some of our results in the special case that R is a constant (corresponding to the case that A always faces the same competitor with a fixed value), since this case simplifies the math and allows for closed-form solutions.

We assume that the probability that the user converts (buys a product from A) is an arbitrary function of the number of times she has seen A's ad. We denote this probability by λ_j, where j is the number of times the user has seen A's ad. Typically, λ_j is unimodal, i.e., it increases at the beginning to reach a peak, and then decreases, although we will not make any such assumption.

Advertiser A's value per conversion is denoted by v. In the next section, we will discuss the problem of optimal allocation of ad space (to A or to the competitor). This can be viewed as the auctioneer's problem when trying to choose between A and its competitor to maximize social welfare, or A's problem when designing a bidding agent to submit a per-impression bid on its behalf each time. As it turns out, these views are equivalent.

The optimal allocation problem is one side of the multiple attribution problem. The other side is the problem of distributing A's payment among the publisher on which A's ad is displayed. This is an important problem when each of these pages is owned by a possibly different publisher, which is a common case in marketplaces like Google's DoubleClick Ad Exchange or Yahoo!'s Right Media Exchange [13]. We will discuss publisher fairness criteria in Section 6.

4   The Ad Allocation Problem

Given the values of the parameters of the model defined in the previous section (i.e., q, λ_i's, and the distribution of R), the goal of the ad allocation problem is to decide when to show A's ad to maximize the expected social welfare. Here the social welfare is the sum of the values that A and its competitor derive. Another way to look at this problem is to assume that at its core, the ad space is allocated through a second-price pay-per-impression auction[1], and conversion-seeking advertisers like A need to participate in the auction through a bidding agent that bids a per-impression value for each auction. The objective of such a bidding agent is the value to A minus its cost, which is equal to R if A wins. The difference between this objective and the objective of social-welfare maximizing auctioneer is an additive term equal to expectation of the sum of the R values. Therefore, the two optimizations are the same. In this section and the next, we solve this optimization problem by modeling it as a Markov Decision Process (MDP) and solving the corresponding Bellman equations [16]. We will also derive the value that A's bidding agent should bid to achieve the optimal outcome.

MDP formulation. We can define an MDP as follows: for each j, where $j - 1$ represents the number of ad views so far, we have three states a_j, b_j, and c_j. The state a_j represents the probabilistic state right before the next time the user views a page on which an ad can be displayed. This state has a transition with probability q to the *quit* state (which is a terminal state), and another with probability $1 - q$ to b_j. At b_j, the value of R is realized and we need to make a decision between not showing the ad, which would give a reward of R and takes us back to the state a_j, or to show the ad, which would take us to the state c_j. This is a probabilistic state with probability of transition of $(1 - \lambda_j)$ to a_{j+1} (corresponding to the non-conversion event) and probability of transition of λ_j to a terminal *convert* state. The reward of this transition is v (the value of conversion) plus the value of the infinite sequence of alternative ads starting from this point. Since the number of page visits follows an exponential distribution, this value is $v + (1 - q)E[R]/q$. The state a_1 is the starting state. Figure 1 illustrates the process.

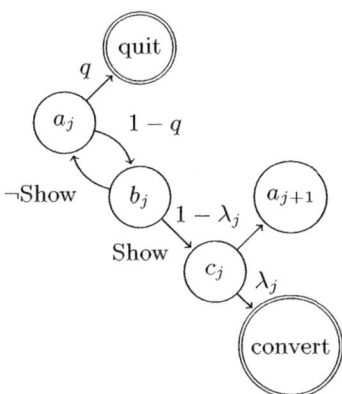

Fig. 1. Multiple Attribution MDP

---

[1] This is the case in marketplaces such as Yahoo!'s Right Media Exchange.

The Bellman Equation. We denote the total social welfare we obtain from this user starting from the state b_j by V_j. At this state, we need to choose between showing the competitor's ad or showing A's ad. In the former case, we immediately get a value of R and with prob. $1 - q$ will be taken back to the state b_j. Therefore, the expected value in this case is $R + (1 - q)V_j$. In the latter case, with probability λ_j a conversion happens, which results in a value of v for the conversion plus $(1 - q)E[R]/q$ for the sequence of competitor ads we can show afterward. With probability $1 - \lambda_j$, we get no conversion and will be taken to the state b_{j+1} with probability $1 - q$. Therefore, the expected value in this case is $\lambda_j(v + (1 - q)r/q) + (1 - \lambda_j)(1 - q)V_{j+1}$, where $r = E[R]$. To summarize:

Proposition 1. *The values V_j of the expected total value starting from the state b_j satisfy the following equation:*

$$V_j = E_R[\max(R + (1 - q)V_j, \lambda_j(v + \frac{(1 - q)r}{q}) + (1 - \lambda_j)(1 - q)V_{j+1})], \quad (1)$$

where $r = E[R]$. The value V_1 indicates the maximum expected social welfare in our model.

5   Computing the Values

In this section, we show how (1) can be simplified to a recurrence relation that can be used to compute V_j's. This recurrence has a simple form, but involves a function that is, in general, non-linear (depending on the distribution of R), and therefore its solution cannot be written in closed form. However, we can do this in the case that R is a constant. Also, we derive the values that a bidding agent that participates in a pay-per-impression auction on behalf of A should bid for each impression.

5.1   The General Recurrence

We start with the Bellman equation (1) and simplify it in each step, eventually writing it in terms of a particular function that captures the effect of q and the distribution of R. First, we rewrite the equation in terms of new variables $W_j := (1 - q)(V_j - r/q)$. Intuitively, W_j is the maximum value starting from the state a_j, minus the value starting from this state without the presence of advertiser A. By replacing V_j's by W_j's in (1) we obtain

$$\frac{W_j}{1 - q} = E[\max(R + W_j, \lambda_j v + (1 - \lambda_j)W_{j+1})] - r. \quad (2)$$

Before simplifying this equation further, notice that this means that in the optimal allocation, the advertiser A wins if and only if $R + W_j \le \lambda_j v + (1 - \lambda_j)W_{j+1}$. Thus,

Proposition 2. *In the optimal allocation, at a point where the user has already seen A's ad $j - 1$ times, the next impression will be allocated to A if and only if the cost of this impression (R) is at most $\lambda_j v + (1 - \lambda_j)W_{j+1} - W_j$.*

To write (2) in a simpler form, we define $h(x) := \mathrm{E}[max(R, x)]$. Clearly, $h(.)$ is a function that only depends on the distribution of R. After subtracting W_j from both sides of (2), we can write this equation as

$$\frac{qW_j}{1-q} = h(\lambda_j v + (1 - \lambda_j)W_{j+1} - W_j) - r. \tag{3}$$

Note that h is by definition a continuous non-decreasing function. For a value $\beta \geq 0$, consider the following equation in terms of the variable x: $qx/(1-q) = h(\beta - x) - r$. At $x = 0$, the right-hand side of this equation is $h(\beta) - r = h(\beta) - h(0) \geq 0$ and the left-hand side is zero. At $x = \beta$, the right-hand side is $h(0) - r = 0$ and the left-hand side is non-negative. Therefore, since the right-hand side of the equation is non-increasing in x, the left-hand side is strictly increasing, and both sides are continuous functions of x, this equation has a unique solution in $[0, \beta]$. We denote the value of this solution by $u(\beta)$.

Proposition 3. *For any value of q and distribution of R, the function $u(.)$ is well-defined, non-decreasing, and continuous, and satisfies $\forall\ \beta :\ u(\beta) \in [0, \beta]$.*

Note that $u(.)$ is defined purely in terms of the distribution of R and the value of q, and in fact, it captures all the information about these parameters that is relevant for the allocation problem. Using this function, (3) can be rewritten as:

$$W_j = u(\lambda_j v + (1 - \lambda_j)W_{j+1}). \tag{4}$$

Obtaining an explicit formula for V_j is only possible if $u(.)$ has a simple form. Unfortunately, this function is often complex and non-linear.[2] However, the above equation gives a straightforward way to compute W_j's numerically: start with a large enough j^* so that $W_{j^*} = 0$, and then move backward to compute W_j for $j = j^* - 1, \ldots, 1$. Such a value of j^* exists in most realistic scenarios; for example, any j^* such that for all $j > j^*$, $\lambda_j v$ is less than the minimum of R (say, the value of the reserve price) suffices. To summarize,

Theorem 1. *Let W_j's be the values computed using (4). Then the optimal allocation can be obtained by submitting a per-impression bid of $bid_j := \lambda_j v + (1 - \lambda_j)W_{j+1} - W_j$ on behalf of A in a state where the user has already seen the ad $j - 1$ times. The social welfare achieved by this mechanism is $r/q + W_1/(1-q)$.*

5.2 Closed-form Solution for Constant R

In the case that R is a constant r, we can significantly simplify the recurrence (4). First, note that by definition, $h(x) = \max(r, x)$. Therefore, $u(\beta)$ is the solution of the equation $qx/(1-q) = \max(\beta - x - r, 0)$. It is easy to see that when $\beta \geq r$, the solution of the above equation is $(1-q)(\beta-r)$, and when $\beta < r$, this solution is zero. Therefore, $u(\beta) = (1 - q)\max(\beta - r, 0)$. This gives

$$W_j = (1 - q)\max(\lambda_j v - r + (1 - \lambda_j)W_{j+1}, 0). \tag{5}$$

---

[2] For the uniform distribution, $u(.)$ is the solution of a quadratic equation; for the exponential distribution $u(.)$ cannot be written in closed form.

To solve this recurrence, we can expand W_{j+1} in the above expression, and iterative. This results in the following explicit expression, which can be easily verified by induction using the above recurrence (5):

$$W_j = (1 - q) \max_{l \geq j-1} \left\{ \sum_{s=j}^{l} (\lambda_s v - r) \psi_s / \psi_j \right\}, \tag{6}$$

where $\psi_i := \prod_{t=1}^{i-1}(1 - q)(1 - \lambda_t)$ is the probability that the user visits at least i times and each time (except possibly the last time) does not convert on A's ad.

In the above expression an empty sum is defined as zero and an empty product is defined as one. So the final solution can be written as follows:

$$V_1 = \frac{r}{q} + \max_{l \geq 0} \left\{ \sum_{s=1}^{l} (\lambda_s v - r) \psi_s \right\}. \tag{7}$$

To summarize:

Theorem 2. *Let l^* be the value of l that achieves the maximum in (7). Then in the optimal allocation, A's ad is shown until the user converts or she sees the ad l^* times. After a conversion happens or this number of ad views is reached, the competitors ad is shown.*

6   Pricing and Publisher Fairness

In the last section, we showed how we can design a bidding agent that translates the advertiser A's values into an effective bid per impression every time there is an advertising opportunity. If this advertiser could pay per impression (we will call this the *pay-per-impression scenario*), this would have been the end of the story: on each auction, we would use the bidding agent to bid, and if A wins based on this bid, she will pay the value of the competitor's bid R. This value would be disbursed to the publisher responsible for that impression. It is not hard to see that this scheme is equivalent to the VCG mechanism from A's perspective (i.e., it allocates the good optimally and charges A the externality she imposes on others), and therefore A has incentive to truthfully report her value per conversion v. Also, the mechanism seems intuitively "fair" for publishers.

However, some advertisers are strict pay-per-conversion advertisers. For these advertisers the payment scheme should satisfy the following property:

Ex-Post Individual Rationality (Ex-Post IR): At any outcome where a conversion has not happened, A does not pay anything. At an outcome where a conversion has happened, A pays at most her value per conversion v.

In addition to the above, we require *Efficiency* (getting the optimal allocation characterized in the last section) and *Incentive Compatibiltiy (IC)*. Note that these two properties imply that in expectation, the amount the advertiser must be charged is the externality it imposes on the others. This is equal to the sum of R on impressions where A's ad is shown. In other words, in expectation, the mechanism should charge the same amount as in the pay-per-impression scenario. The challenge is to implement this while respecting Ex-Post IR.

As we will show in the next subsection, this can be achieved with a simple uniform pricing. This method is simple and works well when there is only one publisher (so there is no issue of fairness). In Section 6.2, we define and study a natural notion of fairness when there are multiple publishers. We will show that there are instances where the uniform pricing method *cannot* result in a fair distribution of payments to publishers. On the positive side, in the case of constant R, we will show that the problem can be formulated as a network flow problem, and will use the maximum-flow minimum-cut theorem to prove that a fair, ex-post IR, and incentive compatible payment rule always exists. As this is a special case of the max-flow min-cut problem, we will also be able to give a simpler and faster algorithm for computing the payments.

6.1   The Uniform Pricing Method

The idea of the uniform pricing method is to charge the same amount for all conversions, regardless of how many ad impressions A gets prior to the conversion. This uniform cost is set at a level to get the advertiser to pay the right amount in expectation. Using the optimality of the allocation, we can show that this scheme satisfies Ex-Post IR. We first illustrate this in the ase of constant R.

First, note that $W_1 \geq 0$. This can be seen directly from the defintion of W_1 and V_1 as the optimal solution of the MDP, or from Equation (6). Let ℓ be the value that maximizes (7). Thus we have $\sum_{s=1}^{\ell}(\lambda_s v - r)\psi_s \geq 0$, or, equivalently:

$$v \geq r \cdot \frac{\sum_{s=1}^{\ell} \psi_s}{\sum_{s=1}^{\ell} \lambda_s \psi_s}. \tag{8}$$

Now consider the expected externality imposed by the advertiser on others. The probability that the ad is shown exactly ℓ times is ψ_ℓ. For some $s < \ell$ the probability that it is shown exactly s times is $\psi_s - \psi_{s+1}$. Therefore, the total expected externality imposed on others by the advertiser is $r \sum_{s=1}^{\ell} s\psi_s - r \sum_{s=1}^{\ell-1} s\psi_{s+1} = r \sum_{s=1}^{\ell} \psi_s$. On the other hand, the probability that the user converts after the i-th view is $\lambda_i \psi_i$. Thus the total probability of conversion is $\sum_{s=1}^{\ell} \lambda_s \psi_s$. Therefore if for each conversion, we charge the advertiser $r \cdot \frac{\sum_{s=1}^{\ell} \psi_s}{\sum_{s=1}^{\ell} \lambda_s \psi_s}$, the expected payment of the advertiser will be equal to the externality it imposes on others (i.e., the IC payment). Also, by Equation (8), the payment per conversion is at most v, and hence Ex-Post IR is also satisfied.

This method can be applied in the general case (when R is not a constant): On any conversion, independent of the history of impressions that lead to this conversion, we charge the advertiser an amount equal to

$$price := \frac{E}{P_{conv}}, \tag{9}$$

where E is the expected total externality that A imposes on the competitors, and P_{conv} is the overall probability of conversion for A. By definition, with this charging scheme in expectation A pays $price \times P_{conv} = E$, which is the incentive compatible payment. To show that the above price satisfies Ex-Post IR, we compare this scenario with the pay-per-impression scenario defined at the beginning of Section 6. It is easy to see that the outcome in both cases is the same and A's payment is also the same in both scenarios in expectation. Therefore, since A's utility in the pay-per-impression scenario is non-negative, it is non-negative here too, implying that $price \leq v$.

6.2   Publisher Fairness

There are two main motivations for studying the multiple attribution problem: the first is to ensure the efficiency of the market outcome, and the second is to ensure that each ad publisher who has contributed in the purchase funnel that has lead to a conversion gets a fair share of the conversion price. So far, we have been concerned with the first aspect: efficiency. In this section we turn to the second aspect: fairness among publishers.

We first need to define the notion of fairness for publishers. Our definition is motivated by the hypothetical pay-per-impression scenario defined at the beginning of Section 6. In this scenario, each publisher who displays A's ad, receives a payment equal to the opportunity cost of this impression. We define fairness in our setting by requiring the same payments *in expectation*:

Publisher Fairness. For each i, the expected value the i'th publisher receives from A is equal to the expected opportunity cost (R) of this publisher conditioned on A winning.

Note that this is a natural property to require, since it is natural for the publisher to request to be paid an amount at least equal to the opportunity cost of the impressions it provides (if this is not satisfied, the publisher could refuse to accept pay-per-conversion advertisers), and since the advertiser's payment is the total externality it imposes on the competitors, no publisher cannot hope to get more than its expected opportunity cost without hurting another publisher.

As we will show below, Publisher Fairness imposes a non-trivial constraint on the payments. In fact, for some payment rules like the uniform scheme defined in Section 6.1, it is not possible to distribute the payment among the publishers in a way that satisfies Publisher Fairness. To illustrate this and prepare for the result of the next section (showing that for constant R, there is a payment rule satisfying Publisher Fairness), we focus on the case of constant R, and introduce some notations.

We number the publishers in the order the user visits ad-bearing pages. Let x_{ij} be the payout to publisher j if the conversion occurs after precisely i views. This quantity is only defined for $i \geq j$, since for $i < j$, the user will either never visit publisher j, or visit this publisher after she is already converted. Also, we only define the variables x_{ij} for $i, j \leq \ell$ where ℓ is the index that maximizes the value in Equation (7), since after this index, A's ad will not be shown.

We can write our desired properties in terms of the $x_{i,j}$ variables. First, we formulate the Publisher Fairness property. For every publisher $j = 1, \ldots, \ell$, conditioned the user visiting j, the probability that it visits exactly i publishers ($i \geq j$) and then it converts is precisely $\psi_i \lambda_i / \psi_j$. Thus, the total expected payment to j, conditioned on the user visiting j can be written as $\sum_{i \geq j} x_{ij} \psi_i \lambda_i / \psi_j$. Therefore the Publisher Fairness property can be written as follows:

$$\forall j : \quad \sum_{i \geq j} x_{ij} \psi_i \lambda_i / \psi_j = r. \tag{10}$$

This property also implies that the payments are incentive compatible: since for each publisher the total payment of A is equal to the externality A imposes on its competitors on this publisher, the total expected payment of A is also equal to the total expected externality it imposes on the competitors. Therefore, all that remains is to formulate the Ex-Post IR property. The total payment of A in case a conversion happens after precisely i impressions is $\sum_{j \leq i} x_{ij}$. Therefore, Ex-Post IR is equivalent to the following.

$$\forall i : \quad \sum_{j \leq i} x_{ij} \leq v. \tag{11}$$

We leave the proof of the following theorem to the full version of the paper.

Theorem 3. *Consider the optimal allocation with the uniform pricing rule defined in the last section. There are instances in this mechanism where there is no way to distribute the advertiser's payment among the publishers in a way that satisfies Publisher Fairness.*

6.3   Fair Payments via Max-Flow Min-Cut

The main result of this section is the existence of a fair payment rule when R is constant. The proof (omitted due to lack of space) is based on formulating the constraints as flow constrains and using the max-flow min-cut theorem.

Theorem 4. *When R is a constant, the optimal allocation rule can be supplemented with a payment scheme that satisfies Incentive Compatibility, Ex-Post Individual Rationality, and Publisher Fairness.*

7   Conclusion

In this work we showed how myopic optimization by the publishers can lead to inefficient allocations in the case when displaying an impression for an advertiser changes the user's conversion probability on subsequent visits. We formulated the optimal allocation problem in this setting as a Markov Decision Process and derived the optimal allocation and a way to translate the advertiser's per-conversion value to bids for each impression. We then studied how the advertiser should be charged in the case of a conversion, and how this charge should be split between publishers in order to achieve incentive compatibility and individula rationality for the advertiser and fairness for the publishers.

Our model is fairly general, yet simple enough to be practical. Perhaps the most important assumption in the model, which is sometimes inaccurate, is that we assumed that the conversion probability depends *only* on the number of views, and *not* on the identity of the publishers that display the ad to the user. One can imagine generalizing this notion, in a manner similar to the separable click-through rate model of sponsored search – that the probability of conversion is a separable function of the number of user visits and the identity of the publisher. Another way to relax this assumption is to assume each publisher has a weight, and the conversion probability of the user at each point is a function of the total weight of the publishers that have shown the ad to the user. When all weights are 1, this model reduces to our identical publisher model. We leave this as an interesting open problem.

References

1. Aggarwal, G., Goel, A., Motwani, R.: Truthful auctions for pricing search keywords. In: Proceedings of the 7th ACM Conference on Electronic Commerce, EC 2006, pp. 1–7. ACM, New York (2006)
2. Backstrom, L., Huttenlocher, D., Kleinberg, J., Lan, X.: Group formation in large social networks: membership, growth, and evolution. In: Proceedings of the 12th ACM SIGKDD International Conference on Knowledge Discovery and Data Mining, pp. 44–54. ACM Press, New York (2006)
3. Barry, T.E.: The development of the hierarchy of effects: An historical perspective. Current Issues and Research in Advertising 10, 251–295 (1987)
4. Danaher, P.J.: Advertising models. In: Wierenga, B. (ed.) Handbook of Marketing Decision Models. International Series in Operations Research and Management Science, vol. 121, pp. 81–106. Springer, Heidelberg (2008)
5. Dreller, J.: In the trenches SEM pre-click & post-click double feature: Conversion attribution and Q&A with analytics guru Eric Peterson. Search Engine Land (November 2008),
   http://searchengineland.com/qa-with-eric-t-peterson-on-web-analytics-for-sem-and-a-look-into-conversion-attribution-15544
6. Dreller, J.: Research brief: Conversion attribution. Fuor Digital (December 2008),
   http://fuor.com/resources/leadership/briefs/conversion-attribution-research-brief
7. Edelman, B., Ostrovsky, M., Schwarz, M.: Internet advertising and the generalized second-price auction: Selling billions of dollars worth of keywords. American Economic Review 97(1), 242–259 (2007)
8. Fain, D.C., Pedersen, J.O.: Sponsored search: A brief history. Bulletin of the American Society for Information Science and Technology 32(2), 12–13 (2006)
9. Farahat, A.: Privacy preserving frequency capping in internet banner advertising. In: Proceedings of the 18th International Conference on World Wide Web, WWW 2009, pp. 1147–1148. ACM, New York (2009)
10. ClearSaleing Inc. Attribution management,
    http://www.attributionmanagement.com/
11. Lewis, R.: Where's the 'wear-out'? Working paper. Yahoo! Research (2011)
12. Mahdian, M., Tomak, K.: Pay-per-action model for online advertising. In: Deng, X., Graham, F.C. (eds.) WINE 2007. LNCS, vol. 4858, pp. 549–557. Springer, Heidelberg (2007)

13. Muthukrishnan, S.: Ad exchanges: Research issues. In: Leonardi, S. (ed.) WINE 2009. LNCS, vol. 5929, pp. 1–12. Springer, Heidelberg (2009)
14. PricewaterhouseCoopers. IAB internet advertising revenue report: 2009 full-year results (2010),
    http://www.iab.net/media/file/IAB-Ad-Revenue-Full-Year-2009.pdf
15. Strong, E.K.: The Psychology of Selling Advertising. McGraw-Hill, New York (1925)
16. White, D.J.: Markov Decision Processes. Wiley, Chichester (1993)

On Communication Protocols That Compute Almost Privately

Marco Comi[1], Bhaskar DasGupta[1],
Michael Schapira[2], and Venkatakumar Srinivasan[1]

[1] Department of Computer Science, University of Illinois at Chicago, IL 60607
ingmarco85@gmail.com, {bdasgup,vsrini7}@uic.edu
[2] Department of Computer Science, Princeton University, Princeton, NJ 08540
ms7@cs.princeton.edu

Abstract. A traditionally desired goal when designing auction mechanisms is *incentive compatibility*, *i.e.*, ensuring that bidders fare best by truthfully reporting their preferences. A *complementary* goal, which has, thus far, received significantly less attention, is to *preserve privacy*, *i.e.*, to ensure that bidders reveal no more information than necessary. We further investigate and generalize the approximate privacy model for two-party communication recently introduced by Feigenbaum *et al.* [8]. We explore the privacy properties of a natural class of communication protocols that we refer to as *"dissection protocols"*. Dissection protocols include, among others, the bisection auction in [9,10] and the bisection protocol for the millionaires problem in [8]. Informally, in a dissection protocol the communicating parties are restricted to answering simple questions of the form *"Is your input between the values α and β (under a pre-defined order over the possible inputs)?"*.

We prove that for a large class of functions called *tiling functions*, which include the 2nd-price Vickrey auction, there *always* exists a dissection protocol that provides a *constant average-case privacy approximation ratio* for uniform or "almost uniform" probability distributions over inputs. To establish this result we present an interesting connection between the approximate privacy framework and basic concepts in computational geometry. We show that such a good privacy approximation ratio for tiling functions does *not*, in general, exist in the *worst case*. We also discuss extensions of the basic setup to more than two parties and to non-tiling functions, and provide calculations of privacy approximation ratios for two functions of interest.

Keywords: Approximate Privacy, Auctions, Communication Protocols.

1 Introduction

Consider the following interaction between two parties, Alice and Bob. Each of the two parties, Alice and Bob, holds a *private* input, x_{bob} and y_{alice} respectively, not known to the other party. The two parties aim to compute a function f of the two private inputs. Alice and Bob alternately query each other to make available a *small* amount of information about their private inputs, *e.g.*, an answer to a range query on their private inputs or a few bits of their private inputs. This

G. Persiano (Ed.): SAGT 2011, LNCS 6982, pp. 44–56, 2011.

process ends when each of them has seen enough information to be able to compute the value of $f(x_{\text{bob}}, y_{\text{alice}})$. The central question that is the focus of this paper is:

Can we design a communication protocol whose execution reveals, to both Alice and Bob, as well as to any eavesdropper, as little information as possible about other the other's private input beyond what is necessary to compute the function value?

Note that there are two conflicting constraints: Alice and Bob need to communicate sufficient information for computing the function value, but would prefer not to communicate too much information about their private inputs. This setting can be generalized in an obvious manner to $d > 1$ parties $\text{party}_1, \text{party}_2, \ldots, \text{party}_d$ computing a d-ary f by querying the parties in round-robin order, allowing each party to broadcast information about its private input (via a public communication channel).

Privacy preserving computational models such as the one described above have become an important research area due to the increasingly widespread usage of sensitive data in networked environments, as evidenced by distributed computing applications, game-theoretic settings (*e.g.*, auctions) and more. Over the years computer scientists have explored many *quantifications* of privacy in computation. Much of this research focused on designing *perfectly* privacy-preserving protocols, *i.e.*, protocols whose execution reveals *no* information about the parties' private inputs beyond that implied by the outcome of the computation. Unfortunately, perfect privacy is often either *impossible*, or *infeasibly costly* to achieve. To overcome this, researchers have also investigated various notions of *approximate privacy* [7,8].

In this paper, we adopt the approximate privacy framework of [8] that quantifies approximate privacy via the *privacy approximation ratios* (PARs) of protocols for computing a deterministic function of two private inputs. *Informally*, PAR captures the objective that an observer of the transcript of the entire protocol will not be able to distinguish the real inputs of the two communicating parties from *as large a set as possible* of other inputs. To capture this intuition, [8] makes use of the machinery of communication-complexity theory to provide a geometric and combinatorial interpretation of protocols. [8] formulates both the worst-case and the average-case version of PARs and studies the tradeoff between privacy preservation and communication complexity for several functions.

1.1 Economic Motivation

The original motivation of this line of research, as explained in [8], comes from privacy concerns in auction theory. A traditionally desired goal when designing an auction mechanism is to ensure that it is *incentive compatible*, *i.e.*, bidders fare best by truthfully reporting their preferences. More recently, attention has also been given to the *complementary* goal of preserving the privacy of the bidders (both with respect to each other and to the auctioneer/mechanism). Take, for example, the famous 2nd-price Vickrey auction of an item. Consider the ascending-price English auction, *i.e.*, the straightforward protocol in which the price of the item is incrementally increased, bidders drop out when their value for the item is exceeded until the identity of winner is determined, and the winner is then charged the second-highest bid. Intuitively, this protocol reveals more information than what is *absolutely necessary* to compute the outcome, *i.e.*, the identity of the winner and the second-highest bid. Specifically, observe under the ascending-price English auction not only will the value of the second-highest bidder be revealed, but so will the values of all other bidders but the winner.

Can we design communication protocols which implement the 2nd-price Vickrey auction in an (approximately) privacy-preserving manner? Can we design such protocols that are computationally- or communication-efficient? These sort of questions motivate our work. We consider a setting that captures applications of the above type, and explore the privacy-preservation and communication-complexity guarantees achievable in this setting.

2   Summary of Our Contributions

Any investigation of approximate privacy for multi-party computation starts by defining how we quantify approximate privacy. In this paper, we use the combinatorial framework of [8] for quantification of approximate privacy for two parties via PARs and present its natural extension to three or more parties. Often, parties' inputs have a natural ordering, *e.g.*, the private input of a party belongs to some range of integers $\{L, L+1, \ldots, M\}$ (as is the case when computing, say, the maximum or minimum of two inputs). When designing protocols for such environments, a natural restriction is to only allow the protocol to ask each party questions of the form *"Is your input between the values α and β (under this natural order over possible inputs)?"*. We refer to this type of protocols as *dissection protocols* and study the privacy properties of this natural class of protocols. We note that the bisection and c-bisection protocols for the millionaires problem and other problems in [8], as well as the bisection auction in [9,10], all fall within this category of protocols. Our findings are summarized below.

Average- and worst-case PARs for tiling functions for two party computation. We first consider a broad class of functions, namely the *tiling functions*, that encompasses several well-studied functions (*e.g.*, Vickrey's second-price auctions). Informally, a two-variable tiling function is a function whose output space can be viewed as a collection of disjoint combinatorial rectangles in the two-dimensional plane, where the function has the same value within each rectangle. A first natural question for investigation is to classify those tiling functions for which there exists a perfectly privacy-preserving dissection protocol. We observe that for every Boolean tiling functions (*i.e.*, tiling functions which output binary values) *this is indeed the case*. In contrast, for tiling functions with a range of just three values, perfectly privacy-preserving computation is no longer necessarily possible (even when not restricted to dissection protocols).

We next turn our attention to PARs. We prove that for *every* tiling function there exists a dissection protocol that achieves a constant PAR in the average case (that is, when the parties' private values are drawn from an uniform or *almost* uniform probability distribution). To establish this result, we make use of results on the binary space partitioning problems studied in the computational geometry literature. We complement this positive result for dissection protocols with the following negative result: *there exist tiling functions for which no dissection protocol can achieve a constant* PAR *in the worst-case.*

Extensions to non-tiling functions and three-party communication. We discuss two extensions of the above results. We explain how our constant average-case PAR result for tiling functions can be extended to a family of "almost" tiling functions. In addition, we consider the case of *more than two* parties. We show that in this setting it is *no longer true* that for every tiling function there exists a dissection protocol that achieves a constant PAR in the average case. Namely, we exhibit a three-dimensional tiling function for which *every* dissection protocol exhibits *exponential* average- and worst-case PARs, *even when an unlimited number of communication steps is allowed.*

PARs for the set covering and equality functions. [8] presents bounds on the average-case and the worst-case PARs of the bisection protocol — a special case of dissection protocols — for several functions. We analyze the PARs of the bisection protocol for two well-studied Boolean functions: the set-covering and equality functions; the equality function provides a useful testbed for evaluating privacy preserving protocols [3] [11, Example 1.21] and set-covering type of functions are useful for studying the differences between deterministic and non-deterministic communication complexities [11]. We show that, for both functions, the bisection protocol *fails to achieve* good PARs in both the average- and the worst-case.

3 Summary of Prior Related Works

3.1 Privacy-Preserving Computation

Privacy-preserving computation has been the subject of extensive research and has been approached from information-theoretic [3], cryptographic [5], statistical [12], communication complexity [13,16], statistical database query [7] and other perspectives [11]. Among these, most relevant to our work is the approximate privacy framework of Feigenbaum *et al.* [8] that presents a metric for quantifying privacy preservation building on the work of Chor and Kushilevitz [6] on characterizing perfectly privately computable computation and on the work of Kushilevitz [13] on the communication complexity of perfectly private computation. The bisection, c-bisection and bounded bisection protocols of [8] fall within our category of dissection protocol since we allow the input space of each party to be divided into two subsets of arbitrary size. There are also some other formulations of perfectly and approximately privacy-preserving computation in the literature, but they are inapplicable in our context. For example, the differential privacy model (see [7]) approaches privacy in a different context via adding noise to the result of a database query in such a way as to preserve the privacy of the individual records but still have the result convey nontrivial information,

3.2   Binary Space Partition (BSP)

BSPs present a way to implement a *geometric divide-and-conquer* strategy and is an extremely popular approach in numerous applications such as hidden surface removal, ray-tracing, visibility problems, solid geometry, motion planning and spatial databases. However, to the best of our knowledge, a connection between BSPs bounds such as in [14,15,2,4] and approximate privacy has not been explored before.

4   The Model and Basic Definitions

4.1   Two-Party Approximate Privacy Model of [8]

We have two parties party_1 and party_2, having binary strings x_1 and x_2 respectively, which represents their private values in some set $\mathcal{U}^{\text{in}}$. The common goal of the two parties is to compute the value $f(x_1, x_2)$ of a given public-knowledge function f. Before a communication protocol P starts, each party_i initializes its *"set of maintained inputs"* $\mathcal{U}_i^{\text{in}}$ to $\mathcal{U}^{\text{in}}$. In one step of communication, one party transmits a bit indicating in which of two parts of its input space its private input lies. The other party then updates its set of maintained inputs accordingly. The very last information transmitted in the protocol P contains the value of of $f(x_1, x_2)$. The final transcript of the protocol is denoted by $s(x_1, x_2)$.

Denoting the domain of outputs by $\mathcal{U}^{\text{out}}$, any function $f : \mathcal{U}^{\text{in}} \times \mathcal{U}^{\text{in}} \mapsto \mathcal{U}^{\text{out}}$ can be visualized as $|\mathcal{U}^{\text{in}}| \times |\mathcal{U}^{\text{in}}|$ matrix with entries from $\mathcal{U}^{\text{out}}$ in which the first dimension represents the possible values of party_1, ordered by some permutation Π_1, while the second dimension represents the possible values of party_2, ordered by some permutation Π_2, and each entry contains the value of f associated with a particular set of inputs from the two parties. This matrix will be denoted by $A_{\Pi_1, \Pi_2}(f)$, or sometimes simply by A. We present the following definitions from [11,8].

Definition 1 (Regions, partitions). *A region of A is any subset of entries in A. A partition of A is a collection of disjoint regions in A whose union is A.*

Definition 2 (Rectangles, tilings, refinements). *A rectangle in A is a sub-matrix of A. A tiling of A is a partition of A into rectangles. A tiling T_1 of A is a refinement of another tiling T_2 of A if every rectangle in T_1 is contained in some rectangle in T_2.*

Definition 3 (Monochromatic, maximal monochromatic and ideal monochromatic partitions). *A region R of A is monochromatic if all entries in R are of the same value. A monochromatic partition of A is a partition with only monochromatic regions. A monochromatic region of A is a maximal monochromatic region if no monochromatic region in A properly contains it. The ideal monochromatic partition of A consists of the maximal monochromatic regions.*

Definition 4 (Perfect privacy). *Protocol P achieves perfect privacy if, for every two sets of inputs (x_1, x_2) and (x_1', x_2') such that $f(x_1, x_2) = f(x_1', x_2')$, it holds that $s(x_1, \ x_2) = s(x'_1, \ x'_2)$. Equivalently, a protocol P for f achieves perfectly privacy if the monochromatic tiling induced by P is the ideal monochromatic partition of $A(f)$.*

Definition 5 (Worst case and average case PAR of a protocol P). *Let $R^P(x_1, x_2)$ be the monochromatic rectangle containing the cell $A(x_1, x_2)$ induced by P, $R^I(x_1, x_2)$ be the monochromatic region containing the cell $A(x_1, y_1)$ in the ideal monochromatic partition of A, and $\mathcal{D}$ be a probability distribution over the space of inputs. Then P has a worst-case PAR of α_{worst} and an average case PAR of $\alpha_{\mathcal{D}}$ under distribution $\mathcal{D}$ provided*[1]

$$\alpha_{\text{worst}} = \max_{(x_1,x_2)\in \mathcal{U}^{\text{in}}\times\mathcal{U}^{\text{in}}} \frac{|\,R^I(x_1,x_2)\,|}{|\,R^P(x_1,x_2)|} \ \ and \ \ \alpha_{\mathcal{D}} = \sum_{(x_1,x_2)\in \mathcal{U}^{\text{in}}\times\mathcal{U}^{\text{in}}} \Pr_{\mathcal{D}}\left[x_1 \ \& \ x_2\right]\frac{|R^I(x_1,x_2)|}{|R^P(x_1,x_2)|}$$

Definition 6 (PAR for a function). *The worst-case (average-case) PAR for a function f is the minimum, over all protocols P for f, of the worst-case (average-case) PAR of P.*

Extension to Multi-party Computation In the multi-party setup, we have $d > 2$ parties $\mathsf{party}_1, \mathsf{party}_2, \ldots, \mathsf{party}_d$ computing a d-ary function $f : (\mathcal{U}^{\text{in}})^d \mapsto \mathcal{U}^{\text{out}}$. Now, f can be visualized as $|\mathcal{U}^{\text{in}}| \times \cdots \times |\mathcal{U}^{\text{in}}|$ matrix $A_{\Pi_1,\ldots,\Pi_d}(f)$ (or, sometimes simply by A) with entries from $\mathcal{U}^{\text{out}}$ in which the ith dimension represents the possible values of party_i ordered by some permutation Π_i, and each entry of A contains the value of f associated with a particular set of inputs from the d parties. Then, all the previous definitions can be naturally adjusted in the obvious manner, *i.e.*, the input space as a d-dimensional space, each party maintains the input partitions of all other $d - 1$ parties, the transcript of the protocol s is a d-ary function, and rectangles are replaced by d-dimensional hyper-rectangles (Cartesian product of d intervals).

4.2 Dissection Protocols and Tiling Functions for 2-Party Computation

Often in a communication complexity settings the input of each party has a natural ordering, *e.g.*, the set of input of a party from $\{0,1\}^k$ can represent the numbers $0, 1, 2, \ldots, 2^k - 1$ (as is the case when computing the maximum/minimum of two inputs, in the millionaires problem, in second-price auctions, and more). When designing protocols for such environments, a natural restriction is to only the allow protocols such that each party asks questions of the form *"Is your input between a and b (in this natural order over possible inputs)?"*, where $a, b \in \{0,1\}^k$. Notice that after applying an appropriate permutation to the inputs, such a protocol divides the input space into two (not necessarily equal) halves. Below, we formalize these types of protocols as *"dissection protocols"*.

---

[1] The notation $\Pr_{\mathcal{D}}[\mathcal{E}]$ denotes the probability of an event $\mathcal{E}$ under distribution $\mathcal{D}$.

Definition 7 (contiguous subset of inputs). *Given a permutation Π of $\{0,1\}^k$, let $\prec_\Pi$ denote the total order over $\{0,1\}^k$ that Π induces, i.e., $\forall a, b \in \{0,1\}^k$, $a \prec_\Pi b$ provided b comes after a in Π. Then, $I \subseteq \{0,1\}^k$ contiguous with respect to Π if $\forall a, b \in I$, $\forall c \in \{0,1\}^k : a \prec_\Pi c \prec_\Pi b \Longrightarrow c \in I$.*

Definition 8 (dissection protocol). *Given a function $f : \{0,1\}^k \times \{0,1\}^k \mapsto \{0,1\}^t$ and permutations Π_1, Π_2 of $\{0,1\}^k$, a protocol for f is a dissection protocol with respect to (Π_1, Π_2) if, at each communication step, the maintained subset of inputs of each party$_i$ is contiguous with respect to Π_i.*

Observe that *every* protocol P can be regarded as a dissection protocol with respect to *some* permutations over inputs by simply constructing the permutation so that it is consistent with the way P updates the maintained sets of inputs. However, *not* every protocol is a dissection protocol with respect to *specific* permutations. Consider, for example, the case that both Π_1 and Π_2 are the permutation over $\{0,1\}^k$ that orders the elements from lowest to highest binary values. Observe that a protocol that is a dissection protocol with respect to these permutations *cannot* ask questions of the form "Is your input odd or even?", for these questions partition the space of inputs into *non-contiguous* subsets with respect to (Π_1, Π_2).

A special case of interest of the dissection protocol is the "bisection type" protocols that have been investigated in the literature in many contexts [8,10].

Definition 9 (bisection, c-bisection and bounded-bisection protocols). *For a constant $c \in \left[\frac{1}{2}, 1\right)$, a dissection protocol with respect to the permutations (Π_1, Π_2) is called a c-bisection protocol provided at each communication step each party$_i$ partitions its input space of size z into two halves of size cz and $(1-c)z$. A bisection protocol is simply a $\frac{1}{2}$-bisection protocol. For an integer valued function $g(k)$ such that $0 \le g(k) \le k$, bounded-bisection$_{g(k)}$ is the protocol that runs a bisection protocol with $g(k)$ bisection operations followed by a protocol (if necessary) in which each party$_i$ repeatedly partitions its input space into two halves one of which is of size exactly one.*

Definition 10 (tiling and non-tiling functions). *A function $f : \{0,1\}^k \times \{0,1\}^k \mapsto \{0,1\}^t$ is called a tiling function with respect to two permutations (Π_1, Π_2) of $\{0,1\}^k$ if the monochromatic regions in $A_{\Pi_1, \Pi_2}(f)$ form a tiling, and the number of monochromatic regions in this tiling is denote by $\mathsf{r}_f(\Pi_1, \Pi_2)$. Conversely, f is a non-tiling function if f is not a tiling function with respect to every pair of permutations (Π_1, Π_2) of $\{0,1\}^k$.*

Note that a function f that is tiling function with respect to permutations (Π_1, Π_2) may not be a tiling function with respect to a different set of permutations (Π_1', Π_2'). Also, a function f can be a tiling function with respect to two distinct permutation pairs (Π_1, Π_2) and (Π_1', Π_2') with a different number of monochromatic regions. Thus, indeed we need Π_1 and Π_2 in the definition of tiling functions and r_f.

Extensions to Multi-party Computation. For the multi-party computation model involving $d > 2$ parties, the d-ary tiling function f has a permutation Π_i of $\{0,1\}^k$ for each ith argument of f (or, equivalently for each party$_i$). A dissection protocol is generalized to a "round robin" dissection protocol in the following manner. In one "mega" round of communications, parties communicate in a fixed order, say party$_1$, party$_2$, ..., party$_d$, and the mega round is repeated if necessary. Any communication by any party is made available to *all* the other parties. Thus, each communication of the dissection protocol partitions a d-dimensional space by an appropriate *set* of $(d-1)$-dimensional hyperplanes, where the missing dimension in the hyperplane correspond to the index of the party communicating.

5   Two-Party Dissection Protocol for Tiling Functions

5.1   Boolean Tiling Functions

Lemma 1. *Any Boolean tiling function* $f\colon \{0,1\}^k \times \{0,1\}^k \mapsto \{0,1\}$ *with respect to some two permutations* (Π_1, Π_2) *can be computed in a perfectly privacy-preserving manner by a dissection protocol with respect to* (Π_1, Π_2).

Remark 1. The claim of Lemma 1 is false if f outputs three values.

5.2   Average and Worst Case PAR for Non-Boolean Tiling Functions

Let $f : \{0,1\}^k \times \{0,1\}^k \mapsto \{0,1\}^t$ be a given tiling function with respect to permutations (Π_1, Π_2). Neither the c-bisection nor the bounded-bisection protocol performs well in terms of average PAR on arbitrary tiling functions. In this section, we show that *any* tiling function f admits a dissection protocol that has a *small constant* average case PAR. Moreover, we show that this result *cannot* be extended to the case of worst-case PARs.

Constant Average-case PAR for Non-Boolean Functions. Let D_u denote the uniform distribution over all input pairs. We define the notion of a c-approximate uniform distribution $\mathsf{D}_u^{\sim c}$; note that $\mathsf{D}_u^{\sim 0} \equiv \mathsf{D}_u$.

Definition 11 (c-approximate uniform distribution). *A c-approximate uniform distribution $\mathsf{D}_u^{\sim c}$ is a distribution in which the probabilities of the input pairs are close to that for the uniform distribution as a linear function of c, namely* $\max_{(\mathbf{x},\mathbf{y}),\,(\mathbf{x}',\mathbf{y}') \in \{0,1\}^k \times \{0,1\}^k} \left| \Pr_{\mathsf{D}_u^{\sim c}}[\mathbf{x} \,\&\, \mathbf{y}] - \Pr_{\mathsf{D}_u^{\sim c}}[\mathbf{x}' \,\&\, \mathbf{y}'] \right| \leq c\,2^{-2k}$.

Theorem 1.
(a) *A tiling function f with respect to permutations (Π_1, Π_2) admits a dissection protocol P with respect to the same permutations (Π_1, Π_2) using at most $4\,\mathsf{r}_f(\Pi_1, \Pi_2)$ communication steps such that* $\alpha_{\mathsf{D}_u^{\sim c}} \leq 4 + 4\,c$.

(b) *For all $0 \leq c < 9/8$, there exists a tiling function $f\colon \{0,1\}^k \times \{0,1\}^k \mapsto \{0,1\}^2$ such that, for any two permutations (Π_1, Π_2) of $\{0,1\}^k$, every dissection protocol with respect to (Π_1, Π_2) using any number of communication steps has* $\alpha_{\mathsf{D}_u^{\sim c}} \geq (11/9) + (2/81)c$.

Proof. We only provide the proof of **(a)**; a proof of the other part can be found in the full version of the paper. Let $\mathcal{S} = \{S_1, S_2, \ldots, S_{r_f}\}$ be the set of $r_f = r_f(\Pi_1, \Pi_2)$ ideal monochromatic rectangles in the tiling of f induced by the permutations (Π_1, Π_2) and consider a protocol P that is a dissection protocol with respect to (Π_1, Π_2). Suppose that the ideal monochromatic rectangle $S_i \in \mathcal{S}$ has y_i elements, and P partitions this rectangle into t_i rectangles $S_{i,1}, \ldots, S_{i,t_i}$ having $z_{i,1}, \ldots, z_{i,t_i}$ elements, respectively. Then, it follows that

$$\alpha_{D_u} = \sum_{(x_1,x_2) \in \mathcal{U} \times \mathcal{U}} \Pr_{D_u}\left[x_1 \& x_2\right] \frac{|R^I(x_1,x_2)|}{|R^P(x_1,x_2)|}$$

$$= \sum_{i=1}^{r_f} \sum_{j=1}^{t_i} \sum_{(x_1,x_2) \in S_{i,j}} \Pr_{D_u}[x_1 \& x_2] \frac{y_i}{z_{i,j}} = \sum_{i=1}^{r_f} \sum_{j=1}^{t_i} \frac{y_i}{2^{2k}} = \sum_{i=1}^{r_f} \frac{t_i\, y_i}{2^{2k}}.$$

Similarly, it follows that

$$\alpha_{D_u^{\tilde{}}c} \leq \sum_{i=1}^{r_f} \sum_{j=1}^{t_i} \sum_{(x_1,x_2) \in S_{i,j}} \frac{1+c}{2^{2k}} \times \frac{y_i}{z_{i,j}} = \sum_{i=1}^{r_f} \sum_{j=1}^{t_i} \frac{(1+c)\, y_i}{2^{2k}} = \sum_{i=1}^{r_f} \frac{(1+c)\, t_i\, y_i}{2^{2k}}.$$

A binary space partition (BSP) for a collection of *disjoint* rectangles in the two-dimensional plane is defined as follows. The plane is divided into two parts by cutting rectangles with a line if necessary. The two resulting parts of the plane are divided recursively in a similar manner; the process continues until at most one fragment of the original rectangles remains in any part of the plane. This division process can be naturally represented as a binary tree (BSP-tree) where a node represents a part of the plane and stores the cut that splits the plane into two parts that its two children represent and each leaf of the BSP-tree represents the final partitioning of the plane by storing at most one fragment of an input rectangle. The *size* of a BSP is the *number of leaves* in the BSP-tree.

Fact 1. [4][2] *Assume that we have a set $\mathcal{S}$ of disjoint axis-parallel rectangles in the plane. Then, there is a* BSP *of $\mathcal{S}$ such that every rectangle in $\mathcal{S}$ is partitioned into at most 4 rectangles.*

Consider the dissection protocol corresponding to the BSP in Fact 1. Then, using $\max_i\{t_i\} \leq 4$ we get $\alpha_{D_u^{\tilde{}}c} \leq \sum_{i=1}^{r_f} \frac{4\,(1+c)\, y_i}{2^{2k}} = 4\,(1+c)$. The number of communication steps in this protocol is the height of the BSP-tree, *i.e.*, $\leq 4r_f$.

Large Worst-case PAR for Non-Boolean Functions. Can one extend the results of the last section to show that every tiling function admits a dissection protocol that achieves a good PAR *even in the worst case*? We answer this question in the negative by presenting a tiling function for which *every* dissection protocol has *large* worst-case PAR.

Theorem 2. *Let $k > 0$ be an even integer. Then, there exists a tiling function $f : \{0,1\}^k \times \{0,1\}^k \mapsto \{0,1\}^3$ with respect to some two permutations (Π_1, Π_2)*

---

[2] The stronger bounds by Berman, DasGupta and Muthukrishnan [2] apply to *average* number of fragments only.

such that, for any two permutations Π_1' and Π_2' of $\{0,1\}^k$, every dissection protocol for f with respect to (Π_1', Π_2') has $\alpha_{\mathrm{worst}} > 2^{k/2} - 1$.

6 Extensions of the Basic Two-Party Setup

6.1 Non-tiling Functions

A natural extension of the class of tiling functions involves relaxing the constraint that each monochromatic region *must* be a rectangle.

Definition 12 (δ-tiling function). *A function $f\colon \{0,1\}^k \times \{0,1\}^k \mapsto \{0,1\}^t$ is a δ-tiling function with respect to permutations (Π_1, Π_2) of $\{0,1\}^k$ if each maximal monochromatic region of $A_{\Pi_1, \Pi_2}(f)$ is an union of at most δ disjoint rectangles.*

Proposition 1. *For any δ-tiling function f with respect to (Π_1, Π_2) with r maximal monochromatic regions, there is a dissection protocol P with respect to (Π_1, Π_2) using at most $4r\delta$ communication steps such that $\alpha_{\mathrm{D}_{\tilde{u}}^c} \leq (4 + 4c)\,\delta$.*

6.2 Multi-party Computation

How good is the average PAR for a dissection protocol on a d-dimensional tiling function? For a general d, it is non-trivial to compute *precise* bounds because each party$_i$ has her/his own permutation Π_i of the input, the tiles are boxes of *full* dimension and hyperplanes corresponding to each step of the dissection protocol is of dimension *exactly $d-1$*. Nonetheless, we show that the average PAR is very high for dissection protocols even for 3 parties and uniform distribution, thereby suggesting that this quantification of privacy may not provide good bounds for three or more parties.

Theorem 3. *There exists a tiling function $f\colon \{0,1\}^k \times \{0,1\}^k \times \{0,1\}^k \mapsto \{0,1\}^{3k}$ such that, for any three permutations Π_1, Π_2, Π_3 of $\{0,1\}^k$, every dissection protocol with respect to (Π_1, Π_2, Π_3) must have $\alpha_{\mathrm{D}_u} = \Omega\left(2^k\right)$.*

Proof. In the sequel, for convenience *we refer to 3-dimensional hyper-rectangles simply by rectangles* and refer to the arguments of function f via *decimal equivalent of the corresponding binary numbers*. The tiling function for this theorem is adopted from an example of the paper by Paterson and Yao [14,15] with appropriate modifications. The three arguments of f are referred to as dimensions 1, 2 and 3, respectively. Define the *volume* of a rectangle $R = [x_1, x_1'] \times [x_2, x_2'] \times [x_3, x_3'] \subseteq \{0, 1, \ldots, 2^k - 1\}^3$ as $\mathsf{Volume}(R) = \max\{0, \Pi_{i=1}^3 (x_i' - x_i + 1)\}$, and let $[*]$ denote the interval $\left[0, 2^k - 1\right]$. We provide the tiling for the function f:

- For each dimension, we have a set of $\Theta\left(2^{2k}\right)$ rectangles; we refer to these rectangles as *non-trivial* rectangles for this dimension.
  - For dimension 1, these rectangles are of the form $[*] \times [2y, 2y] \times [2z, 2z]$ for every integral value of $0 \leq 2y, 2z < 2^k$.

- For dimension 2, these rectangles are of the form $[2x, 2x] \times [*] \times [2z + 1, 2z + 1]$ for every integral value of $0 \leq 2x, 2z + 1 < 2^k$.
- For dimension 3, these rectangles are of the form $[2x + 1, 2x + 1] \times [2y + 1, 2y + 1] \times [*]$ for every integral value of $0 \leq 2x + 1, 2y + 1 < 2^k$.
- The remaining "trivial" rectangles are each of unit volume such that they together cover the remaining input space.

Let $\mathcal{S}_{\mathrm{non-trivial}}$ be the set of all non-trivial rectangles. Observe that:

- Rectangles in $\mathcal{S}_{\mathrm{non-trivial}}$ are mutually disjoint since any two of them do not intersect in at least one dimension.
- *Each* rectangle in $\mathcal{S}_{\mathrm{non-trivial}}$ has a volume of 2^k and thus the sum of their volumes is $\Theta\left(2^{3k}\right)$.

It now follows that the number of monochromatic regions is $O\left(2^{3k}\right)$. Suppose that a dissection protocol partitions, for $i = 1, 2, \ldots, |\mathcal{S}_{\mathrm{non-trivial}}|$, the ith non-trivial rectangle $R_i \in \mathcal{S}_{\mathrm{non-trivial}}$ into t_i rectangles $R_{i,1}, R_{i,2}, \ldots, R_{i,t_i}$. Then,

$$\alpha_{D_u} \overset{\text{def}}{=} \sum_{\substack{(x,y,z) \in \\ \{0,1\}^k \times \{0,1\}^k \times \{0,1\}^k}} \Pr_{D_u}[x \& y \& z] \frac{|R^I(x,y,z)|}{|R^P(x,y,z)|} \geq \sum_{i=1}^{|\mathcal{S}_{\mathrm{non-trivial}}|} \sum_{j=1}^{t_i} \sum_{(x,y,z) \in R_{i,j}} \Pr_{D_u}[x \& y \& z] \frac{\text{Volume}(R_i)}{\text{Volume}(R_{i,j})}$$

$$= \sum_{i=1}^{|\mathcal{S}_{\mathrm{non-trivial}}|} \sum_{j=1}^{t_i} \frac{2^k}{2^{3k}} = \sum_{i=1}^{|\mathcal{S}_{\mathrm{non-trivial}}|} \left(t_i / 2^{2k}\right)$$

Thus, it suffices to show that $\sum_{i=1}^{|\mathcal{S}_{\mathrm{non-trivial}}|} t_i = \Omega\left(2^{3k}\right)$. Let $\mathcal{Q}$ be the set of maximal *monochromatic* rectangles produced the partitioning of the entire protocol. Consider the two entries $p_{x,y,z} = (2x + 1, 2y, 2z + 1)$ and $p'_{x,y,z} = (2x, 2y, 2z)$. Note that $p_{x,y,z}$ belongs to a trivial rectangle since their third, first and second coordinate does not lie within *any* non-trivial rectangle of dimension 1, 2 and 3, respectively, whereas $p'_{x,y,z}$ belongs to the non-trivial rectangle $[*] \times [2 \times (8y), 2 \times (8y)] \times [2 \times (8z), 2 \times (8z)]$ of dimension 1. Thus, $p_{x,y,z}$ and $p'_{x,y,z}$ cannot belong to the same rectangle in $\mathcal{Q}$. Let $T = \bigcup \left\{ \{p_{8x,8y,8z}, p'_{8x,8y,8z}\} \mid 64 < 16x, 16y, 16z < 2^k - 64 \right\}$. Clearly, $|T| = \Theta\left(2^{3k}\right)$. For an entry (x_1, x_2, x_3), let its neighborhood be defined by the ball $\mathsf{Nbr}(x_1, x_2, x_3) = \left\{ (x'_1, x'_2, x'_3) \mid \forall i : |x_i - x'_i| \leq 4 \right\}$. Note that $\mathsf{Nbr}(p_{8x,8y,8z}) \cap \mathsf{Nbr}(p_{8x',8y',8z'}) = \emptyset$ provided $(x, y, z) \neq (x', y', z')$. Next, we show that, to ensure that the two entries $p_{8x,8y,8z}$ and $p'_{8x,8y,8z}$ are in two different rectangles in $\mathcal{Q}$, the protocol must produce an *additional* fragment of one of the non-trivial rectangles in the neighborhood $\mathsf{Nbr}(p_{8x,8y,8z})$; this would directly imply $\sum_i t_i = \Omega\left(2^{3k}\right)$.

Consider the step of the protocol *before* which $p_{8x,8y,8z}$ and $p'_{8x,8y,8z}$ were contained inside the same rectangle, namely a rectangle Q that includes the rectangle $[16x, 16x + 1] \times [16y, 16y] \times [16z, 16z + 1]$, but after which they are in two different rectangles $Q_1 = [a'_1, b'_1] \times [a'_2, b'_2] \times [a'_3, b'_3]$ and $Q_2 = [a''_1, b''_1] \times [a''_2, b''_2] \times [a''_3, b''_3]$. Remember that both Q_1 and Q_2 must have the same two dimensions and these two dimensions must be the same as the corresponding dimensions of Q. The following cases arise.

Case 1(split via the 1st coordinate): $[a_2', b_2'] = [a_2'', b_2''] \supseteq [16y, 16y]$, $[a_3', b_3'] = [a_3'', b_3''] \supseteq [16z, 16z+1]$, $b_1' = 16x$ and $a_1'' = 16x+1$. Then, a new fragment of a non-trivial rectangle of dimension 2 is produced at $[16x, 16y, 16z] \in \text{Nbr}(p_{8x,8y,8z})$.

Case 2(split via the 2nd coordinate): $[a_1', b_1'] = [a_1'', b_1''] \supseteq [16x, 16x+1]$ and $[a_3', b_3'] = [a_3'', b_3''] \supseteq [16z, 16z+1]$. This case is not possible.

Case 3(split via the 3rd coordinate): $[a_1', b_1'] = [a_1'', b_1''] \supseteq [16x, 16x+1]$, $[a_2', b_2'] = [a_2'', b_2''] \supseteq [16y, 16y]$, $b_3' = 16z$ and $a_3'' = 16z+1$. Then, a new fragment of a non-trivial rectangle of dimension 1 is produced at $[16x, 16y, 16z] \in \text{Nbr}(p_{8x,8y,8z})$.

7   Analysis of the Bisection Protocol for Two Functions

Let $\mathbf{x} = (x_1, x_2, \ldots, x_n) \in \{0, 1\}^k$ and $\mathbf{y} = (y_1, y_2, \ldots, y_n) \in \{0, 1\}^k$. The functions that we consider are the following:

set-covering: $f_{\wedge, \vee}(\mathbf{x}, \mathbf{y}) = \bigwedge_{i=1}^{n} (x_i \vee y_i)$. To interpret this as a set-covering function, suppose that the universe $\mathcal{U}$ consists of n elements $e_1, e_2, \ldots, e_n$ and the vectors $\mathbf{x}$ and $\mathbf{y}$ encode membership of the elements in two sets $S_{\mathbf{x}}$ and $S_{\mathbf{y}}$, i.e., x_i (respectively, y_i) is 1 if and only if $e_i \in S_{\mathbf{x}}$ (respectively, $e_i \in S_{\mathbf{y}}$). Then, $f_{\wedge, \vee}(\mathbf{x}, \mathbf{y}) = 1$ if and only if $S_{\mathbf{x}} \cup S_{\mathbf{y}} = \mathcal{U}$.
equality: $f_=(\mathbf{x}, \mathbf{y}) = 1$ if $x_i = y_i$ for all $1 \leq i \leq k$, and $f_=(\mathbf{x}, \mathbf{y}) = 0$ otherwise.

A summary of our bounds is as follows: for $f_{\wedge, \vee}$, $\alpha_{\text{worst}} \geq \alpha_{\mathsf{D}_u} \geq \left(\frac{3}{2}\right)^{2k}$; for $f_=$, $\alpha_{\mathsf{D}_u} = 2^k - 2 + 2^{1-k}$, and $\alpha_{\text{worst}} = 2^{2k-1} - 2^{k-1}$.

References

1. Ghosh, A., Roughgarden, T., Sundararajan, M.: Universally utility-maximizing privacy mechanisms. In: 41st ACM Symp. on Theory of Computing, pp. 351–360 (2009)
2. Berman, P., DasGupta, B., Muthukrishnan, S.: On the Exact Size of the Binary Space Partitioning of Sets of Isothetic Rectangles with Applications. SIAM Journal of Discrete Mathematics 15(2), 252–267 (2002)
3. Bar-Yehuda, R., Chor, B., Kushilevitz, E., Orlitsky, A.: Privacy, additional information, and communication. IEEE Trans. on Inform. Theory 39, 55–65 (1993)
4. d'Amore, F., Franciosa, P.G.: On the optimal binary plane partition for sets of isothetic rectangles. Information Processing Letters 44, 255–259 (1992)
5. Chaum, D., Crépeau, C., Damgaard, I.: Multiparty, unconditionally secure protocols. In: 22th ACM Symposium on Theory of Computing, pp. 11–19 (1988)
6. Chor, B., Kushilevitz, E.: A zero-one law for boolean privacy. SIAM Journal of Discrete Mathematics 4, 36–47 (1991)
7. Dwork, C.: Differential privacy. In: Bugliesi, M., Preneel, B., Sassone, V., Wegener, I. (eds.) ICALP 2006. LNCS, vol. 4052, pp. 1–12. Springer, Heidelberg (2006)
8. Feigenbaum, J., Jaggard, A., Schapira, M.: Approximate Privacy: Foundations and Quantification. In: ACM Conference on Electronic Commerce, pp. 167–178 (2010)
9. Grigorievaa, E., Heringsb, P.J.-J., Müllera, R., Vermeulena, D.: The communication complexity of private value single-item auctions. Operations Research Letters 34, 491–498 (2006)

10. Grigorievaa, E., Heringsb, P.J.-J., Müllera, R., Vermeulena, D.: The private value single item bisection auction. Economic Theory 30, 107–118 (2007)
11. Kushilevitz, E., Nisan, N.: Communication Complexity. Cambridge University Press, Cambridge (1997)
12. Kifer, D., Lin, B.-R.: An Axiomatic View of Statistical Privacy and Utility. Journal of Privacy and Confidentiality (to appear)
13. Kushilevitz, E.: Privacy and communication complexity. SIAM Journal of Discrete Mathematics 5(2), 273–284 (1992)
14. Paterson, M., Yao, F.F.: Efficient binary space partitions for hidden-surface removal and solid modeling. Discrete & Computational Geometry 5(1), 485–503 (1990)
15. Paterson, M., Yao, F.F.: Optimal binary space partitions for orthogonal objects. Journal of Algorithms 13, 99–113 (1992)
16. Yao, A.C.: Some complexity questions related to distributive computing. In: 11th ACM Symposium on Theory of Computing, pp. 209–213 (1979)

Dynamic Inefficiency: Anarchy without Stability

Noam Berger[1], Michal Feldman[2,*], Ofer Neiman[3], and Mishael Rosenthal[4]

[1] Einstein Institute of Mathematics, Hebrew University of Jerusalem
berger@math.huji.ac.il
[2] School of Business Administration and Center for Rationality,
Hebrew University of Jerusalem
mfeldman@huji.ac.il
[3] Princeton university and Center for Computational Intractability
oneiman@princeton.edu
[4] School of Engineering and Computer Science, Hebrew University of Jerusalem
mishael@cs.huji.ac.il

Abstract. The price of anarchy [16] is by now a standard measure for quantifying the inefficiency introduced in games due to selfish behavior, and is defined as the ratio between the optimal outcome and the worst Nash equilibrium. However, this notion is well defined only for games that always possess a Nash equilibrium (NE). We propose the *dynamic inefficiency* measure, which is roughly defined as the average inefficiency in an infinite best-response dynamic. Both the price of anarchy [16] and the price of sinking [9] can be obtained as special cases of the dynamic inefficiency measure. We consider three natural best-response dynamic rules — *Random Walk* (RW), *Round Robin* (RR) and *Best Improvement* (BI) — which are distinguished according to the order in which players apply best-response moves.

In order to make the above concrete, we use the proposed measure to study the job scheduling setting introduced in [3], and in particular the scheduling policy introduced there. While the proposed policy achieves the best possible price of anarchy with respect to a pure NE, the game induced by the proposed policy may admit no pure NE, thus the *dynamic inefficiency* measure reflects the worst case inefficiency better. We show that the dynamic inefficiency may be arbitrarily higher than the price of anarchy, in any of the three dynamic rules. As the dynamic inefficiency of the RW dynamic coincides with the *price of sinking*, this result resolves an open question raised in [3].

We further use the proposed measure to study the inefficiency of the Hotelling game and the facility location game. We find that using different dynamic rules may yield diverse inefficiency outcomes; moreover, it seems that no single dynamic rule is superior to another.

* The author is supported in part by the Google Inter-university center for Electronic Markets and Auctions and by the Israel Science Foundation (grant number 1219/09) and by the Leon Recanati Fund of the Jerusalem school of business administration.

G. Persiano (Ed.): SAGT 2011, LNCS 6982, pp. 57–68, 2011.
© Springer-Verlag Berlin Heidelberg 2011

1 Introduction

Best-response dynamics are central in the theory of games. The celebrated Nash equilibrium solution concept is implicitly based on the assumption that players follow best-response dynamics until they reach a state from which no player can improve her utility. Best-response dynamics give rise to many interesting questions which have been extensively studied in the literature. Most of the focus concerning best response dynamics has been devoted to convergence issues, such as whether best-response dynamics converge to a Nash equilibrium and what is the rate of convergence.

Best-response dynamics are essentially a large family of dynamics, which differ from each other in the order in which turns are assigned to players [1]. It is well known that the order of the players' moves is crucial to various aspects, such as convergence rate to a Nash equilibrium [6]. Our main goal is to study the effect of the players' order on the obtained (in)efficiency of the outcome.

The most established measure of inefficiency of games is the Price of Anarchy (PoA) [13,16], which is a worst-case measure, defined as the ratio between the worst Nash equilibrium (NE) and the social optimum (with respect to a well-defined social objective function), usually defined with respect to pure strategies. The PoA essentially measures how much the society suffers from players who maximize their individual welfare rather than the social good. The PoA has been evaluated in many settings, such as selfish routing [19,18], job scheduling [13,5,8], network formation [7,1,2], facility location [20], and more. However, this notion is well defined only in settings that are guaranteed to admit a NE.

One approach that has been taken with respect to this challenge, in cases where agents are assumed to use pure strategies, is the introduction of a *sink equilibrium* [9]. A sink equilibrium is a strongly connected component with no outgoing edges in the *configuration graph* associated with a game. The configuration graph has a vertex set associated with the set of pure strategy profiles, and its edges correspond to best-response moves. Unlike pure strategy Nash equilibria, sink equilibria are guaranteed to exist. The social value associated with a sink equilibrium is the expected social value of the stationary distribution of a random walk on the states of the sink. The *price of sinking* is the equivalence of the price of anarchy measure with respect to sink equilibria.

Indeed, the notion of best response lies at the heart of many of the proposed solution concepts, even if just implicitly. The implicit assumption that underlies the notion of social value associated with a sink equilibrium is that in each turn a player is chosen uniformly at random to perform her best response. However, there could be other natural best-response dynamics that arise in different settings.

In this paper, we focus on the following three natural dynamic rules: (i) **random walk (RW)**, where a player is chosen uniformly at random; (ii) **round robin (RR)**, where players play in a cyclic manner according to a pre-defined

---

[1] In fact, best-response dynamics may be asynchronous, but in this paper we restrict attention to synchronous dynamics.

order, and (iii) **best improvement (BI)**, where the player with the current highest (multiplicative) improvement factor plays. Our goal is to study the effect of the players' order on the obtained (in)efficiency of the outcome.

To this end, we introduce the concept of *dynamic inefficiency* as an equivalent measure to the price of anarchy in games that may not admit a Nash equilibrium or in games in which best-response dynamics are not guaranteed to converge to a Nash equilibrium. Every dynamic rule D chooses (deterministically or randomly) a player that performs her best response in each time period. Given a dynamic D and an initial configuration u, one can compute the expected social value obtained by following the rules of dynamic D starting from u. The dynamic inefficiency in a particular game is defined as the expected social welfare with respect to the worst initial configuration. Similarly, the dynamic inefficiency of a particular family of games is defined as the worst dynamic inefficiency over all games in the family. Note that the above definition coincides with the original price of anarchy measure for games in which best-response dynamics always converge to a Nash equilibrium (e.g., in congestion [17] and potential [14] games) for every dynamic rule. Similarly, the dynamic inefficiency of the RW dynamic coincides with the definition of the price of sinking. Thus, we find the dynamic inefficiency a natural generalization of well-established inefficiency measures.

1.1   Our Results

We evaluate the dynamic inefficiency with respect to the three dynamic rules specified above, and in three different applications, namely non-preemptive job scheduling on unrelated machines [3], the Hotelling model [10], and facility location [15]. Our contribution is conceptual as well as technical. First, we introduce a measure which allows us to evaluate the inefficiency of a particular dynamic even if it does not lead to a Nash equilibrium. Second, we develop proof techniques for providing lower bounds for the three dynamic rules. In what follows we present our results in the specific models.

Job scheduling (Section 3). We consider job scheduling on unrelated machines, where each of the n players controls a single job and selects a machine among a set of m machines. The assignment of job i on machine j is associated with a processing time that is denoted by $p_{i,j}$. Each machine schedules its jobs sequentially according to some non-preemptive scheduling policy (i.e., jobs are processed without interference, and no delay is introduced between two consecutive jobs), and the cost of each job in a given profile is its completion time on its machine. The social cost of a given profile is the maximal completion time of any job (known as the makespan objective).

Machines' ordering policies may be local or strongly local. A local policy considers only the parameters of the jobs assigned to it, while a strongly local policy considers only the processing time of the jobs assigned to it on itself (without knowing the processing time of its jobs on other machines). Azar et. al. [3] showed that the PoA of any local policy is $\Omega(\log m)$ and that the PoA of any strongly local policy is $\Omega(m)$ (if a Nash equilibrium exists). Ibarra and Kim [11] showed that the shortest-first (strongly local) policy exhibits a matching $O(m)$ bound, and Azar et. al. [3] showed that the inefficiency-based (local) policy (defined in Section 3) exhibits a matching $O(\log m)$ bound.

We claim that there is a fundamental difference between the last two results. The shortest-first policy induces a potential game [12,14]; thus best-response dynamics always converge to a pure Nash equilibrium, and the PoA is an appropriate measure. In contrast, the inefficiency-based policy induces a game that does not necessarily admit a pure Nash equilibrium [3], and even if a Nash equilibrium exists, not every best-response dynamic converges to a Nash equilibrium. Consequently, the realized inefficiency of the last policy may be much higher than the bound provided by the price of anarchy measure.

We study the dynamic inefficiency of the inefficiency based policy with respect to our three dynamic rules. We show a lower bound of $\Omega(\log \log \mathbf{n})$ for the dynamic inefficiency of the RW rule. This bound may be arbitrarily higher[2] than the price of anarchy, which is bounded by $O(\log \mathbf{m})$. This resolves an open question raised in [3]. For the BI and RR rules, we show in the full version even higher lower bounds of $\Omega(\sqrt{n})$ and $\Omega(n)$, respectively.

Hotelling model Hotelling [10] devised a model where customers are distributed evenly along a line, and there are m strategic players, each choosing a location on the line, with the objective of maximizing the number of customers whose location is closer to her than to any other player. It is well known that this game admits a pure Nash equilibrium if and only if the number of players is different than three. This motivates the evaluation of the dynamic inefficiency measure in settings with three players. The social objective function we consider here is the minimal utility over all players, i.e., we wish to maximize the minimal number of customers one attracts.

We show in the full version that the dynamic inefficiency of the BI rule is upper bounded by a universal constant, while the dynamic inefficiency of the RW and RR rules is lower bounded by $\Omega(n)$, where n is the number of possible locations of players. Thus, the BI dynamics and the RW and RR dynamics exhibit the best possible and worst possible inefficiencies, respectively (up to a constant factor). In contrast to the BI dynamics, the RW and RR dynamics are *configuration-oblivious* (i.e., the next move is determined independently of the current configuration).

Facility location. In facility location games a central designer decides where to locate a public facility on a line, and each player has a single point representing her ideal location. Suppose that the cost associated with each player is the squared distance of her ideal location to the actual placement of the facility, and that we wish to minimize the average cost of the players. Under this objective function the optimal location is the mean of all the points. However, for any chosen location, there will be a player who can decrease her distance from the

---

[2] Note the parameter n; i.e., number of players, versus the parameter m; i.e., number of machines.

chosen location by reporting a false ideal location. Moreover, it is easy to see that if the players know in advance that the mean point of the reported locations is chosen, then every player who is given a turn can actually transfer the location to be exactly at her ideal point. Thus, unless all players are located at exactly the same point, there will be no Nash equilibrium. Our results, that appear in the full version, indicate that the dynamic inefficiency of the RR and RW rules is exactly 2, while that of the BI rule is $\Theta(n)$.

2   Preliminaries

In our analysis it will be convenient to use the following graph-theoretic notation: we think of the configuration set (i.e., pure strategy profiles) as the vertex set of a *configuration graph* $G = (V, E)$. The configuration graph is a directed graph in which there is a directed edge $e = (u, v) \in E$ if and only if there is a player whose best response given the configuration u leads to the configuration v. We assume that each player has a unique best response for each vertex. A *sink* is a vertex with no outgoing edges. A *Nash Equilibrium* (NE) is a configuration in which each player is best responding to the actions of the other players. Thus, a NE is a sink in the configuration graph.

A *social value* of $S(v)$ is associated with each vertex $v \in V$. Two examples of social value functions are the social welfare function, defined as the sum of the players' utilities, and the max-min function, defined as the minimum utility of any player.

A best-response dynamic rule is a function $D : V \times \mathbb{N} \to [n]$, mapping each point in time, possibly depending on the current configuration, to a player $i \in [n]$ who is the next player to apply her best-response strategy. The function D may be deterministic or non-deterministic.

Let $P = \langle u_1, \dots, u_T \rangle$ denote a finite path in the configuration graph (where u_i may equal u_j for some $i \neq j$). The *average social value* associated with a path P is defined as $S(P) = \frac{1}{T} \sum_{t=1}^{T} S(u_t)$. Given a tuple $\langle u, D \rangle$ of a vertex $u \in V$ and a dynamic rule D, let $\mathcal{P}_T(u, D)$ denote the distribution over the paths of length T initiated at vertex u under the dynamic rule D. The social value of the dynamic rule D initiated at vertex u is defined as

$$S(u, D) = \lim_{T \to \infty} \mathbb{E}_{P \sim \mathcal{P}_T(u, D)}[S(P)]. \tag{1}$$

While the expression above is not always well defined, in the full version of the paper we demonstrate that it is always well defined for the dynamic rules considered in this paper.

With this, we are ready to define the notion of *dynamic inefficiency*. Given a finite configuration graph $G = (V, E)$ and a dynamic rule D, the dynamic inefficiency (DI) of G with respect to D is defined as

$$\mathrm{DI}(D, G) = \max_{u \in V} \frac{\mathrm{OPT}}{S(u, D)},$$

where OPT $= \max_{u \in V} S(u)$. That is, DI measures the ratio between the optimal outcome and the social value obtained by a dynamic rule D under the worst possible initial vertex. Finally, for a family of games $\mathcal{G}$, we define the dynamic inefficiency of a dynamic rule D as the dynamic inefficiency of the worst possible $G \in \mathcal{G}$. This is given by

$$\mathrm{DI}(D) = \sup_{G \in \mathcal{G}} \{\mathrm{DI}(D, G)\} \ .$$

In some of the settings, social costs are considered rather than social value. In these cases, the necessary obvious adjustments should be made. In particular, $S(u, D)$ will denote the social *cost*, OPT will be defined as $\min_{u \in V} S(u)$, and the dynamic inefficiency of some dynamic D will be defined as $\mathrm{DI}(D) = \max_{u \in V} \frac{S(u,D)}{\mathrm{OPT}}$. We consider both cases in the sequel.

An important observation is that both the price of anarchy and the price of sinking are obtained as special cases of the dynamic inefficiency. In games for which every best-response dynamic converges to a Nash equilibrium (e.g., potential games [14]), the dynamic inefficiency is independent of the dynamic and is equivalent to the price of anarchy. The price of sinking [9] is equivalent to the dynamic inefficiency with respect to the RW dynamic rule.

3   Dynamic Inefficiency in Job Scheduling

Consider a non-preemptive job scheduling setting on unrelated machines, as described in the Introduction. Define the efficiency of a job i on machine j as

$$\mathrm{eff}(i, j) = \frac{p_{ij}}{\min_{k \in [m]} p_{ik}}.$$

The efficiency-based policy of a machine (proposed by [3]) orders its jobs according to their efficiency, from low to high efficiency, where ties are broken arbitrarily in a pre-defined way.

A configuration of a job scheduling game is a mapping $u : [n] \to [m]$ that maps each job to a machine. The processing time of machine j in configuration u is $\mathrm{time}_u(j) = \sum_{i \in u^{-1}(j)} p_{ij}$, and the social value function we are interested in is the *makespan* — the longest processing time on any machine, i.e., $S(u) = \max_{j \in [m]} \mathrm{time}_u(j)$.

The players are the jobs to be processed, their actions are the machines they choose to run on, and the cost of a job is its own completion time.

3.1   Random Walk Dynamic

In this section we consider the RW dynamic, where in each turn a player is chosen uniformly at random to play. The main result of this section is the establishment of a lower bound of $\Omega(\log \log n)$ for the dynamic inefficiency of job scheduling under the efficiency-based policy. This means that the inefficiency may tend to infinity with the number of jobs, even though the number of machines is constant. This result should be contrasted with the $O(\log m)$ upper bound on the price of anarchy, established by [3]. The main result is cast in the following theorem.

Theorem 1. *There exists a family of instances G_n of machine scheduling on a constant number of machines, such that*

$$\text{DI}(RW, G_n) \geq \Omega(\log \log n) \ ,$$

where n is the number of jobs. In particular the dynamic inefficiency is not bounded with respect to the number of machines.

Remark: The definition of the dynamic inefficiency with respect to the RW dynamic rule coincides with the definition of the *price of sinking*. Thus, the last result can be interpreted as a lower bound on the price of sinking.

The assertion of Theorem 1 is established in the following sections.

The Construction. Let us begin with an informal description of the example. As the base for our construction we use an instance, given in [3], with a constant number of machines and jobs, that admits no Nash equilibrium. Then we add to it n jobs indexed by $1, \ldots, n$ and one machine, such that the n additional jobs have an incentive to run on two machines. On the first machine, denoted by W, the total processing time of all the n jobs is smaller than 2, while on the second machine T the processing time of any job is ≈ 1. Each of these jobs has an incentive to move from machine W to machine T if it has the minimal index on T, thus increasing the processing time on T. We show that the expected number of jobs on T, and hence also the expected makespan, is at least $\Omega(\log \log n)$, while the optimum is some universal constant.

Formally, there will be 5 machines, denoted by A, B, C, T, W, and $n + 5$ jobs denoted by $0, 1, 2, \ldots, n$ and $\alpha, \beta, \gamma, \delta$. The following table shows the processing time for the jobs on the machines:

	A	B	C	T	W
0	4	24	3.95	25	∞
α	2	12	1.98	∞	∞
β	5	28	4.9	∞	∞
γ	20	∞	∞	∞	∞
δ	∞	∞	50	∞	∞
i	∞	∞	$\frac{1}{50i^3}$	$\sum_{j=1}^{i} \frac{1}{j^2} - \epsilon$	$\frac{1}{i^2}$

In the last row, i stands for any job $1, 2, \ldots, n$, and $\epsilon = \frac{1}{10 \cdot 2^n}$.

Useful Properties

Proposition 1. *The inefficiency policy induces the following order on the machines:*

- *On machine A the order is $(\gamma, \alpha, 0, \beta)$.*
- *On machine B the order is $(\beta, \alpha, 0)$.*
- *On machine C the order is $(\delta, \alpha, \beta, 0, 1, 2, \ldots, n)$.*
- *On machine T the order is $(0, 1, 2, \ldots, n)$.*
- *On machine W the order is $(1, 2, \ldots, n)$.*

Proof. On machine C every job has efficiency 1; hence given any tie-breaking rule between jobs of equal efficiency we let δ be the one that runs first[3] (the rest of the ordering is arbitrary). On the other machines, the order follows from straightforward calculations.

Note that no job except δ has an incentive to move to machine C, and job γ will always be in machine A. The possible configurations for jobs $0, \alpha, \beta$ are denoted by XYZ; for instance, ABB means that job 0 is on machine A and jobs α, β are on machine B. We shall only consider configurations in G that have incoming edges, and in this example there are 8 such configurations among the possible configurations for jobs $0, \alpha, \beta$. The transitions are

$$BAA \to BBA \to ABA \to ABB \to AAB \to TAB \to TAA$$

From state TAA we can go either to TBA or back to BAA. From TBA the only possible transition will take us back to the state ABA. These are all the possible transitions; hence there is no stable state. At any time a job i for $i > 0$ has an incentive to be in machine T only if it has the *minimal* index from all the jobs in T. We want to show that in the single (non-trivial) strongly connected component of G, the expected number of jobs in machine T is at least $\Omega(\log \log n)$.

Dynamic Inefficiency - Lower Bound. It can be checked that any configuration for jobs $1, 2, \ldots, n$ is possible among machines T, W. Consider the stationary distribution π over this strongly connected component in G. Let $T(i) \subseteq V$ be the set of configurations in which job i is scheduled on machine T, and let $p_n(i) = \sum_{v \in T(i)} \pi(v)$ denote the probability that job i is on machine T. Let $p_n(\emptyset)$ be the probability that no job is on machine T.

Proposition 2. *For any $n > m \geq i$, $p_n(i) = p_m(i)$. This is because the incentives for jobs $\alpha, \beta, \gamma, \delta, 0, 1, \ldots, i$ are not affected by the presence of any job j for $j > i$.*

Using this proposition we shall omit the subscript and write only $p(i)$. The following claim suggests we should focus our attention on how often machine T is empty.

Claim. For any $n \geq 1$, $p_n(\emptyset) \leq p(n)$.

Proof. Job n has an incentive to move to T if and *only if* the configuration is such that T is empty. The probability of job n to get a turn to move is $1/(n+5)$. Hence the probability that job n is in T at some time t is equal to the probability that for some $i \geq 0$ job n entered T at time $t - i$ (i.e., machine T was empty and

---

[3] We can also handle tie-breaking rules that consider the length of the job. For instance, if shorter jobs were scheduled first in case of a tie, we would split job δ into many small jobs.

n got a turn to play) and stayed there for i rounds. The probability that job n stayed in machine T for i rounds is at least $p_i = \left(1 - \frac{1}{n+5}\right)^i$. We conclude that

$$p(n) \geq \frac{p_n(\emptyset)}{n+5} \sum_{i=0}^{\infty} p_i = p_n(\emptyset).$$

The main technical lemma is the following:

Lemma 1. *There exists a universal constant c such that for any $n > 1$, $p_n(\emptyset) \geq \frac{c}{n \log n}$.*

Let us first show that given this lemma we can easily prove the main theorem:

Proof (Proof of Theorem 1). In the single (non-trivial) strongly connected component of G, the expected number of jobs in machine T (with respect to a RW) is at least

$$\sum_{i=1}^{n} p(i) \geq c \sum_{i=2}^{n} \frac{1}{i \log i} \geq (c/2) \log \log n \ .$$

However, there is a configuration in which every job completes execution in time at most 50;[4] hence $\mathrm{OPT}(G) \leq 50$. We conclude that the dynamic inefficiency for G is at least $\Omega(\log \log n)$.

In what follows, we establish the assertion of Lemma 1.

Proof (Proof of Lemma 1). Let $Y \subseteq V$ be the set of configurations in which T is empty, and let t be the expected number of steps between visits to configurations in Y. We have that $p_n(\emptyset) = \frac{1}{t}$, and need to prove that $t \leq O(n \log n)$. We start with a claim on rapidly decreasing integer random variables.

Claim. Fix some $n \in \mathbb{N}$, $n > 1$. Let $x_1, x_2, \ldots$ be random variables getting non-increasing values in $\mathbb{N}$, such that $\mathbb{E}[x_1] \leq n/2$ and for every $i > 0$, $\mathbb{E}[x_{i+1} \mid x_i = k] \leq k/2$; then if we let s be a random variable which is the minimal index such that $x_s = 0$, then $\mathbb{E}[s] \leq \log n + 2$.

Proof. First we prove by induction on i that $\mathbb{E}[x_i] \leq \frac{n}{2^i}$. This holds for $i = 1$; assume it is true for i and then prove for $i + 1$. By the rule of conditional probability,

$$\mathbb{E}[x_{i+1}] = \sum_{j \geq 0} \Pr[x_i = j] \cdot \mathbb{E}[x_{i+1} \mid x_i = j]$$

$$\leq \sum_{j \geq 0} \Pr[x_i = j] \cdot (j/2) = \mathbb{E}[x_i]/2 \leq \frac{n}{2^{i+1}} \ .$$

---

[4] For instance, if all jobs $1, 2, \ldots, n$ are on machine W, then job i will finish in time $\sum_{j=1}^{i} \frac{1}{j^2} < 2$

Note that if $x_i = 0$ for some i, then it must be that $x_j = 0$ for all $j > i$. Now for any integer $i > 0$,

$$
\begin{aligned}
\Pr[s = \log n + i] &\leq \Pr[s > \log n + i - 1] \\
&= \Pr[x_{\log n + i - 1} \geq 1] \\
&\leq \mathbb{E}[x_{\log n + i - 1}] \leq 1/2^{i-1}
\end{aligned}
$$

the second inequality is a Markov inequality. We conclude that

$$
\mathbb{E}[s] = \sum_{i=1}^{\log n} i \cdot \Pr[s = i] + \sum_{i=\log n + 1}^{\infty} i \cdot \Pr[s = i]
$$

$$
\leq \log n + \sum_{i=\log n + 1}^{\infty} i/2^{i-1} \leq \log n + 2 .
$$

Claim. Assume that we are in a configuration u in which job $i \in \{0, 1, \ldots, n\}$ is in machine T. The expected time until we reach a configuration in which job i is not in machine T is at most $O(n)$.

Proof. First note that $p(0) = c'$ for some constant $0 < c' < 1$, this is because jobs $0, \alpha, \beta, \gamma, \delta$ are not affected at all by the location of any job from $1, 2, \ldots, n$, and hence when they get to play they will follow one of the two cycles shown earlier, which implies that in a constant fraction c' of the time job 0 will be in machine T, and in the other $1 - c'$ fraction it will be on another machine.

Note that job i will have incentive to leave T if job 0 is in machine T when i gets its turn to play. Let $q(i)$ be the event that i gets a turn to play (which is independent of the current configuration), $T(i)$ denotes the event that job i is in machine T, then we have that the probability that job i will leave machine T is at least

$$
\begin{aligned}
\Pr[q(i) \wedge T(0) \mid T(i)] &= \Pr[q(i) \mid T(0) \wedge T(i)] \cdot \Pr[T(0) \mid T(i)] \\
&= Pr[q(i)] \cdot \Pr[T(0)] = \frac{c'}{n+5}
\end{aligned}
$$

Now the expected time until job i will leave is at most $\frac{1}{\Pr[q(i) \wedge T(0) \mid T(i)]} \leq O(n)$.

Claim. Let $\ell = \ell(t)$ be a random variable that is the minimal job in T at time t, and let x be the random variable that is the next job that enters T. Then $\mathbb{E}[x \mid \ell = m] \leq m/2$.

Proof. The jobs that have an incentive to enter machine T are $0, 1, \ldots, m-1$. It is easy to see that for any job $i \in \{1, \ldots, m-1\}$, $\Pr[x = i \mid \ell = m] \leq \frac{1}{m-1}$ (note that job 0 has also some small probability to enter T, but it does not contribute to the expectation). Now

$$
\mathbb{E}[x \mid \ell = m] = \sum_{i=1}^{m-1} i \cdot \Pr[x = i \mid \ell = m] \leq m/2 .
$$

Let $y \in Y$ be any configuration in which T is empty. We define a series of random variables $x_1, x_2, \ldots$ as follows. Let x_1 be the index of the first job to enter T, and let x_i be the maximal index of a job in T when x_{i-1} leaves T (and 0 if T is empty). Note that when $x_{i-1} = k$ is the maximal index of a job in T, no job with index larger than k has an incentive to move to T; hence given that $x_{i-1} = k$ it must be that $x_i \leq k$. Let s be the minimal index such that $x_s = 0$; then either machine T is empty or by Claim 3.1 is expected to become empty in $O(n)$ steps. It is easy to see that $\mathbb{E}[x_1] \leq n/2$, and the tricky part is to bound the expected value of x_i.

Claim. $\mathbb{E}[x_i | x_{i-1} = k] \leq k/2$.

Proof. Fix some k such that $x_{i-1} = k$. Consider the time $\bar{t}$ in which job k became the maximal job in T, and consider the time $t' < \bar{t}$ in which job k moved to machine T and did not leave until time $\bar{t}$. In time t' it must be that no job i, for $i < k$, was in machine T, since job k had an incentive to move to T. In particular, x_i is not in T at time t'

Consider the time $t'' > t'$ in which x_i enters T, and stays until job k leaves. In time t'' the minimal job in T is at most k; hence by Claim 3.1 we have that $\mathbb{E}[x_i \mid x_{i-1} = k] \leq k/2$.

Consider the random variables $x_1, x_2, \ldots, x_s$. By Claim 3.1 they satisfy the conditions of Claim 3.1; hence $\mathbb{E}[s] \leq \log n + 2$. By Claim 3.1 we have that the expected time until the maximal job leaves T is at most $O(n)$. We conclude that in expectation after $O(n \log n)$ steps the maximal job in T will be 0, and within additional $O(n)$ steps machine T is expected to become empty. This concludes the proof.

4   Conclusion

We study the notion of dynamic inefficiency, which generalizes well-studied notions of inefficiency such as the price of anarchy and the price of sinking, and quantify it in three different applications. In games where best-response dynamics are not guaranteed to converge to an equilibrium, dynamic inefficiency reflects better the inefficiency that may arise. It would be of interest to quantify the dynamic inefficiency in additional applications. It is of a particular interest to study whether there exist families of games for which one dynamic rule is always superior to another.

In the job scheduling realm, our work demonstrates that the inefficiency based policy suggested by [3] suffers from an extremely high price of sinking. A natural open question arises: is there a local policy that always admits a Nash equilibrium and exhibits a PoA of $o(m)$? (recall that m is the number of machines). Alternatively, is there a local policy that exhibits a dynamic inefficiency of $o(m)$ for some best response dynamic rule? Recently, Caragiannis [4] found a *preemptive* local policy that always admits a Nash equilibrium and has a PoA of $O(\log m)$. However, we are primarily interested in non-preemptive policies.

References

1. Albers, S., Elits, S., Even-Dar, E., Mansour, Y., Roditty, L.: On Nash equilibria for a network creation game. In: Seventeenth Annual ACM-SIAM Symposium on Discrete Algorithms (2006)
2. Anshelevich, E., Dasgupta, A., Tardos, É., Wexler, T.: Near-Optimal Network Design with Selfish Agents. In: STOC 2003 (2003)
3. Azar, Y., Jain, K., Mirrokni, V.: (almost) optimal coordination mechanisms for unrelated machine scheduling. In: Proceedings of the Nineteenth Annual ACM-SIAM Symposium on Discrete Algorithms, SODA 2008, pp. 323–332. Society for Industrial and Applied Mathematics, Philadelphia (2008)
4. Caragiannis, I.: Efficient coordination mechanisms for unrelated machine scheduling. In: Proceedings of the Twentieth Annual ACM-SIAM Symposium on Discrete Algorithms, SODA 2009, pp. 815–824. Society for Industrial and Applied Mathematics, Philadelphia (2009)
5. Czumaj, A., Vöcking, B.: Tight bounds for worst-case equilibria. In: SODA, pp. 413–420 (2002)
6. Even-Dar, E., Kesselman, A., Mansour, Y.: Convergence time to nash equilibria. In: Baeten, J.C.M., Lenstra, J.K., Parrow, J., Woeginger, G.J. (eds.) ICALP 2003. LNCS, vol. 2719, pp. 502–513. Springer, Heidelberg (2003)
7. Fabrikant, A., Luthra, A., Maneva, E., Papadimitriou, C., Shenker, S.: On a network creation game. In: ACM Symposium on Principles of Distributed Computing, PODC (2003)
8. Feldman, M., Tamir, T.: Conflicting congestion effects in resource allocation games. In: Papadimitriou, C., Zhang, S. (eds.) WINE 2008. LNCS, vol. 5385, pp. 109–117. Springer, Heidelberg (2008)
9. Goemans, M., Mirrokni, V., Vetta, A.: Sink equilibria and convergence. In: FOCS 2005: Proceedings of the 46th Annual IEEE Symposium on Foundations of Computer Science, pp. 142–154. IEEE Computer Society, Washington, DC (2005)
10. Hotelling, H.: Stability in competition. Economic Journal 39(53), 41–57 (1929)
11. Ibarra, O.H., Kim, C.E.: Heuristic algorithms for scheduling independent tasks on nonidentical processors. Journal of the ACM 24, 280–289 (1977)
12. Immorlica, N., Li, L., Mirrokni, V.S., Schulz, A.S.: Coordination mechanisms for selfish scheduling. Theor. Comput. Sci. 410, 1589–1598 (2009)
13. Koutsoupias, E., Papadimitriou, C.: Worst-case equilibria. In: Meinel, C., Tison, S. (eds.) STACS 1999. LNCS, vol. 1563, pp. 404–413. Springer, Heidelberg (1999)
14. Monderer, D., Shapley, L.S.: Potential Games. Games and Economic Behavior 14, 124–143 (1996)
15. Moulin, H.: On strategy-proofness and single-peakedness. Public Choice 35, 437–455 (1980)
16. Papadimitriou, C.: Algorithms, games, and the Internet. In: Proceedings of 33rd STOC, pp. 749–753 (2001)
17. Rosenthal, R.W.: A class of games possessing pure-strategy Nash equilibria. International Journal of Game Theory 2, 65–67 (1973)
18. Roughgarden, T.: The price of anarchy is independent of the network topology. In: STOC 2002, pp. 428–437 (2002)
19. Roughgarden, T., Tardos, E.: How bad is selfish routing? Journal of the ACM 49(2), 236–259 (2002)
20. Vetta, A.R.: Nash equilibria in competitive societies with applications to facility location, traffic routing and auctions. In: Symposium on the Foundations of Computer Science (FOCS), pp. 416–425 (2002)

Throw One's Cake — and Eat It Too

Orit Arzi, Yonatan Aumann, and Yair Dombb

Department of Computer Science, Bar-Ilan University, Ramat Gan, Israel
{oritarzi1,yair.biu}@gmail.com, aumann@cs.biu.ac.il

"Envy, desire, and the pursuit of honor drive a person from the world."

Chapters of Our Fathers 4:27

Abstract. We consider the problem of fairly dividing a heterogeneous cake between a number of players with different tastes. In this setting, it is known that fairness requirements may result in a suboptimal division from the social welfare standpoint. Here, we show that in some cases, discarding some of the cake and fairly dividing only the remainder may be socially preferable to any fair division of the entire cake. We study this phenomenon, providing asymptotically-tight bounds on the social improvement achievable by such discarding.

1 Introduction

Cake cutting is a standard metaphor used for modeling the problem of fair division of goods among multiple players. "Fairness" can be defined in several different ways, with *envy-freeness* being one of the more prominent ones. A division is *envy-free* if no player prefers getting a piece given to someone else.

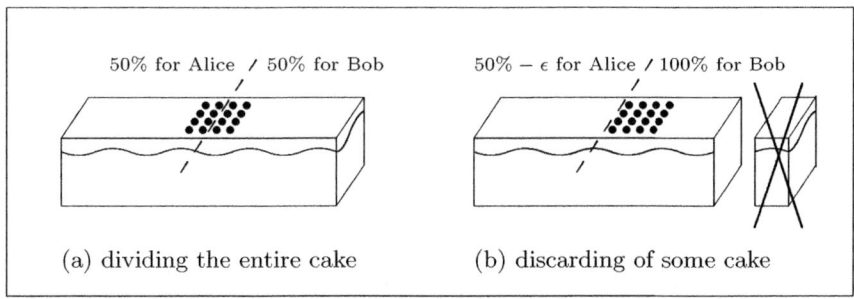

50% for Alice / 50% for Bob   50% − ε for Alice / 100% for Bob

(a) dividing the entire cake   (b) discarding of some cake

Fig. 1. Discarding of part of the cake allows for a socially-preferable envy-free division

Consider the rectangular cake depicted in Figure 1(a). It is a chocolate cake sprinkled with candies right along the middle. Suppose you have two kids: Alice and Bob. Alice likes the base of the cake, but is indifferent to the candies; Bob is the opposite: he cares only for the candies. It is easy to see that if each of the children must get one consecutive piece of the cake, then splitting the cake along the middle is the only possible envy-free division. Any other split would result in one child getting less than 50% (by his or her valuation) and envying the other. But this division is rather wasteful: if Bob could only get a small additional fraction from Alice (small in her view), he would be doubly as happy. Is there any possible way to make this happen — without introducing envy?

Interestingly, the answer is in the affirmative. By discarding a small piece from the right-end of the cake, we can now place the cut to the left of candies, giving the right piece to Bob and the left to Alice (see Figure 1(b)). Alice would no longer envy Bob, as he gets the same amount of the cake as she does. The overall happiness level would substantially increase: Bob is doubly happy, and Alice is only ϵ less happy.

The above is a particular example of what we call the *dumping effect* — the phenomenon in which one can increase the social welfare of envy-free divisions by discarding (=*dumping*) some of the cake. The above provides an example of a *utilitarian $1.5 - \epsilon$ dumping effect* (i.e. the utilitarian welfare, defined as the sum of individual utilities, increases by a factor of $1.5 - \epsilon$). In this paper we analyze the dumping effect: under what circumstances may it arise? what social welfare can it improve? and by how much? Interestingly, we show that at times much can be gained by such discarding of some of the cake. We show:

- With regards to utilitarian welfare, the dumping effect with n players can be as high as $\Theta(\sqrt{n})$; i.e. there are cases where discarding some of the cake allows for an envy-free division that is $\Theta(\sqrt{n})$ better (from the utilitarian standpoint) than any envy-free division of the entire cake. This bound is asymptotically tight.
- With regards to egalitarian welfare, the dumping effect with n players can be as high as $\frac{n}{3}$. Egalitarian welfare is defined as the utility obtained by the least-happy player. In particular, we show a case where discarding some cake allows us to improve from an allocation in which at least one player gets no more than $1/n$ to an allocation in which everybody gets at least $\approx 1/3$ (!). Our construction almost matches the upper bound of $\frac{n}{2}$ following from [1]; for $n \leq 4$ we show that the bound of $\frac{n}{2}$ can actually be obtained.
- With regards to Pareto efficiency, there are instances in which discarding some cake allows for an envy-free division that Pareto dominates every envy-free division of the entire cake. We show that by discarding even one piece of the cake it may be possible to *double* the utility of all but two players without harming these remaining two players.

All of our results are for divisions that require that each player get one consecutive piece of the cake. For divisions that allow players to get arbitrarily many pieces of the cake we show that no dumping effect is possible.

Related work. The problem of fair division has been studied in many different fields and settings. Modern mathematical treatment of fair division via the cake cutting abstraction started in the 1940s [15]. Since then, many works presented algorithms or protocols for fair division [16,9,2,6], as well as theorems establishing the existence of fair divisions (under different interpretations of fairness) in different settings [7,16]. Starting from the mid 1990s, several books appeared on the subject [3,13,11], and much attention was given to the question of finding bounds on the number of steps required for dividing a cake fairly [10,14,8,12].

A more recent work by Caragiannis et al. [4] added the issue of social welfare into the framework of cake cutting. In particular, Caragiannis et al. aimed at showing bounds on the loss of social welfare caused by fairness requirements, by defining and analyzing the Price of Fairness (defined for different fairness criteria). The work in [4] considered fair division of divisible and indivisible goods, as well as divisible and indivisible chores; for each of these settings, it has provided bounds on the highest possible degradation in utilitarian welfare caused by three prominent fairness requirements — proportionality, envy-freeness, and equitability. Following this line of work, a recent work by a subset of the authors here [1] analyzed the utilitarian and egalitarian Price of Fairness in the setting of cake cutting where each piece is required to be a single connected interval. (This is in contrast to the work of [4] that allowed a piece in the division to be comprised of any union of intervals.)

Finally, the concept of partial divisions of a cake (in which not all the cake is allotted to the players) has also been considered in [6] and very recently in [5]. Interestingly, in each of these works, discarding of some of the cake serves a different purpose. In [6], the authors present a proportional, envy-free and truthful cake cutting algorithm for players with valuation functions of a restricted form. In that work, disposing of some of the cake is what ensures that the players have no incentive to lie to the protocol. In [5], a restricted case of non-additive valuation functions is considered, with one of the results being an approximately-proportional, envy-free protocol for two players. In that protocol, some cake is discarded in order to guarantee envy-freeness. Here, we show that leaving some cake unallocated can also increase social welfare.

2   Definitions and Preliminaries

As customary, we assume a 1-dimensional cake that is represented by the interval $[0, 1]$. We denote the set of players by $[n]$ (where $[n] = \{1, \dots, n\}$), and assume that each player i has a nonatomic (additive) measure v_i mapping each interval of the cake to its value for player i, and having $v_i([0, 1]) = 1$. Let x be some division of the cake between the players; we denote the value player i assigns to player j's piece in x by $u_i(x, j)$. We say that a division x is *complete* if it leaves no cake unallocated; otherwise, we say that the division is *partial*.

Definition 1. *We say that a cake instance with n players exhibits an α-dumping effect (with $\alpha > 1$ and with respect to some social welfare function $w(\cdot)$) if there exists a partial division y such that*

1. *y is envy-free; i.e. $u_i(y, i) \geq u_i(y, j)$ for all $i, j \in [n]$, and*
2. *for every envy-free complete division x, $w(y) \geq \alpha \cdot w(x)$.*

In this work, we consider two prominent social welfare functions: utilitarian and egalitarian. The utilitarian welfare of a division x is the sum of the players' utilities; formally, we write $u(x) = \sum_{i \in [n]} u_i(x, i)$. The egalitarian welfare of a division x is the utility of the worst-off player, i.e. $eg(x) = \min_{i \in [n]} u_i(x, i)$.

From this point forward, we restrict the discussion to divisions in which every player gets a single connected interval of the cake. The first reason for this restriction is that giving the players such connected pieces seem more "natural", and is in many scenarios more desirable than giving pieces composed of unions of intervals. The second reason is captured by the following simple result:

Proposition 1. *If players are allowed to get non-connected pieces (that are composed of unions of intervals), there can be no utilitarian or egalitarian dumping effect. In addition, in this setting no envy-free partial division can Pareto dominate all envy-free complete divisions.*

Proof. We prove for utilitarian welfare; the proof for egalitarian welfare and Pareto domination is analogous. Suppose that we allow such non-connected divisions, and assume that there is a utilitarian α-dumping effect, with $\alpha > 1$. Then there exists an envy-free partial division y such that for every envy-free complete division x, $\sum_{i \in [n]} u_i(y, i) \geq \alpha \cdot \sum_{i \in [n]} u_i(x, i)$. Let $U \subseteq [0, 1]$ be the part of the cake that was not allocated to the players in y; it is known (e.g. [7]) that U itself has a complete envy-free division y'. Note that giving each player i her part from y' in addition to her original piece from y yields again an envy-free division; call this division z. Clearly, z is a complete division of $[0, 1]$ having $u_i(z, i) \geq u_i(y, i)$ for all $i \in [n]$. It follows that for every envy-free division x

$$\sum_{i \in [n]} u_i(z, i) \geq \sum_{i \in [n]} u_i(y, i) \geq \alpha \cdot \sum_{i \in [n]} u_i(x, i) > \sum_{i \in [n]} u_i(x, i) \; ;$$

a contradiction.

We thus formally define a connected division of a cake to n players simply as a sequence of n non-intersecting open intervals[1]; the first interval is given to the first player, the second to the second player, etc. We will say that such a division is complete if the union of these intervals (including their endpoints) equals the entire cake; otherwise, we will say that the division is partial. Note that a partial division may leave several disjoint intervals unallocated.

Finally, we give the definition of the Price of Envy-Freeness, first defined in [4], which aims to measure the highest degradation in social welfare that may be necessary to achieve envy-freeness. As we show next, the Price of Envy-Freeness for a cake instance gives an upper bound on the dumping effect for the same instance and welfare function.

Definition 2. *Let I be a cake instance, X the set of all complete divisions of I, and X_{EF} the set of all complete envy-free divisions of I. The* Price of Envy-Freeness *of the cake instance I, with respect to a social welfare function $w(\cdot)$, is defined as the ratio*

---

[1] Since we assume that the valuation functions of all players are nonatomic, open and closed intervals always have the same value.

$$\frac{\max_{x \in X} w(x)}{\max_{y \in X_{EF}} w(y)} \ .$$

Proposition 2. *The utilitarian (resp. egalitarian) dumping effect is bounded from above by the utilitarian (resp. egalitarian) Price of Envy-Freeness.*

Proof. We again prove only for utilitarian welfare. Assume, by contradiction, that there exists a cake cutting instance with n players and with utilitarian dumping effect of β, while the utilitarian Price of Envy-Freeness for these n players is $\alpha < \beta$. Then there exists a partial division y such that for every envy-free complete division x, $\sum_{i \in [n]} u_i(y, i) \geq \beta \cdot \sum_{i \in [n]} u_i(x, i)$. Note that every inclusion-maximal unalloted interval is adjacent to at least one interval that is given to a player. Therefore, consider the complete division z which allocates each player her interval as in y, and in addition adds the previously-unalloted intervals to the piece of one of the adjacent players (chosen arbitrarily). This is clearly a (not necessarily envy-free) complete division in which $u_i(z, i) \geq u_i(y, i)$ for every $i \in [n]$. We get that for every envy-free division x

$$\sum_{i \in [n]} u_i(z, i) \geq \sum_{i \in [n]} u_i(y, i) \geq \beta \cdot \sum_{i \in [n]} u_i(x, i) > \alpha \cdot \sum_{i \in [n]} u_i(x, i) \ ,$$

contradicting our bound on the utilitarian Price of Envy-Freeness.

3   Utilitarian Welfare

Theorem 1. *The utilitarian dumping effect with n players may be as high as $\Theta(\sqrt{n})$, and this bound is asymptotically tight.*

We show that for every $k, t \in \mathbb{N}$, there exists a cake cutting instance with $n = 2k(3t - 2)$ players in which throwing away $(t - 1)k$ intervals can improve the utilitarian welfare of the best envy-free division by a factor exceeding $\frac{kt+2t}{k+3t}$. In particular, choosing $t = \Theta(k)$ yields an improvement of $\Theta(\sqrt{n})$. The matching upper bound follows from Proposition 2, combined with Theorem 1 of [1], which shows an upper bound of $\frac{\sqrt{n}}{2} + 1 - o(1)$ on the utilitarian Price of Envy-Freeness.

For the lower bound, we construct a cake with three parts: a "common" part, a "high-values" part, and a "compensation" part. Furthermore, each of the latter two parts is itself divided into k identical subparts. However, due to space constrains, we are unable to provide here the full details of the construction (which can be found in the full version of the paper). Instead, we illustrate a key structure used in the construction, reason about its properties, and explain in general lines how it is used in the construction of the entire cake.

Let us consider a subset of $3t-2$ players, comprised of $2t-2$ "Type A" players, $t - 1$ "Type B" players, and one "chosen" player C. For $0 \leq i \leq t - 2$ we will have the players $3i + 1$ and $3i + 2$ be of Type A, and the player $3i + 3$ be of Type B; we will say that players $3i + 1$ and $3i + 2$ are "neighbors". In the "high values" part of the cake, the chosen player C has t intervals she desires, each of them being of value $\frac{1}{t}$ to her. Between every two consecutive pieces desired by C there are two more pieces, desired by a pair of neighbors of Type A. Each neighbor desires one of these pieces, and considers that piece to be worth $\frac{4}{n}$ (as can be seen on the top part of Figure 2). In addition, each pair of neighbors desires two more pieces, located in the "compensation" part of the cake. Namely, for every two neighbors $3i+1$ and $3i+2$, we have two pieces desired by $3i+1$ followed by two pieces desired by $3i+2$; each of these pieces is worth $\frac{2}{n}$ to the corresponding player. In between these four pieces, there are three pieces desired by the player $3i+3$ of Type B: the first and third pieces have each a value of $\frac{3}{2n}$ to that player, and the second has value of $\frac{1}{n}$. The reader is again referred to Figure 2 for a graphical representation; note that while the preferences of the chosen player C are completely described, this is not so for the other players, as the pieces we have described do not add up to a value of 1 for these players.

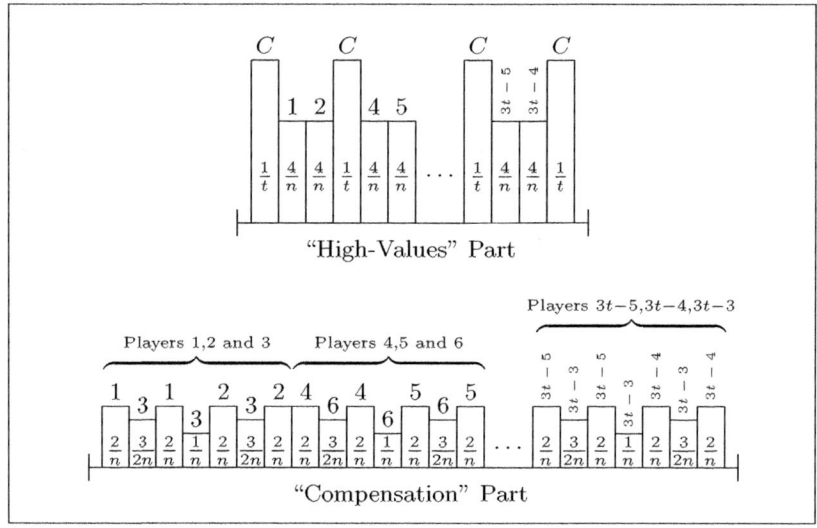

Fig. 2. The (incomplete) preferences of one set of players. The number above each column denotes which player has that valuation.

Lemma 1. *Assume a cake and a set of players with preferences as above for the "high-values" and "compensation" parts (in addition to some preferences on the remainder of the cake). Assume, in addition, that the following properties hold:*

(P1) If an envy-free division gives some Type A player a piece of value $\geq \frac{4}{n}$, this piece must either intersect that player's desired piece from the "high values" part, or contain both of her desired pieces from the "compensation" part.

(P2) If an envy-free division gives some Type B player a piece of value $\geq \frac{2}{n}$, this piece must be completely contained in the "compensation" part.

Then no envy free division gives the chosen player C a piece of value $> \frac{1}{t}$.

Proof. Suppose that C does get such a piece in some envy-free division. It must be that this piece intersects at least two of C's desired pieces; in particular, there are two neighbors of Type A such that C completely devours a desired piece of each of them. Let these players be $3i + 1$ and $3i + 2$; these players consider C's piece as worth at least $\frac{4}{n}$ and thus must each get a piece of at least this value to avoid envy. By the property (P1), the only way to do that is to give each of them their two desired pieces from the compensation part of the cake. Recall that each of the two players has two desired pieces in the compensation part: denote them (from left to right) A_1, A_2, A_3 and A_4. In between those pieces, there are three pieces which we will denote by B_1, B_2 and B_3, desired by the Type B player $3i + 3$. In order to give each of these Type A players a piece of value $\frac{4}{n}$, we must give player $3i + 1$ an interval containing A_1, B_1 and A_2, and player $3i + 2$ an interval containing A_3, B_3 and A_4. Each of these intervals is worth at least $\frac{3}{2n}$ to player $3i + 3$ who thus cannot be satisfied with the piece B_2 (worth to her only $\frac{1}{n}$), and must therefore get her share from another part of the cake. Since no other players have any value for the piece B_2, it must be shared between players $3i + 1$ and $3i + 2$ whose pieces are the closest to it. At least one of these players will get at least half of B_2, and the piece of this player will be worth at least $\frac{2}{n}$ to player $3i + 3$; by (P2), this must cause envy.

In the full construction we have k sets of players, each identical to the set described above. This sums up to k chosen players, $k(2t - 2)$ Type A players, and $k(t-1)$ Type B players, totaling in $k(3t - 2) = \frac{n}{2}$ players. The remaining players are all "common players", who are all just interested in the "common part" of the cake (which for all other players has the missing value needed for the entire value of the cake to sum up to 1). The cake is composed of this "common part", k copies of the "high-values" part, and k copies of the "compensation part" (one copy of each part for each set of players). We then show that, as in Lemma 1, a complete division cannot give any chosen player more than $\frac{1}{t}$. However, throwing away the pieces alluded to as $B2$ in the proof of the lemma allows us to obtain an envy-free division giving each chosen player a value of 1, with only a minor decrease in the value we give to the other players compared to any complete envy-free division.

4   Egalitarian Welfare

Theorem 2. *The egalitarian dumping effect with n players may get arbitrarily close to $\frac{n}{3}$, and this bound is asymptotically tight.*

We show that for every $k \in \mathbb{N}$ there is a cake cutting instance with $n = 3k + 1$ players in which throwing away k intervals can improve the egalitarian welfare of the best envy-free division by a factor arbitrarily close to $\frac{n}{3}$. The matching upper bound follows from Proposition 2, combined with Theorem 5 of [1], which shows an upper bound of $\frac{n}{2}$ on the egalitarian Price of Envy-Freeness.

To illustrate the main ideas of the lower bound (whose full proof is also deferred to the full version of the paper), we consider the simple case of $k = 1$

76   O. Arzi, Y. Aumann, and Y. Dombb

(i.e. $n = 4$). Fix some small $\epsilon > 0$. We create a cake with two parts: the "main part" and the "last player" part. In the main part, we have two "blocks" of four intervals: in both blocks, the first interval is of value $\frac{1}{4}$ to player 4 and the third interval is of value $\frac{1-\epsilon}{3}$ to player 3. The remaining intervals (second and fourth) of the first block are each of value $\frac{1+\epsilon}{4}$ to player 1, while those of the second block are each of value $\frac{1+\epsilon}{4}$ to player 2. The first block is followed by an interval of value ϵ to player 3; we denote this interval by I. The second block is followed by an interval of value $\frac{1-\epsilon}{3}$ to player 3. In the "last player" part we have two intervals of value $\frac{1}{4}$ to player 4; between these intervals there are two more intervals, one considered by player 1 as worth $\frac{1-\epsilon}{2}$, and the other considered by player 2 as worth $\frac{1-\epsilon}{2}$. Figure 3 illustrates these preferences graphically.

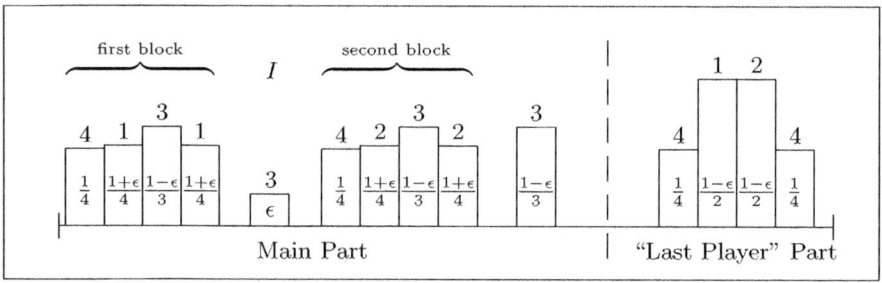

Fig. 3. Preferences of all players for the case of $n = 4$

Lemma 2. *In any complete envy-free division of the above cake, player 4 has utility $\leq \frac{1}{4}$. In particular, this is also a bound on the egalitarian welfare of such divisions.*

Proof. Suppose otherwise, then it has to be that the piece of player 4 intersects at least two of her four desired intervals. If it intersects two of her first three desired pieces, she will completely devour at least one block; however, each block is worth strictly more than $\frac{1}{2}$ to some player, who will then envy player 4.

Thus, player 4's interval must be contained in the "last player" section, and intersect player 4's third and fourth desired intervals. However, in this case, player 4's piece is worth $\frac{1-\epsilon}{2}$ to both players 1 and 2; to ensure envy-freeness, they both must also get a piece of at least that value. If ϵ is small enough, player 1 must get a piece containing the third interval of the first block, and player 2 must get a piece containing the third interval of the second block. Each of these pieces are worth $\frac{1-\epsilon}{3}$ to player 3; this forces player 3 to get the rightmost of her desired intervals in order to avoid envy. The interval I must therefore be split between players 1 and 2, making at least one of them end up with a piece worth more than $\frac{1}{3}$ to player 3, which will then be envious; a contradiction.

Lemma 3. *In the above cake, discarding the interval I allows for an envy-free division with egalitarian welfare of $\frac{1-\epsilon}{3}$.*

Proof. Suppose we discard the piece I. We can now allocate the entire "last player" part to player 4, giving her utility $\frac{1}{2}$. In the main part, we give player 1 the entire first block, and player 2 the entire second block. Finally we give player 3 the interval following the second block. It is easy to verify that this division is indeed envy-free, and that its egalitarian welfare is $\frac{1-\epsilon}{3}$.

We have shown a dumping effect of $\frac{4(1-\epsilon)}{3}$ for the case of $n = 4$ players. The proof of Theorem 2 uses a generalization of this construction, with $3k$ players taking the role of players 1-3, and a single player taking the role of player 4. However, due to space constraints, we omit the full details and proof of this generalization from this extended abstract.

Finally, we state the following theorem, whose proof is also deferred to the full version of the paper, showing that the bound of $\frac{n}{2}$ is indeed tight for $n \leq 4$.

Theorem 3. *For $n \leq 4$ players, there are examples in which the egalitarian dumping effect is arbitrarily close to $\frac{n}{2}$.*

5 Pareto-Dominant Partial Divisions

A division x is said to *Pareto dominate* another division y if for all i, $u_i(x, i) \geq u_i(y, i)$, and at least one of these inequalities is strict; in other words, if at least one player does better in x than in y, and no one does worse. x *strictly Pareto dominates* y if for all i, $u_i(x, i) > u_i(y, i)$, i.e. if *everyone* is doing better in x.

We first show that starting from *any* envy-free complete division it is impossible to strictly improve the utility of all players simultaneously.

Theorem 4. *Let x be an envy-free complete division. Then there is no other division, partial or complete, that strictly Pareto dominates x.*

Proof. Our proof hinges on the following observation, due to [1]:

> *Let y be a division such that $u_i(y, i) > u_i(x, i)$ for some $i \in [n]$. Since i values any other piece in the division x at most as much as her own, it has to be that in y, i gets an interval that intersects pieces that were given to at least two different players in x (possibly including i herself).*

In other words, in order for a player i to get a piece worth more than her piece in x, she must get at least one "boundary" (between two consecutive pieces) from x. Thus, a (partial or complete) division that strictly Pareto dominates x must give (at least) one such boundary to each player. However, since x is a connected division it contains only $n - 1$ boundaries, one less than the number of players.

It it thus interesting that there *do* exist instances in which an envy-free partial division (non-strictly) Pareto dominates *every* envy-free complete division. Moreover, in some cases the partial division improves the utility of almost all the players, and by a significant (constant) factor.

Theorem 5. *For every $n > 2$, there exists a cake cutting instance with n players and an envy-free partial division giving $n - 2$ players* twice *the value they would get in any envy-free complete division, while giving the remaining two players at least as much as they would get in any envy-free complete division.*

Proof. Let $n > 2$, and fix some $0 < \epsilon < \frac{1}{n(n+1)}$. Consider the following valuations:

- Each player $1 \le i \le n - 1$ ("focused players") desires only the interval $(\frac{i}{n} - \epsilon, \frac{i}{n} + \epsilon)$, and considers it to be of value 1.
- Player n assigns a uniform valuation to the entire cake.

Now, for envy-free complete division of the above cake, it must be that:

1. *Player n gets a piece of physical size $\ge \frac{1}{n}$.*
   Since we give away all the cake, some player must get a piece of physical size at least $\frac{1}{n}$; if player n does not get such a piece, she will envy that player.
2. *Player n cannot get any piece containing some neighborhood of a point $\frac{i}{n}$ for $i \in [n-1]$.*
   Because of the previous observation, if player n gets such a piece then her piece contains the interval $(\frac{i}{n} - \delta, \frac{i+1}{n} - \delta)$ for some $0 < \delta < 1$. Such a piece is always worth strictly more than $\frac{1}{2}$ to player i, and will make her envious. Therefore, *player n must get a piece of the form $(\frac{i-1}{n}, \frac{i}{n})$ for some $i \in [n]$.*
3. *Every "focused player" has to get a piece of physical size exactly $\frac{1}{n}$.*
   First, it is clear that if some player $i \in [n-1]$ gets a piece of size $> \frac{1}{n}$, player n will envy that player. On the other hand, we have that the players $[n-1]$ have to share cake of total physical size $\frac{n-1}{n}$; since none of them can get a piece of size larger than $\frac{1}{n}$, each of them must get a piece of size exactly $\frac{1}{n}$.

From these observations we obtain that in any envy-free division, all the cuts are at points $\frac{i}{n}$ with $i \in [n-1]$; in such a division, player n always has utility $\frac{1}{n}$, and every other player has utility $\frac{1}{2}$.

Now consider the following partial division of the cake. First, give player n the piece $(0, \frac{1}{n})$, and player 1 the piece $(\frac{1}{n}, \frac{2}{n} - 2\epsilon)$. Then give each player $2 \le i \le n-1$ the next piece of size $\frac{1}{n} - \epsilon$, which is the interval $\left(i \cdot (\frac{1}{n} - \epsilon), (i+1) \cdot (\frac{1}{n} - \epsilon)\right)$. Finally, we throw away the (non-allocated) remainder.

In this division, player n has value $\frac{1}{n}$, which is just as good as in any complete division. Similarly, player 1 has value $\frac{1}{2}$, which again is as good as she can get in any complete division. Players 2 through $n - 1$, on the other hand, get each her entire desired interval; otherwise, the position of the right boundary of player $(n - 1)$'s piece must be to the left of the point $\frac{n-1}{n} + \epsilon$. However, since we took $\epsilon < \frac{1}{n(n+1)}$, we have that the right boundary of player $(n - 1)$'s piece is at

$$\left(\frac{1}{n} - \epsilon\right) \cdot n > \left(\frac{1}{n} - \frac{1}{n(n+1)}\right) \cdot n = \frac{n}{n+1} = \frac{n^2 - 1}{n(n+1)} + \frac{1}{n(n+1)} > \frac{n-1}{n} + \epsilon.$$

Therefore, each of these players gets a piece of value 1, which is twice what they could get in any complete division.

Finally, it is clear that none of the players feel envy: Players 2 through $n-1$ feel no envy (having gotten all they desire in the cake). Player n feels no envy since she receives the physically-largest piece in the division. Player 1 also feels no envy as her piece has value $\frac{1}{2}$, and so no other player could have gotten a piece with a larger value for her.

We note that the construction above can be also used to show a utilitarian dumping effect arbitrarily close to $\frac{2(n-1)+\frac{2}{n}}{(n-1)+\frac{2}{n}}$ (one need only move the leftmost boundary in the partial division to $\frac{1}{n} - \epsilon$). While this is asymptotically inferior to the bound shown in Theorem 1, this construction is much simpler, and works for as few as two players. In fact, for $n=2$ this construction coincides with the example given in the introduction, and moreover gives a provably tight lower bound: the dumping effect of $\frac{3}{2} - \epsilon$ we obtain in this case matches the $n - \frac{1}{2}$ upper bound on the utilitarian Price of Envy-Freeness given in [4].

6 Discussion and Open Problems

In this work, we have studied the dumping effect and its possible magnitude. We have shown that the increase in welfare when discarding some of the cake can be substantial, moving from $1/n$ to $\Theta(1)$ for egalitarian welfare and from $\Theta(1)$ to $\Theta(\sqrt{n})$ for utilitarian welfare, and have shown a Pareto improvement that improves by a factor of two all but two players. In fact, in some cases discarding some of the cake can essentially eliminate the social cost associated with fair division. It is interesting to note that all of our lower bound constructions have an additional nice property — no player desires any discarded piece more than her own piece. Thus, not only do players not envy each other, but they also do not feel much loss with any discarded piece.

Several problems remain open. First, we note that while our bounds for the utilitarian and egalitarian welfare functions are asymptotically tight, there are still constant gaps which await closure. With regards to Pareto improvement, we provided a construction where all but two players improve their utility by a factor of two. An interesting open problem is to see whether a stronger Pareto effect can be obtained. Before we do so, however, we must first define the exact criteria by which we evaluate Pareto improvements. Possible criteria include: the number of players that increase their utility, the largest utility increase by any player, and the total utility increase of the players (= utilitarian welfare).

More important, perhaps, is that all of our results are existential in nature, but do not provide guidance on what to do in specific cases. It is thus of interest to develop algorithms to determine what, if any, parts of the cake it is best to discard in order to gain the most social welfare, for the different welfare functions.

Finally, our work joins other recent works [6,5] that imply that leaving some cake unallocated may be a useful technique in fair division algorithms. Following this direction, it may be interesting to see if discarding of some cake may also help in finding socially-efficient envy-free connected divisions. Generalizing beyond fair division, it would be be interesting to see if such an approach, of intentionally forgoing or discarding some of the available goods, can also benefit other social interaction settings.

References

1. Aumann, Y., Dombb, Y.: The efficiency of fair division with connected pieces. In: Saberi, A. (ed.) WINE 2010. LNCS, vol. 6484, pp. 26–37. Springer, Heidelberg (2010)
2. Brams, S.J., Taylor, A.D.: An envy-free cake division protocol. The American Mathematical Monthly 102(1), 9–18 (1995)
3. Brams, S.J., Taylor, A.D.: Fair Division: From cake cutting to dispute resolution. Cambridge University Press, New York (1996)
4. Caragiannis, I., Kaklamanis, C., Kanellopoulos, P., Kyropoulou, M.: The efficiency of fair division. In: Leonardi, S. (ed.) WINE 2009. LNCS, vol. 5929, pp. 475–482. Springer, Heidelberg (2009)
5. Caragiannis, I., Lai, J., Procaccia, A.: Towards more expressive cake cutting. In: IJCAI: International Joint Conferences on Artificial Intelligence (2011)
6. Chen, Y., Lai, J., Parkes, D.C., Procaccia, A.D.: Truth, justice, and cake cutting. In: AAAI (2010)
7. Dubins, L.E., Spanier, E.H.: How to cut a cake fairly. The American Mathematical Monthly 68(1), 1–17 (1961)
8. Edmonds, J., Pruhs, K.: Cake cutting really is not a piece of cake. In: SODA 2006: Proceedings of the Seventeenth Annual ACM-SIAM Symposium on Discrete Algorithm, pp. 271–278. ACM, New York (2006)
9. Even, S., Paz, A.: A note on cake cutting. Discrete Applied Mathematics 7(3), 285–296 (1984)
10. Magdon-Ismail, M., Busch, C., Krishnamoorthy, M.S.: Cake-cutting is not a piece of cake. In: Alt, H., Habib, M. (eds.) STACS 2003. LNCS, vol. 2607, pp. 596–607. Springer, Heidelberg (2003)
11. Moulin, H.J.: Fair Division and Collective Welfare. Number 0262633116 in MIT Press Books. The MIT Press (2004)
12. Procaccia, A.D.: Thou shalt covet thy neighbor's cake. In: IJCAI, pp. 239–244 (2009)
13. Robertson, J., Webb, W.: Cake-cutting algorithms: Be fair if you can. A K Peters, Ltd., Natick (1998)
14. Sgall, J., Woeginger, G.J.: A lower bound for cake cutting. In: Di Battista, G., Zwick, U. (eds.) ESA 2003. LNCS, vol. 2832, pp. 459–469. Springer, Heidelberg (2003)
15. Steinhaus, H.: Sur la division pragmatique. Econometrica 17(Supplement: Report of the Washington Meeting), 315–319 (1949)
16. Stromquist, W.: How to cut a cake fairly. The American Mathematical Monthly 87(8), 640–644 (1980)

The Price of Optimum in a Matching Game*

Bruno Escoffier[1,2], Laurent Gourvès[2,1], and Jérôme Monnot[2,1]

[1] Université de Paris-Dauphine, LAMSADE, 75775 Paris, France
[2] CNRS, UMR 7243, 75775 Paris, France

Abstract. Due to the lack of coordination, it is unlikely that the selfish players of a strategic game reach a socially good state. Using Stackelberg strategies is a popular way to improve the system's performance. Stackelberg strategies consist of controlling the action of a fraction α of the players. However compelling an agent can be costly, unpopular or just hard to implement. It is then natural to ask for the least costly way to reach a desired state. This paper deals with a simple strategic game which has a high price of anarchy: the nodes of a simple graph are independent agents who try to form pairs. We analyse the optimization problem where the action of a *minimum* number of players shall be fixed and any possible equilibrium of the modified game must be a social optimum (a maximum matching).

For this problem, deciding whether a solution is feasible or not is not straitforward, but we prove that it can be done in polynomial time. In addition the problem is shown to be APX-hard, since its restriction to graphs admitting a vertex cover is equivalent, from the approximability point of view, to VERTEX COVER in general graphs.

1 Introduction

We propose analyzing the following non cooperative game. The input is a simple graph $G = (V, E)$ where every vertex is controlled by a player whose strategy set is his neighborhood in G. If a vertex v selects a neighbor u while u selects v then the two nodes are matched and they both have utility 1. If a vertex v selects a neighbor u but u does not select v then v is unmatched and its utility is 0. Each player aims at maximizing its own utility.

Matchings in graphs are a model for many practical situations where nodes may represent autonomous entities (e.g. the *stable marriage problem* [1] and the *assignment game* [2]). For instance, suppose that each node is a tennis player searching for a partner. An edge between two players means that they are available at the same time, or just that they know each other. As another example, consider a set of companies on one side, each offering a job, and on the other side a set of applicants. There is an edge if the worker is qualified for and interested in the job.

Taking the number of matched nodes as the social welfare associated with a strategy profile (a maximum cardinality matching is then a social optimum),

---

* This work is supported by ANR, project COCA, ANR-09-JCJC-0066.

G. Persiano (Ed.): SAGT 2011, LNCS 6982, pp. 81–92, 2011.

we can rapidly observe that the game has a high *price of anarchy*. The system needs regulation because the uncoordinated and selfish behavior of the players deteriorates its performance. How can we do this regulation? One can enforce a maximum matching but forcing some nodes' strategy may be costly, unpopular or simply hard to implement. When both cases (complete freedom and total regulation) are not satisfactory, it is necessary to make a tradeoff. In this paper we propose to fix the strategy of some nodes; the other players are free to make their choice. The only requirement is that every equilibrium of the modified game is a social optimum (a maximum matching). Because it is unpopular/costly, the number of forced players should be *minimum*. We call the optimization problem MFV for *minimum forced vertices*. The challenging task is to identify the nodes which play a central role in the graph. As we will see, the problematic is known as the *price of optimum* [3,4] in the well established framework of *Stackelberg games*.

1.1   Related Work

There is a great interest in how uncoordinated and selfish agents make use of a common resource [5,6]. A popular way of modeling the problem is by means of a noncooperative game and by viewing its Nash equilibria as outcomes of selfish behavior. In this context, the *price of anarchy* (PoA) [6,7], defined as the value of the worst Nash equilibrium relative to the social optimum, is a well established measure of the performance deterioration. A game with a high PoA needs regulation and several ways to improve the system performance exist in the literature, including *coordination mechanisms* [8] and *Stackelberg strategies* [3,4,9,10,11,12,13,14]. This paper deals with the latter.

In [9] Roughgarden studies a nonatomic scheduling problem where a rate of flow r should be to assigned to a set of parallel machines with load dependent latencies. There are two kinds of players: a leader controlling a fraction α of r and a set of followers, with everyone of them handling an infinitesimal part of $(1 - \alpha)r$. The leader, interested in optimizing the total latency, plays first (i.e. assigns αr to the machines) and keeps his strategy fixed. The followers react independently and selfishly to the leaders strategy, optimizing their own latency. The author gives helpful properties of the game: (i) Nash assignments exist and are essentially unique, (ii) there exists an assignment induced by a Stackelberg strategy and any two such assignments have equal cost. He provides an algorithm for computing a leader strategy that induces an equilibrium with total latency no more than $1/\alpha$ times that of the optimal assignment of jobs to machines. This is the best possible approximation ratio but the algorithm does not always use *at best* the amount of players that it can compel. Indeed Roughgarden shows that it is NP-hard to compute such an optimal Stackelberg strategy.

Further extensions and improvements on the nonatomic scheduling problem can be found in [10,11,14].

In [13] Sharma and Williamson introduce the Stackelberg threshold (also called the *value of altruism*) which is the minimum amount of centrally controlled flow such that the cost of the resulting equilibrium solution is strictly less than the cost of the Nash equilibrium (where no fraction of the flow is controlled). The approach is considered for the nonatomic scheduling game.

In [3] Kaporis and Spirakis study the nonatomic scheduling game with the aim of computing the least portion of flow α^* that the leader must control in order to enforce the overall optimum cost. α^* is then called the *price of optimum*. They provide an optimal algorithm which works for single-commodity instances – instances which are more general than parallel instances discussed in [9]. As mentioned in [3] the concept of price of optimum dates back to the work of Korilis, Lazar and Orda [4] who study a different system environment.

Finally one can mention the work of Fotakis [12] who proposed an atomic (discrete) version of Roughgarden's approach.

1.2   Motivation, Organization and Results

The matching game studied in this article captures practical situations where uncoordinated agents try to form pairs. We focus on the Nash equilibria because this solution concept is arguably the most important concept in game theory in order to capture possible outcomes of strategic games. As stated in Theorem 1 (next section), the game has a high price of anarchy so it is relevant to study Stackelberg strategies, the Stackelberg threshold and the price of optimum. This paper is devoted to the last approach because MFV is exactly the price of optimum of the matching game. Recall that in MFV, a leader interested in the social welfare fixes the strategy of a *minimum* number of nodes so that any equilibrium reached by the unforced nodes creates a *maximum* number of pairs.

At this point one can stress important differences between the nonatomic scheduling game [9,3] and the matching game: the matching game is atomic, there may be several completely different optima, and two equilibra induced by a common Stackelberg strategy may have significantly differing social utility. Due to the last observation, it is not trivial to decide whether a Stackelberg strategy induces a social optimum or not. In other words, separating feasible and infeasible solutions to the MFV problem is not direct.

We first give a formal definition of the noncooperative game and show that it has a high price of anarchy (Section 2). The MFV problem is then introduced. In Section 3 we show that we can decide in polynomial time whether a solution is feasible or not. In particular one can detect graphs for which any pure Nash equilibrium corresponds to a maximum matching though no vertex is forced.

Next we investigate the complexity and the approximability of MFV. Our result is that MFV in graphs admitting a perfect matching is, from the approximability point of view, equivalent to VERTEX COVER. Hence MFV is APX-hard in general graphs and there exists a 2-approximation algorithm in graphs admitting a perfect matching. Concluding remarks and future works are given in Section 5. Due to space limitations, some proofs are omitted.

2   The Strategic Game and the Optimization Problem

We are given a simple connected graph $G = (V, E)$. Every vertex is controlled by a player so we interchangeably mention a vertex and the player who controls it. The strategy set of every player i is his neighborhood in G, denoted by $\mathcal{N}_G(i)$. Then the strategy set of a leaf in G is a singleton. Throughout the article S_i designates the action/strategy of player i. A player is matched if the neighbor that he selects also selects him. Then i is matched under S if $S_{S_i} = i$. A player has *utility* 1 when he is matched, otherwise his utility is 0. The utility of player i under state S is denoted by $u_i(S)$.

The *social welfare* is defined as the number of matched nodes. We focus on the pure strategy Nash equilibria, considering them as the possible outcomes of the game. It is not difficult to see that every instance admits a pure Nash equilibrium. In addition, the players converge to a Nash equilibrium after at most $|V|/2$ rounds.

Interestingly there are some graphs for which the players always reach a social optimum: paths of length 1, 2 and 4; cycles of length 3 and 5; stars, etc. However the social welfare can be very far from the social optimum in many instances as the following result states.

Theorem 1. *The PoA is* $\max\{2/|V|, 1/\Delta\}$ *where* Δ *denotes the maximum degree of a node.*

Theorem 1 indicates that the system needs regulation to achieve an acceptable state where a maximum number of players are matched. That is why we introduce a related optimization problem, called MFV for *minimum forced vertices*.

For a graph $G = (V, E)$, instance of the MFV problem, a solution is a pair $\langle T, Q \rangle$ where Q is a subset of players and every player i in Q is forced to select node $T_i \in \mathcal{N}_G(i)$ (i.e. $T = (T_i)_{i \in Q}$). In the following $\langle T, Q \rangle$ is called a *Stackelberg strategy* or simply a *solution*. A state S is a *Stackelberg equilibrium* resulting from the Stackelberg strategy $\langle T, Q \rangle$ if $S_i = T_i$ for every $i \in Q$, and $\forall i \in V \setminus Q$, $\forall j \in \mathcal{N}_G(i)$, $u_i(S) \geq u_i(S_{-i}, j)$. Here (S_{-i}, j) denotes S where S_i is set to j.

A solution $\langle T, Q \rangle$ to the MFV problem is said feasible if every Stackelberg equilibrium is a social optimum. The value of $\langle T, Q \rangle$ is $|Q|$. This value is to be minimized.

Now let us introduce some notions that we use throughout the article. The matching induced by a strategy profile S is denoted by $\mathcal{M}^S$ and defined as $\{(u, v) \in E : S_u = v \text{ and } S_v = u\}$. We also define three useful notions of compatibility:

- A matching M and a state S are compatible if $M = \mathcal{M}^S$;
- A state S and a solution $\langle T, Q \rangle$ are compatible if $T_i = S_i$ for all $i \in Q$;
- A matching M and a solution $\langle T, Q \rangle$ are compatible if there exists a state S compatible with both M and $\langle T, Q \rangle$.

We sometimes write that a matching is *induced* by a solution if they are compatible.

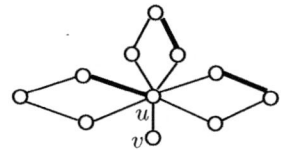

Fig. 1. Example with a non optimum matching M (bold edges)

Let us consider the example depicted in Figure 1. If no node is forced, then one can easily find an equilibrium inducing the non optimum matching M (bold edges). On the other hand, if we force u to play v, then any Stackelberg equilibrium induces a maximum matching. Hence, forcing this node is an optimal solution to the MFV problem.

3   Feasible Solutions

It is easy to produce a feasible solution to any instance of the MFV problem: simply compute a maximum matching and force each matched node toward its mate in the matching. However, deciding whether a given solution, even the empty one, is feasible or not is not straightforward. This is due to the fact that the social welfares of two Stackelberg equilibria compatible with a solution $\langle T, Q \rangle$ may significantly differ.

Let $\mathcal{M}^*$ be a maximum matching compatible with a solution $\langle T, Q \rangle$. Then it is easy to see that there exists a Stackelberg equilibrium S^* compatible with both $\mathcal{M}^*$ and $\langle T, Q \rangle$ (if v is forced then it plays T_v, if it is matched in $\mathcal{M}^*$ then it plays its mate, otherwise it plays any vertex in his neighborhood). By definition, if $\mathcal{M}^*$ is not maximum in G, then $\langle T, Q \rangle$ is not feasible. However, the reverse is not true: another Stackelberg equilibrium compatible with $\langle T, Q \rangle$ may induce a matching which is not maximum in G. In this section we show how to determine in polynomial time if a solution $\langle T, Q \rangle$ is feasible or not.

In the sequel, $\mathcal{M}^*$ denotes a matching compatible with $\langle T, Q \rangle$ and we assume that $\mathcal{M}^*$ is maximum in G. In addition S^* denotes a Stackelberg equilibrium compatible with $\langle T, Q \rangle$ and $\mathcal{M}^*$. Note that a maximum matching compatible with a solution $\langle T, Q \rangle$ can be computed in polynomial time: start from G, remove every edge (u, v) such that u is forced to play a node w different from v, and compute a maximum matching in the resulting graph.

We will resort to patterns called *diminishing configurations*. As we will prove, a solution is not feasible if and only if it contains a diminishing configuration with respect to *any* compatible maximum matching. In fact the presence of a diminishing configuration is an opportunity for the players to reach a stable, but non optimal, state. The next key point is that one can detect diminishing configurations in polynomial time. It is trivial for all but one diminishing configuration; the difficult case is reduced to a known result due to Jack Edmonds and quoted in [15]. The diminishing configurations are of three kinds: long, short and average.

Definition 1. — $\mathcal{M}^*$ and $\langle T, Q \rangle$ possess a *long diminishing configuration* if
there are $2r$ vertices $v_1, \ldots, v_{2r}$ arranged in a path as on Figure 2 (a) (r is
an integer such that $r \geq 2$) and a strategy profile S^* satisfying
  - S^* is a Stackelberg equilibrium compatible with $\langle T, Q \rangle$ and $\mathcal{M}^*$
  - if $v_1 \notin Q$ then there is no $v \in V \setminus \{v_1, \ldots, v_{2r}\}$ such that $S_v^* = v_1$
  - if $v_{2r} \notin Q$ then there is no $v \in V \setminus \{v_1, \ldots, v_{2r}\}$ such that $S_v^* = v_{2r}$
- $\mathcal{M}^*$ and $\langle T, Q \rangle$ possess a *short diminishing configuration* if there exists one
  pattern among those depicted on Figure 2 (b) to (g) and a strategy profile S^*
  satisfying
  - S^* is a Stackelberg equilibrium compatible with $\langle T, Q \rangle$ and $\mathcal{M}^*$
  - there is no node $v \in V \setminus \{v_1, v_2\}$ such that $S_v^* \in \{v_1, v_2\}$
- $\mathcal{M}^*$ and $\langle T, Q \rangle$ possess an *average diminishing configuration* if there exists
  one pattern among those depicted on Figure 2 (h) to (j) and a strategy profile
  S^* satisfying
  - S^* is a Stackelberg equilibrium compatible with $\langle T, Q \rangle$ and $\mathcal{M}^*$
  - there is no $v \in V \setminus \{v_1, v_2, y\}$ such that $S_v^* \in \{v_1, v_2, z\}$

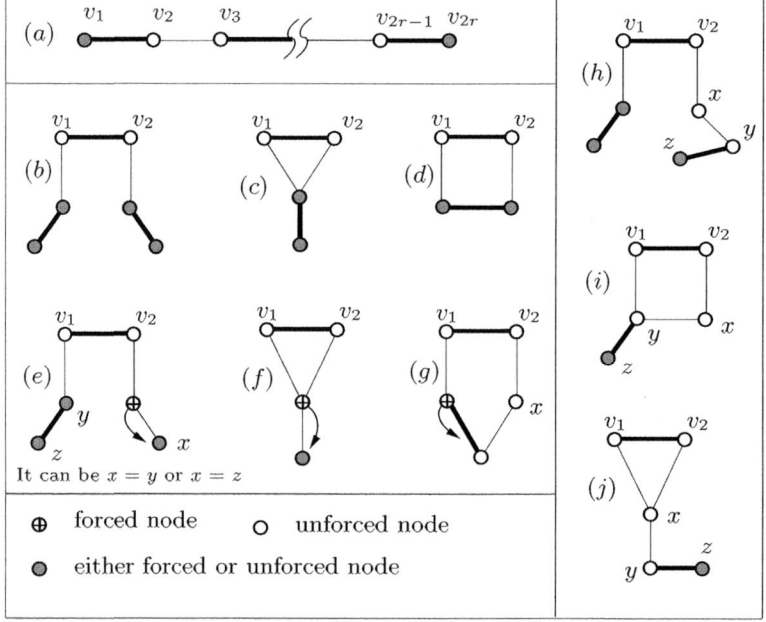

Fig. 2. The ten diminishing configurations. Every bold edge belongs to $\mathcal{M}^*$, every thin
edge belongs to $E \setminus \mathcal{M}^*$. A white node is not in Q while crossed node must belong to
Q. Grey nodes can be in Q or not. For the case (e), nodes x and y (resp. z) can be the
same.

In the following Lemma, we assume that $\mathcal{M}^*$, the maximum matching compatible with $\langle T, Q \rangle$, is also maximum in G.

Lemma 1. $\langle T, Q \rangle$ *is not feasible if and only if* $\mathcal{M}^*$ *and* $\langle T, Q \rangle$ *possess a diminishing configuration.*

Proof. (sketch)

($\Leftarrow$) Suppose that $\mathcal{M}^*$ and $\langle T, Q \rangle$ possess a diminishing configuration. One can slightly modify S^*, as done on Figure 3 for each case, such that the strategy profile remains a Stackelberg equilibrium and the corresponding matching has decreased by one unit. Therefore $\langle T, Q \rangle$ is not feasible.

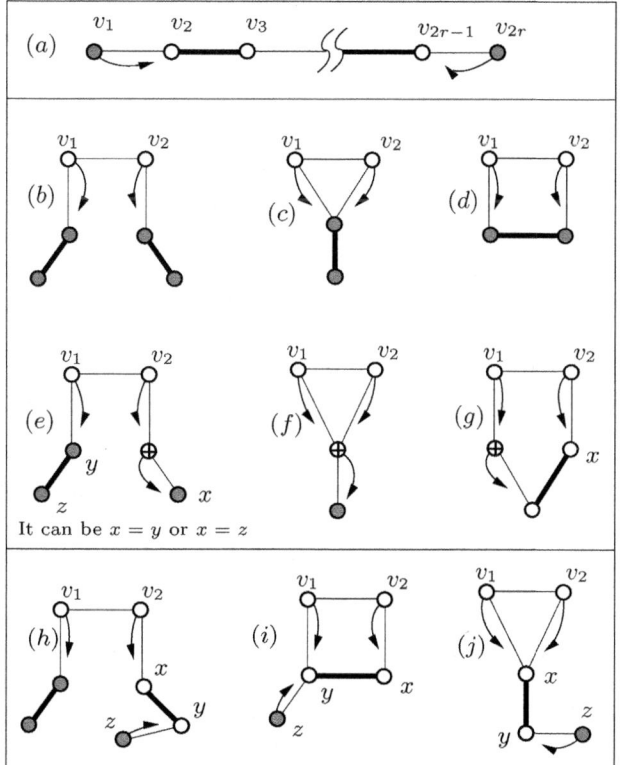

Fig. 3. For each configuration of Figure 2 there is a way to decrease the matching by one unit. The corresponding strategy profile remains a Stackelberg equilibrium compatible with $\langle T, Q \rangle$.

($\Rightarrow$) Let S' be a Stackelberg equilibrium compatible with $\langle T, Q \rangle$ such that its associated matching $\mathcal{M}'$ is not maximum in G. Consider the symmetric difference $\mathcal{M}' \Delta \mathcal{M}^*$. Its connected components are of four kinds:

 - a path which starts with an edge of $\mathcal{M}'$, alternates edges of $\mathcal{M}^*$ and $\mathcal{M}'$, and ends with an edge of $\mathcal{M}'$ (see case 1 in Figure 4)
 - a path which starts with an edge of $\mathcal{M}'$, alternates edges of $\mathcal{M}^*$ and $\mathcal{M}'$, and ends with an edge of $\mathcal{M}^*$ (see case 2 in Figure 4)

- a path which starts with an edge of $\mathcal{M}^*$, alternates edges of $\mathcal{M}^*$ and $\mathcal{M}'$, and ends with an edge of $\mathcal{M}^*$ (see case 3 in Figure 4)
- an even cycle which alternates edges of $\mathcal{M}^*$ and $\mathcal{M}'$ (see case 4 in Figure 4)

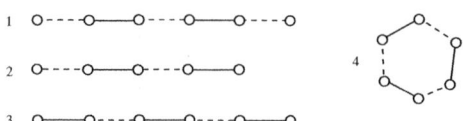

Fig. 4. The four cases for the connected components of $\mathcal{M}'\Delta\mathcal{M}^*$. Edges of $\mathcal{M}^*$ and $\mathcal{M}'$ are respectively solid and dashed.

Since $|\mathcal{M}^*| > |\mathcal{M}'|$ there must be one component of the third kind because this is the only case which contains more edges of $\mathcal{M}^*$ than edges of $\mathcal{M}'$. Notice that the two nodes on the extremities of this component are unmatched in $\mathcal{M}'$. We consider two cases, whether this component contains at least two edges of $\mathcal{M}^*$ (Case A), or just one (Case B).
Due to space limitation, the proof of Case B is omitted.

Case A: Let us denote by $v_1, \ldots, v_{2r}$ the nodes of the component. Since $S'_v \neq S^*_v$ holds for every $v \in \{v_2, \cdots, v_{2r-1}\}$, we deduce that no node of $\{v_2, \cdots, v_{2r-1}\}$ is forced. If v_1 and v_{2r} belong to Q then a long diminishing configuration is found. Suppose that neither v_1 nor v_{2r} belong to Q. If there is no vertex $v \in V \setminus \{v_1, \ldots, v_{2r}\}$ such that $S^*_v \in \{v_1, v_{2r}\}$ then a long diminishing configuration is found. If there is a node $v \in Q$ such that $T_v \in \{v_1, v_{2r}\}$ then S' is not a Stackelberg equilibrium (v_1 and v_{2r} are unforced and unmatched in $\mathcal{M}'$ so one of them can play v and be matched), contradiction. If there is a leaf v adjacent to v_1 or v_{2r} then S' is not a Stackelberg equilibrium (v_1 and v_{2r} are unforced and unmatched in $\mathcal{M}'$ so one of them can play v and be matched), contradiction.

If there is a node $v \notin Q \cup \{v_2, v_{2r-1}\}$ such that $S^*_v \in \{v_1, v_{2r}\}$ then v is unmatched in $\mathcal{M}^*$, all its neighbors are matched because $\mathcal{M}^*$ is maximum, and $S'_v \notin \{v_1, v_{2r}\}$ because S' is a Stackelberg equilibrium. One can set $S^*_v \leftarrow S'_v$ every time this case happens and deduce that a long diminishing configuration exists. Indeed S^* though modified remains a Stackelberg equilibrium compatible with $\mathcal{M}^*$ and $\langle T, Q \rangle$.

The last case is when $v_1 \notin Q$ while $v_{2r} \in Q$ (the case $v_{2r} \notin Q$ while $v_1 \in Q$ is completely symmetric). If there is no node $v \in V \setminus \{v_1, \cdots, v_{2r}\}$ such that $S^*_v = v_1$ then a long diminishing configuration exists. If there is a node $v \in Q$ such that $T_v = v_1$ then S' is not a Stackelberg equilibrium, contradiction. If there is a node $v \notin Q$ such that $S^*_v = v_1$ then $S'_v \neq v_1$ because S' is a Stackelberg equilibrium. One can set $S^*_v \leftarrow S'_v$ every time this case happens and deduce that a long diminishing configuration exists. $\quad\square$

Notice that Lemma 1 is obtained with *any* maximum matching $\mathcal{M}^*$ compatible with $\langle T, Q \rangle$.

Theorem 2. *One can decide in polynomial time whether a solution $\langle T, Q \rangle$ is feasible.*

Proof. Compute a maximum matching $\mathcal{M}^*$ compatible with $\langle T, Q \rangle$. If $\mathcal{M}^*$ is not optimum in G then $\langle T, Q \rangle$ is not feasible. From now on suppose that $\mathcal{M}^*$ is optimum. Let S^* be any Stackelberg equilibrium compatible with both $\mathcal{M}^*$ and $\langle T, Q \rangle$. Using Lemma 1, $\langle T, Q \rangle$ is not feasible iff $\mathcal{M}^*$ and $\langle T, Q \rangle$ possess a diminishing configuration. Short and average diminishing configurations contain a constant number of nodes so we can easily check their existence in polynomial time.

For every pair of distinct nodes $\{a, b\}$ such that a and b are matched in $\mathcal{M}^*$ but not together, we check whether a long diminishing configuration with extremities a and b exists. Suppose that a (resp. b) is matched with a' (resp. b'). If there is a node $v \in V \setminus \{a', b'\}$ such that $S_v^* \in \{a, b\}$ in every Stackelberg equilibrium S^* compatible with $\langle T, Q \rangle$ then the answer is no (it happens when $v \in Q$ and $T_v \in \{a, b\}$, or $v \in Q$ and $\mathcal{N}_G(v) \subseteq \{a, b\}$). Otherwise every unmatched neighbor v of a or b can play a strategy $S_v^* \notin \{a, b\}$ and S^* remains a Stackelberg equilibrium compatible with $\langle T, Q \rangle$. If $\{a', b'\} \cap Q \neq \emptyset$ then the answer is also no. Now consider the graph $G' = G[V \setminus (Q \cup \{a, b\})]$ to which we add the edges (a, a') and (b, b'). Deciding whether there exists an $a - b$ path in G' which alternates edges of $\mathcal{M}^*$ and edges not in $\mathcal{M}^*$, and such that the first and last edge of this path are respectively (a, a') and (b, b'), can be done in $O(n^{2.5})$ steps. This result is due to J. Edmonds and a sketch of proof can be found in [15] (Lemma 1.1). This problem is equivalent to checking whether a long diminishing configuration with extremities a and b exists in G. □

We have mentioned in the previous section that for some graphs, forcing no node is the optimal Stackelberg strategy, leading to an optimal solution with value 0. Such a particular case can be detected in polynomial time by Theorem 2. In the next section, we focus on instances for which the strategy of at least one node must be fixed.

4   Complexity and Approximation

Let $\mathcal{G}$ be the class of graphs admitting a perfect matching. We focus on this important class of graphs to show that MFV is APX-hard.

Theorem 3. *For any $\rho \geq 1$, MFV restricted to graphs in $\mathcal{G}$ is ρ-approximable in polynomial time if and only if the minimum VERTEX COVER problem (in general graphs) is ρ-approximable in polynomial time.*

Proof. The proof will be done in two steps. In the first step, we will give a polynomial time reduction preserving approximation from the minimum VERTEX COVER problem to MFV restricted to graphs in $\mathcal{G}$, while in the second step we will produce a polynomial time reduction preserving approximation from MFV restricted to graphs in $\mathcal{G}$ to the minimum VERTEX COVER problem.

• First step. Let G be a simple graph, instance of the minimum VERTEX COVER problem. We suppose that $V(G) = \{v_1, \cdots, v_n\}$ and $E(G) = \{e_1, \cdots, e_m\}$. Let us build a simple graph G', instance of MFV, as follows: take G, add a copy of every vertex and link every vertex to its copy. More formally we set $V(G') = \{v_1, \cdots, v_n\} \cup \{v'_1, \cdots, v'_n\}$ and $E(G') = \{e_1, \cdots, e_m\} \cup \{(v_i, v'_i) : i \in \{1, \ldots, n\}\}$. Remark that G' admits a unique perfect matching made of all edges (v_i, v'_i). Then $G \in \mathcal{G}$. We claim that G admits a vertex cover C of size at most k iff G' admits a feasible solution of size at most k.

$(\Rightarrow)$ Consider the solution $\langle T, Q \rangle$ where $Q = C$ and for every $v_i \in C$, set $T_{v_i} = v'_i$. Since C is a vertex cover, there is no pair of nodes $v_i, v_j \in V(G) \setminus C$ such that $(v_i, v_j) \in E(G)$. Hence v_i (resp. v_j) can only be matched with v'_i (resp. v'_j), and $\langle T, Q \rangle$ is a feasible solution of value $|Q| = |C|$.

$(\Leftarrow)$ Suppose that there are two nodes $v_i, v_j \in V(G) \setminus Q$ such that $(v_i, v_j) \in E(G)$. These nodes can match together because they are not forced, contradicting that $\langle T, Q \rangle$ is a feasible solution (v_i and v_j must be matched with v'_i and v'_j respectively). Therefore $Q \cap V(G)$ is a vertex cover of G, of size at most $|Q|$.

• Second step. Let $G = (V, E)$ be a graph admitting a perfect matching, i.e., $G \in \mathcal{G}$. Let G' be a graph defined as $V(G') = \{v \in V(G) : d_G(v) > 1\}$ and $E(G') = \{(x, y) \in E(G) : x, y \in V(G')\}$. We claim that there is a vertex cover of size at most k in G' iff MFV has a solution of value at most k in G.

$(\Rightarrow)$ Let C be a vertex cover of size k in G'. Compute a maximum matching $\mathcal{M}$ of G. Build a solution $\langle T, Q \rangle$ to the MFV problem as follows: force every node of C to follow the matching $\mathcal{M}$. The matching being perfect, this is always possible. It is clear that k nodes are forced.

Let us prove that every Stackelberg equilibrium S compatible with $\langle T, Q \rangle$ induces the optimal matching $\mathcal{M}$. Take an edge $(u, v) \in \mathcal{M}$. If both u and v are forced then $T_u = v$ and $T_v = u$, by construction. Suppose that only u is forced. We have $T_u = v$ and there is no node $w \neq u$ such that $T_w = v$, by construction. If there is an unforced node $w \in \mathcal{N}_G(v)$ then either $w \in V(G')$, which contradicts the fact that C is a valid vertex cover of G', or $v \in V(G) \setminus V(G')$, contradicts the fact that $\mathcal{M}$ is a perfect matching of G. Now suppose that neither u nor v is forced. At least one of them, say u, has degree 1 since otherwise C is not a valid vertex cover. As previously an unforced node $w \in \mathcal{N}_G(v)$ would contradict that C is a valid vertex cover. Then u can only play v and v's rational behavior is to play u.

$(\Leftarrow)$ Take a solution $\langle T, Q \rangle$ with $|Q| = k$ and build a vertex cover $C := V(G') \cap Q$. It is clear that C has size at most k. We can observe that C is not a vertex cover in G' iff there exists an edge (u, v) with $d_G(u) > 1$, $d_G(v) > 1$ and $\{u, v\} \cap Q = \emptyset$. Let $\mathcal{M}$ be an optimal (and perfect) matching induced by a Stackelberg equilibrium compatible with $\langle T, Q \rangle$. If u and v are matched together in $\mathcal{M}$ then u (resp. v) has a matched neighbor $u' \neq v$

(resp. $v' \neq u$). If u (resp. v) plays u' (resp. v') then we get an equilibrium which contradicts the fact that $\langle T, Q \rangle$ is a feasible solution. If u and v are not matched together in $\mathcal{M}$ then suppose that u is matched with u' while v is matched with v' (the matching is perfect). If we remove (u, u'), (v, v') and add (u, v) then the state is an equilibrium (neither u' nor v' can deviate and be matched with a node since $\mathcal{M}$ is perfect) but the resulting matching is not optimal.   □

Based on known results for VERTEX COVER [16,17] we deduce that MFV is APX-hard and even NP-hard to approximate within ratio 1.36 (in general), but it is 2-approximable in polynomial time in $\mathcal{G}$.

5 Conclusion

To summarize, one can decide in polynomial time whether a solution is feasible or infeasible for *any* instance of MFV. MFV is APX-hard in general and a 2-approximation algorithm exists for the case of graphs admitting a perfect matching ($\mathcal{G}$). The approximability of this problem in general graphs is worth being considered. In particular, is there a way to generalize the 2-approximation algorithm to general graphs? In a preliminary version of this article [18] we achieved an approximation ratio of 6 thanks to a long and tedious analysis.

The model studied for the MVF problem can be extended in many directions. First, forcing a node v may cost $c(v)$ instead of a unit cost as it is supposed in the present article. Another possible extension would be to require a feasible Stackelberg strategy to reach only an approximation of the social optimum.

Finally, our study focuses on Nash equilibria, i.e. states resilient to deviations by any single player. An interesting extension is to deal with simultaneous deviations by several players. In particular, simultaneous deviations by two players is considered in the stable marriage problem [1]. In our setting, a state is called a k-strong equilibrium if it is resilient to deviations by at most k players (a strong equilibrium is resilient to deviations by any number of players). Then the k-strong price of anarchy is defined as the price of anarchy but for k-strong equilibria [19].

Dealing with this last issue, we can show that the notions of 2-strong, k-strong and strong equilibria coincide for the matching game (proof omitted), and that states resilient to deviations by several players are much better in term of social welfare than Nash equilibria.

Proposition 1. *The strong price of anarchy is* $1/2$.

Proof. Let S be a 2-strong equilibrium whereas S^* is an optimum state. For every edge $(i, j) \in E$, we have $\max\{u_i(S), u_j(S)\} \geq 1$. Take a maximum cardinality matching $\mathcal{M}^*$ and use the previous inequality to get that $\mathcal{SW}(S) = \sum_{i \in V} u_i(S) \geq \sum_{\{i,j\} \in \mathcal{M}^*} u_i(S) + u_j(S) \geq \sum_{\{i,j\} \in \mathcal{M}^*} \max\{u_i(S), u_j(S)\}$. It follows that $\mathcal{SW}(S) \geq |\mathcal{M}^*| = \mathcal{SW}(S^*)/2$. Take a path of length 3 as a tight example.   □

Considering the MFV problem for strong equilibria is an interesting topic that is worth being considered in some future works.

References

1. Gale, D., Shapley, L.S.: College admissions and the stability of marriage. The American Mathematical Monthly 69(1), 9–15 (1962)
2. Shapley, L.S., Shubik, M.: The assignment game i: The core. International Journal of Game Theory 1, 111–130 (1972)
3. Kaporis, A.C., Spirakis, P.G.: Stackelberg games: The price of optimum. In: Kao, M.Y. (ed.) Encyclopedia of Algorithms. Springer, Heidelberg (2008)
4. Korilis, Y.A., Lazar, A.A., Orda, A.: Achieving network optima using stackelberg routing strategies. IEEE/ACM Trans. Netw. 5(1), 161–173 (1997)
5. Roughgarden, T., Tardos, É.: How bad is selfish routing? J. ACM 49(2), 236–259 (2002)
6. Koutsoupias, E., Papadimitriou, C.H.: Worst-case equilibria. Computer Science Review 3(2), 65–69 (2009)
7. Papadimitriou, C.H.: Algorithms, games, and the internet. In: STOC, pp. 749–753 (2001)
8. Christodoulou, G., Koutsoupias, E., Nanavati, A.: Coordination mechanisms. Theor. Comput. Sci. 410, 3327–3336 (2009)
9. Roughgarden, T.: Stackelberg scheduling strategies. SIAM J. Comput. 33(2), 332–350 (2004)
10. Kumar, V.S.A., Marathe, M.V.: Improved results for stackelberg scheduling strategies. In: Widmayer, P., Triguero, F., Morales, R., Hennessy, M., Eidenbenz, S., Conejo, R. (eds.) ICALP 2002. LNCS, vol. 2380, pp. 776–787. Springer, Heidelberg (2002)
11. Swamy, C.: The effectiveness of stackelberg strategies and tolls for network congestion games. In: SODA, pp. 1133–1142. SIAM, Philadelphia (2007)
12. Fotakis, D.: Stackelberg strategies for atomic congestion games. Theory Comput. Syst. 47(1), 218–249 (2010)
13. Sharma, Y., Williamson, D.P.: Stackelberg thresholds in network routing games or the value of altruism. Games and Economic Behavior 67, 174–190 (2009)
14. Bonifaci, V., Harks, T., Schäfer, G.: Stackelberg routing in arbitrary networks. Math. Oper. Res. 35(2), 330–346 (2010)
15. Manoussakis, Y.: Alternating paths in edge-colored complete graphs. Discrete Applied Mathematics 56(2-3), 297–309 (1995)
16. Hochbaum, D.S.: Approximation Algorithms for NP-Hard Problems. PWS Publishing Company, Boston (1996)
17. Dinur, I., Safra, S.: On the hardness of approximating minimum vertex cover. Annals of Mathematics 162(1), 439–485 (2005)
18. Escoffier, B., Gourvès, L., Monnot, J.: Minimum regulation of uncoordinated matchings. CoRR abs/1012.3889 (2010)
19. Andelman, N., Feldman, M., Mansour, Y.: Strong price of anarchy. Games and Economic Behavior 65, 289–317 (2009)

Pareto Optimality in Coalition Formation

Haris Aziz, Felix Brandt, and Paul Harrenstein

Institut für Informatik, Technische Universität München,
80538 München, Germany
{aziz,brandtf,harrenst}@in.tum.de

Abstract. A minimal requirement on allocative efficiency in the social sciences is Pareto optimality. In this paper, we identify a far-reaching structural connection between Pareto optimal and perfect partitions that has various algorithmic consequences for coalition formation. In particular, we show that computing and verifying Pareto optimal partitions in general hedonic games and B-hedonic games is intractable while both problems are tractable for roommate games and W-hedonic games. The latter two positive results are obtained by reductions to maximum weight matching and clique packing, respectively.

1 Introduction

Topics concerning coalitions and coalition formation have come under increasing scrutiny of computer scientists. The reason for this may be obvious. For the proper operation of distributed and multiagent systems, cooperation may be required. At the same time, collaboration in very large groups may also lead to unnecessary overhead, which may even exceed the positive effects of cooperation. To model such situations formally, concepts from the social and economic sciences have proved to be very helpful and thus provide the mathematical basis for a better understanding of the issues involved.

Coalition formation games, which were first formalized by Drèze and Greenberg [9], model coalition formation in settings in which utility is *non-transferable*. In many such situations it is natural to assume that a player's appreciation of a coalition structure only depends on the coalition he is a member of and not on how the remaining players are grouped. Initiated by Banerjee et al. [3] and Bogomolnaia and Jackson [4], much of the work on coalition formation now concentrates on these so-called *hedonic games*. In this paper, we focus on Pareto optimality and individual rationality in this rich class of coalition formation games.

The main question in coalition formation games is which coalitions one may reasonably expect to form. To get a proper formal grasp of this issue, a number of stability concepts have been proposed for hedonic games—such as the core or Nash stability—and much research concentrates on conditions for existence, the structure, and computation of stable and efficient partitions. *Pareto optimality*—which holds if no coalition structure is strictly better for some player without being strictly worse for another—and *individual rationality*—which holds if every player is satisfied in the sense that no player would rather be on his own—are commonly considered minimal requirements for any reasonable partition.

G. Persiano (Ed.): SAGT 2011, LNCS 6982, pp. 93–104, 2011.

Another reason to investigate Pareto optimal partitions algorithmically is that, in contrast to other stability concepts like the core, they are guaranteed to exist. This even holds if we additionally require individual rationality. Moreover, even though the *Gale-Shapley algorithm* returns a core stable matching for marriage games, it is already NP-hard to check whether the core is empty in various classes and representations of hedonic games, such as roommate games [12], general hedonic games [2], and games with $\mathscr{B}$- and $\mathscr{W}$-preferences [7, 6]. Interestingly, when the status-quo partition cannot be changed without the mutual consent of all players, Pareto optimality can be seen as a stability notion [11].

We investigate both the problem of finding a Pareto optimal and individually rational partition and the problem of deciding whether a partition is Pareto optimal. In particular, our results concern *general hedonic games*, B-hedonic and W-hedonic games (two classes of games in which each player's preferences over coalitions are based on his most preferred and least preferred player in his coalition, respectively), and *roommate games*.

Many of our results, both positive and negative, rely on the concept of *perfection* and how it relates to Pareto optimality. A *perfect* partition is one that is most desirable for every player. We find (*a*) that under extremely mild conditions, NP-hardness of finding a perfect partition implies NP-hardness of finding a Pareto optimal partition (Lemma 1), and (*b*) that under stronger but equally well-specified circumstances, feasibility of finding a perfect partition implies feasibility of finding a Pareto optimal partition (Lemma 2). The latter we show via a Turing reduction to the problem of computing a perfect partition. At the heart of this algorithm, which we refer to as the *Preference Refinement Algorithm (PRA)*, lies a fundamental insight of how perfection and Pareto optimality are related. It turns out that a partition is Pareto optimal for a particular preference profile if and only if the partition is perfect for another but related profile (Theorem 1). In this way PRA is also applicable to any other discrete allocation setting.

For general allocation problems, *serial dictatorship*—which chooses subsequently the most preferred allocation for a player given a fixed ranking of all players—is well-established as a procedure for finding Pareto optimal solutions (see, e.g., [1]). However, it is only guaranteed to do so, if the players' preferences over outcomes are strict, which is not feasible in many compact representations. Moreover, when applied to coalition formation games, there may be Pareto optimal partitions that serial dictatorship is unable to find, which may have serious repercussions if also other considerations, like fairness, are taken into account. By contrast, PRA handles weak preferences well and is complete in the sense that it may return any Pareto optimal partition, provided that the subroutine that calculates perfect partitions can compute any perfect partition (Theorem 2).

2 Preliminaries

In this section, we review the terminology and notation used in this paper.

Hedonic games. Let N be a set of n players. A *coalition* is any non-empty subset of N. By $\mathcal{N}_i$ we denote the set of coalitions player i belongs to, i.e., $\mathcal{N}_i = \{S \subseteq N : i \in S\}$. A *coalition structure*, or simply a *partition*, is a partition π of the players N into coalitions, where $\pi(i)$ is the coalition player i belongs to.

A *hedonic game* is a pair (N, R), where $R = (R_1, \ldots, R_n)$ is a *preference profile* specifying the preferences of each player i as a binary, complete, reflexive, and transitive *preference relation* R_i over $\mathcal{N}_i$. If R_i is also anti-symmetric we say that i's preferences are *strict*. We adopt the conventions of social choice theory by writing $S\, P_i\, T$ if $S\, R_i\, T$ but not $T\, R_i\, S$—i.e., if *i strictly prefers* S to T—and $S\, I_i\, T$ if both $S\, R_i\, T$ and $T\, R_i\, S$— i.e., if i is *indifferent* between S and T.

For a player i, a coalition S in $\mathcal{N}_i$ is *acceptable* if for i being in S is at least preferable as being alone—i.e., if $S\, R_i\, \{i\}$—and *unacceptable* otherwise.

In a similar fashion, for X a subset of $\mathcal{N}_i$, a coalition S in X is said to be *most preferred in X by i* if $S\, R_i\, T$ for all $T \in X$ and *least preferred in X by i* if $T\, R_i\, S$ for all $T \in X$. In case $X = \mathcal{N}_i$ we generally omit the reference to X. The sets of most and least preferred coalitions in X by i, we denote by $\max_{R_i}(X)$ and $\min_{R_i}(X)$, respectively.

In hedonic games, players are only interested in the coalition they are in. Accordingly, preferences over coalitions naturally extend to preferences over partitions and we write $\pi\, R_i\, \pi'$ if $\pi(i)\, R_i\, \pi'(i)$. We also say that partition π is *acceptable* or *unacceptable* to a player i according to whether $\pi(i)$ is acceptable or unacceptable to i, respectively. Moreover, π is *individually rational* if π is acceptable to all players. A partition π is *Pareto optimal for R* if there is no partition π' with $\pi'\, R_j\, \pi$ for all players j and $\pi'\, P_i\, \pi$ for at least one player i. Partition π is, moreover, said to be *weakly Pareto optimal for R_i* if there is no π' with $\pi'\, P_i\, \pi$ for all players i.

Classes of hedonic games. The number of potential coalitions grows exponentially in the number of players. In this sense, hedonic games are relatively large objects and for algorithmic purposes it is often useful to look at classes of games that allow for concise representations.

For *general hedonic games*, we will assume that each player expresses his preferences only over his acceptable coalitions. This representation is also known as *Representation by Individually Rational Lists of Coalitions* [2].

We now describe classes of hedonic games in which the players' preferences over coalitions are induced by their preferences over the other players. For R_i such preferences of player i over players, we say that a player j is *acceptable* to i if $j\, R_i\, i$ and *unacceptable* otherwise. Any coalition containing an unacceptable player is unacceptable to player i.

Roommate games. The class of *roommate games*, which are well-known from the literature on matching theory, can be defined as those hedonic games in which only coalitions of size one or two are acceptable and preferences R_i over other players are extended naturally over preferences over coalitions in the following way: $\{i, j\}\, R_i\, \{i, k\}$ if and only if $j\, R_i\, k$ for all $j, k \in N$.

B-hedonic and W-hedonic games. For a subset J of players, we denote by $\max_{R_i}(J)$ and $\min_{R_i}(J)$ the sets of the most and least preferred players in J by i, respectively. We will assume that $\max_{R_i}(\emptyset) = \min_{R_i}(\emptyset) = \{i\}$. In a *B-hedonic* game the preferences R_i of a player i over players extend to preferences over coalitions in such a way that, for all coalitions S and T in $\mathcal{N}_i$, we have $S\ R_i\ T$ if and only if $j\ R_i\ k$ for all $j \in \max_{R_i}(S \setminus \{i\})$ and $k \in \max_{R_i}(T \setminus \{i\})$ or some j in T is unacceptable to i. Analogously, in a *W-hedonic game* (N, R), we have $S\ R_i\ T$ if and only if $j\ R_i\ k$ for all $j \in \min_{R_i}(S \setminus \{i\})$ and $k \in \min_{R_i}(T \setminus \{i\})$ or some j in T is unacceptable to i.[1]

3   Perfection and Pareto Optimality

Pareto optimality constitutes rather a minimal efficiency requirement on partitions. A much stronger condition is that of *perfection*. We say that a partition π is *perfect* if $\pi(i)$ is a most preferred coalition for all players i. Thus, every perfect partition is Pareto optimal but not necessarily the other way round. Perfect partitions are obviously very desirable, but, in contrast to Pareto optimal ones, they are unfortunately not guaranteed to exist. Nevertheless, there exists a strong structural connection between the two concepts, which we exploit in our algorithm for finding Pareto optimal partitions in Section 4.

The problem of finding a perfect partition (PerfectPartition) we formally specify as follows.

> PerfectPartition
> *Instance*: A preference profile R
> *Question*: Find a perfect partition for R.
>   If no perfect partition exists, output $\emptyset$.

We will later see that the complexity of PerfectPartition depends on the specific class of hedonic games that is being considered. By contrast, the related problem of *checking* whether a partition is perfect is an almost trivial problem for virtually all reasonable classes of games. If perfect partitions exist, they clearly coincide with the Pareto optimal ones. Hence, an oracle to compute a Pareto optimal partition can be used to solve PerfectPartition. If this Pareto optimal partition is perfect we are done, if it is not, no perfect partitions exist. Thus, we obtain the following simple lemma, which we will invoke in our hardness proofs for computing Pareto optimal partitions.

Lemma 1. *For every class of hedonic games for which it can be checked in polynomial time whether a given partition is perfect, NP-hardness of* PerfectPartition *implies NP-hardness of computing a Pareto optimal partition.*

It might be less obvious that a procedure solving PerfectPartition can also be deployed as an oracle for an algorithm to compute Pareto optimal partitions. To do so, we first give a characterization of Pareto optimal partitions in terms of perfect partitions, which forms the mathematical heart of the Preference Refinement Algorithm to be presented in the next section.

The connection between perfection and Pareto optimality can intuitively be explained as follows. If all players are indifferent among all coalitions, all partitions are

---

[1] W-hedonic games are equivalent to hedonic games with $\mathcal{W}$-preferences if individually rational outcomes are assumed. Unlike hedonic games with $\mathcal{B}$-preferences, B-hedonic games are defined in analogy to W-hedonic games and the preferences are not based on coalition sizes (cf. [7]).

perfect. It follows that the players can always relax their preferences up to a point where perfect partitions become possible. We find that, if a partition is perfect for a minimally relaxed preference profile—in the sense that, if any one player relaxes his preferences only slightly less, no perfect partition is possible anymore—, this partition is Pareto optimal for the original unrelaxed preference profile. To see this, assume π is perfect in some minimally relaxed preference profile and that some player i reasserts some strict preferences he had previously relaxed, thus rendering π no longer perfect. Now, π does not rank among i's most preferred partitions anymore. By assumption, none of i's most preferred partitions is also most preferred by all other players. Hence, it is impossible to find a partition π' that is better for i than π, without some other player strictly preferring π to π'. It follows that π is Pareto optimal.

To make this argumentation precise, we introduce the concept of a coarsening of a preference profile and the lattices these coarsenings define. Let $R = (R_1, \ldots, R_n)$ and $R' = (R'_1, \ldots, R'_n)$ be preference profiles over a set X and let i be a player. We write $R_i \leq_i R'_i$ if

$$R_i|_{\{x,y\}} = R'_i|_{\{x,y\}} \text{ for all } x \in X \text{ and all } y \in X \setminus \max_{R_i}(X).$$

Accordingly, R_i is exactly like R'_i, except that in R'_i player i may have strict preferences among some of his most preferred coalitions in R_i. Thus, R'_i is finer than R_i. For instance, let R_i, R'_i, R''_i, and R'''_i be such that $\pi_1 P_i \pi_2 P_i \pi_3$, $\pi_1 I'_i \pi_2 P'_i \pi_3$, $\pi_1 P''_i \pi_2 I''_i \pi_3$, and $\pi_1 I'''_i \pi_2 I'''_i \pi_3$, then $R'''_i \leq_i R'_i \leq_i R_i$ and $R'''_i \leq_i R''_i$, but not $R''_i \leq_i R_i$. It can easily be established that $\leq_i$ is a linear order for each player i.

We say that a preference profile $R = (R_1, \ldots, R_n)$ over X is a *coarsening of* or *coarsens* another preference profile $R' = (R'_1, \ldots, R'_n)$ over X whenever $R_i \leq_i R'_i$ for every player i. In that case we also say that R' *refines* R and write $R \leq R'$. Moreover, we write $R < R'$ if $R \leq R'$ but not $R' \leq R$. Thus, if R' refines R, i.e., if $R \leq R'$, then for each i and all coalitions S and T we have that $S R'_i T$ implies $S R_i T$, but not necessarily the other way round. It is worth observing that, if a partition is perfect for some preference profile R, then it is also perfect for any coarsening of R. The same holds for Pareto optimal partitions.

For preference profiles R and R' with $R \leq R'$, let $[R, R']$ denote the set $\{R'' : R \leq R'' \leq R'\}$, i.e., the set of all coarsenings of R' that also refine R. $([R, R'], \leq)$ is a complete lattice with R and R' as bottom and top element, respectively. We say that R *covers* R' if R is a minimal refinement of R' with $R' \neq R$, i.e., if $R' < R$ and there is no R'' such that $R' < R'' < R$. Observe that, if R covers R', R and R' coincide for all but one player, say i, for whom R_i is the unique minimal refinement of R'_i such that $R'_i \neq R_i$. We also denote R_i by $\texttt{Cover}(R'_i)$. If no cover of R'_i exists, $\texttt{Cover}(R'_i)$ returns the empty set.

We are now in a position to prove the following theorem, which characterizes Pareto optimal partitions for a preference profile R as those that are perfect for particular coarsenings R' of R. These R' are such that no perfect partitions exist for any preference profile that covers R'.

Theorem 1. *Let $(N, R^\top)$ and $(N, R^\perp)$ be hedonic games such that $R^\perp \leq R^\top$ and π a perfect partition for $R^\perp$. Then, π is Pareto optimal for $R^\top$ if and only if there is some $R \in [R^\perp, R^\top]$ such that (i) π is a perfect partition for R and (ii) there is no perfect partition for any $R' \in [R^\perp, R^\top]$ that covers R.*

Proof. For the if-direction, assume there is some $R \in [R^\perp, R^\top]$ such that π is perfect for R and there is no perfect partition for any $R' \in [R^\perp, R^\top]$ that covers R. For contradiction, also assume π is not Pareto optimal for $R^\top$. Then, there is some π' such that $\pi' R_j^\top \pi$ for all j and $\pi' P_i^\top \pi$ for some i. By $R \leq R^\top$ and π being perfect for R, it follows that π' is a perfect partition for R as well. Hence, $\pi' I_i \pi$. It follows that there is some $R' = (R_1, \ldots, R_{i-1}, R_i', R_{i+1}, \ldots, R_n)$ in $([R_\perp, R^\top], \leq)$ that covers R. Also observe that, because R_i' is the unique minimal refinement of R_i such that $R_i <_i R_i'$, and $\pi' P_i^\top \pi$ even if $\pi' I_i \pi$, π' is still perfect for R', a contradiction.

For the only-if direction assume that π is Pareto optimal for $R^\top$. Let R be the finest coarsening of $R^\top$ in $[R^\top, R^\perp]$ for which π is perfect. Observe that $R = (R_1, \ldots, R_n)$ can be defined such that $R_i = R_i^\top \cup \{(X, Y) : X R_i^\top \pi \text{ and } Y R_i^\top \pi\}$ for all i. Since π is perfect for $R^\perp$, we have $R^\perp \leq R$. If $R = R^\top$, we are done immediately. Otherwise, consider an arbitrary $R' \in [R^\perp, R^\top]$ that covers R and assume for contradiction that some perfect partition π' exists for R'. Then, in particular, $\pi' R_k' \pi$ for all k. Since R' covers R, there is exactly one i with $R_i' \neq R_i$, whereas $R_j' = R_j$ for all $j \neq i$. As π is perfect for R, we also have $\pi R_j' \pi'$ for all $j \neq i$. Since R' is a finer coarsening of $R^\top$ than R, π is not perfect for R' by assumption. It follows that $\pi' P_i' \pi$. Hence, π is not Pareto optimal for R'. As $R' \leq R^\top$, we may conclude that π is not Pareto optimal for $R^\top$, a contradiction.   □

4   The Preference Refinement Algorithm

In this section, we present the *Preference Refinement Algorithm (PRA)*, a general algorithm to compute Pareto optimal and individually rational partitions. The algorithm invokes an oracle solving `PerfectPartition` and is based on the formal connection between Pareto optimality and perfection made explicit in Theorem 1. We define `PerfectPartition` to return ∅ if $R_i = \emptyset$ for some i.

The idea underlying the algorithm is as follows. To calculate a Pareto optimal and individually rational partition for a hedonic game (N, R), first find that coarsening R' of R in which each player is indifferent among all his acceptable coalitions and his preferences among unacceptable coalitions are as in R. In this coarsening, a perfect and individually rational partition is guaranteed to exist. Then, we search the lattice $([R', R], \leq)$ for a preference profile that allows for a perfect partition but none of the profiles covering it do. By virtue of Theorem 1, every perfect partition for such a preference profile will be a Pareto optimal partition for R. By only refining the preferences of one player at a time, we can use divide-and-conquer to conduct the search. A formal specification of PRA is given in Algorithm 1. `Refine`$(Q_i^\perp, Q_i^\top)$ returns a refinement $Q_i' \in (Q_i^\perp, Q_i^\top]$, i.e., Q_i' is a refinement of $Q_i^\perp$ but not a refinement of $Q_i^\top$. `Refine`$(Q_i^\perp, Q_i^\top)$ can be defined in at least three fundamental ways:

(*i*) `Refine`$(Q_i^\perp, Q_i^\top) = Q_i'$ such that the number of refinements from $Q_i^\perp$ to Q_i' is half of the number of refinements from $Q_i^\perp$ to $Q_i^\top$ (default divide-and-conquer setting);

(*ii*) `Refine`$(Q_i^\perp, Q_i^\top) = Q_i^\top$ (serial dictator setting); and

(*iii*) `Refine`$(Q_i^\perp, Q_i^\top) = $ `Cover`$(Q_i^\perp)$.

Algorithm 1. Preference Refinement Algorithm (PRA)

Input: Hedonic game (N, R)
Output: Pareto optimal and individually rational partition

1 $Q_i^\top \leftarrow R_i$, for each $i \in N$
2 $Q_i^\perp \leftarrow R_i \cup \{(X, Y) : X\, R_i\, \{i\} \text{ and } Y\, R_i\, \{i\}\}$, for each $i \in N$
3 $J \leftarrow N$
4 **while** $J \neq \emptyset$ **do**
5      $i \in J$
6   **if** $\texttt{PerfectPartition}(N, (Q_1^\perp, \ldots, Q_{i-1}^\perp, \texttt{Cover}(Q_i^\perp), Q_{i+1}^\perp, \ldots, Q_n^\perp)) = \emptyset$ **then**
7          $J \leftarrow J \setminus \{i\}$
8   **else**
9          $Q_i' \leftarrow \texttt{Refine}(Q_i^\perp, Q_i^\top)$
10   **if** $\texttt{PerfectPartition}(N, (Q_1^\perp, \ldots, Q_{i-1}^\perp, Q_i', Q_{i+1}^\perp, \ldots, Q_n^\perp)) \neq \emptyset$ **then**
11             $Q_i^\perp \leftarrow Q_i'$
12   **else**
13             $Q_i^\top \leftarrow Q_i''$ where $\texttt{Cover}(Q_i'') = Q_i'$
14   **end if**
15   **end if**
16 **end while**
17 **return** $\texttt{PerfectPartition}(N, Q^\perp)$

The following theorem shows the correctness and completeness of PRA.

Theorem 2. *For any hedonic game (N, R),*

   (i) *PRA returns an individually rational and Pareto optimal partition.*
   (ii) *For every individually rational and Pareto optimal partition π', there is an execution of PRA that returns a partition π such that $\pi\, I_i\, \pi'$ for all i in N.*

Proof. For *(i)*, we prove that during an execution of PRA, for each assignment of $Q^\perp$, there exists a perfect partition π for that assignment. This claim certainly holds for the first assignment of $Q^\perp$, the coarsest acceptable coarsening of R. Furthermore, $Q^\perp$ is only refined (Step 9) if there exists a perfect partition for a refinement of $Q^\perp$. Let Q^* be the final assignment of $Q^\perp$. Then, we argue that the partition π returned by PRA is Pareto optimal and individually rational. By Theorem 1, if π were not Pareto optimal, there would exist a covering of Q^* for which a perfect partition still exists and Q^* would not be the final assignment of $Q^\perp$. Since, each player at least gets one of his acceptable coalitions, π is also individually rational.

For *(ii)*, first observe that, by Theorem 1, for each Pareto optimal and individually rational partition π for a preference profile R there is some coarsening Q^* of R where π is perfect and no perfect partitions exist for any covering of Q^*. By individual rationality of π, it follows that Q^* is a refinement of the initial assignment of $Q^\perp$. An appropriate number of coverings of the initial assignment of $Q^\perp$ with respect to each player results in a final assignment Q^* of $Q^\perp$. The perfect partition for Q^* that is returned by PRA is then such that $\pi\, I_i\, \pi'$ for all i in N.   □

We now specify the conditions under which PRA runs in polynomial time.

Lemma 2. *For any class of hedonic games for which any coarsening and* PerfectPartition *can be computed in polynomial time, PRA runs in polynomial time.*

Furthermore, if for a given preference profile R and partition π, a coarsening of R for which π is perfect can be computed in polynomial time, it can also be verified in polynomial time whether π is Pareto optimal.

Proof (Sketch). Under the given conditions, we prove that PRA runs in polynomial time. We first prove that the while-loop in PRA iterates a polynomial number of times. In each iteration of the while-loop, either a player i which cannot be further improved is removed from J (Step 7) or we enter the first else condition. In the first else, either $Q^\perp$ is set to Q_i' or $Q_i^\top$ is set to Q_i'' where $\text{Cover}(Q_i'') = Q_i'$. In either case, we discard from future consideration, half of the refinements of $Q_i^\perp$ due to the default divide-and-conquer definition of Refine in order to find a suitable refinement of the current $Q^\perp$ with respect to i. Therefore, even if the representation of (N, R) may be such that each player differentiates between an exponential number of coalitions, divide-and-conquer ensures that PRA iterates a polynomial number of times. As the crucial subroutine PerfectPartition takes polynomial time, PRA runs in polynomial time.

For the second part of the lemma, we run PRA to find a Pareto optimal partition that Pareto dominates π. We therefore modify Step 2 by setting $Q_i^\perp$ to a coarsening of R for which π is a perfect partition. Since such a coarsening can be computed in polynomial time as stated by the condition in the lemma, Step 2 takes polynomial time. Since an initial perfect partition exists for $Q_i^\perp$, we run PRA as usual after Step 2.   □

PRA applies not only to general hedonic games but to many natural classes of hedonic games in which equivalence classes (of possibly exponentially many coalitions) for each player are implicitly defined.[2] In fact PRA runs in polynomial time even if there are an exponential number of equivalence classes. Note that the lattice $[Q^\perp, R]$ can be of exponential height and doubly-exponential width. PRA traverses though this lattice in an orderly way to compute a Pareto optimal partition.

Serial dictatorship is a well-studied mechanism in resource allocation, in which an arbitrary player is chosen as the 'dictator' who is then given his most favored allocation and the process is repeated until all players or resources have been dealt with. In the context of coalition formation, *serial dictatorship* is well-defined only if in every iteration, the dictator has a *unique* most preferred coalition.

Proposition 1. *For general hedonic games, W-hedonic games, and roommate games, a Pareto optimal partition can be computed in polynomial time when preferences are strict.*

Proposition 1 follows from the application of serial dictatorship to hedonic games with strict preferences over the coalitions. If the preferences over coalitions are not strict,

---

[2] For example, in W-hedonic games, $\max_{R_i}(N)$ specifies the set of favorite players of player i but can also implicitly represent all those coalitions S such that the least preferred player in S is also a favorite player for i.

then the decision to assign one of the favorite coalitions to the dictator may be sub-optimal. Even if players expresses strict preferences over other players, serial dictator-ship may not work if the preferences induced over coalitions admit ties.

We see that if serial dictatorship works properly and efficiently in some setting, then so can PRA by simulating serial dictatorship. If in each iteration in Algorithm 1, the same player is chosen in Step 5 (until it is deleted from J) and $Q_i^\top$ is chosen in Step 9, then PRA can simulate serial dictatorship. Therefore PRA can also achieve the positive results of Proposition 1.

PRA has another advantage over serial dictatorship. Abdulkadiroğlu and Sönmez [1] showed that in the case of strict preferences and house allocation settings, every Pareto optimal allocation can be achieved by serial dictatorship. In the case of coalition forma-tion, however, it is easy to construct a four-player hedonic game with strict preferences for which there is a Pareto optimal partition that serial dictatorship cannot return.

5   Computational Results

In this section, we consider the problem of VERIFICATION (verifying whether a given partition is Pareto optimal) and COMPUTATION (computing a Pareto optimal partition) for the classes and representations of hedonic games mentioned in the preliminaries.

5.1   General Hedonic Games

As shown in Proposition 1, Pareto optimal partitions can be found efficiently for general hedonic games with strict preferences. If preferences are not strict, the problem turns out to be NP-hard. We prove this statement by utilizing Lemma 1 and showing that PerfectPartition is NP-hard by a reduction from ExactCoverBy3Sets (X3C).

Theorem 3. *For a general hedonic game, computing a Pareto optimal partition is NP-hard even when each player has a maximum of four acceptable coalitions and the max-imum size of each coalition is three.*

Interestingly, *verifying* Pareto optimality is coNP-complete even for strict preferences.[3]

Theorem 4. *For a general hedonic game, verifying whether a partition π is Pareto op-timal and whether π is weakly Pareto optimal is coNP-complete even when preferences are strict and π consists of the grand coalition of all players.*

5.2   Roommate Games

For the class of roommate games, we obtain more positive results.

Theorem 5. *For roommate games, an individually rational and Pareto optimal coali-tion can be computed in polynomial time.*

---

[3] Theorem 4 contrasts with the general observation that *"in the area of matching theory usually ties are 'responsible' for NP-completeness"* [5].

Proof (Sketch). We utilize Lemma 1. It is sufficient to show that PerfectPartition can be solved in time $O(n^3)$.

We say that $j \in F(i)$ if and only if j is a favorite player in i's preference list. Construct an undirected graph $G = (V, E)$ where $V = N \cup (N \times \{0\})$, $E = \{\{i, j\} : i \neq j \wedge i \in F(j) \wedge j \in F(j)\} \cup \{\{i, (i, 0)\} : i \in F(i)\}$.

Then the claim is that there exists a perfect partition for (N, R) if and only if there exists a matching of size n in graph G. It is clear that in a matching of size n, each $v \in N$ is matched. If there exists a perfect partition, then each player in N is matched to a player $j \neq i$ such that $j \in F(i)$ or i is unmatched but $i \in F(i)$. In either case there exists a matching M in which i is matched. In the first case, i is matched to j in a matching M in G. In the second case, i is matched to $(i, 0)$.

Now assume that there exists a matching M of size n in G. Then, each $i \in N$ is matched to $j \neq i$ or $(i, 0)$. If i is matched to j, then we know $\{i, j\} \in E$ and therefore $j \in F(i)$. If i is matched to $(i, 0)$, then we know $\{i, (i, 0)\} \in E$ and therefore $i \in F(i)$. Thus, there exists a perfect partition. □

By utilizing the second part of Lemma 1, it can be seen that there exists an algorithm to compute a Pareto optimal improvement of a given roommate matching which takes time $O(n^3) \cdot O(n \log(n)) = O(n^4 \log(n))$. As a corollary we get the following.

Theorem 6. *For roommate games, it can be checked in polynomial time whether a partition is Pareto optimal.*

We can devise a tailor-made algorithm for roommate games which finds a Pareto optimal Pareto improvement of a given matching in $O(n^3)$—the same asymptotic complexity required by the algorithm of Morrill [11] for the restricted case of strict preferences.

5.3 W-Hedonic Games

We now turn to Pareto optimality in W-hedonic games.

Theorem 7. *For W-hedonic games, a partition that is both individually rational and Pareto optimal can be computed in polynomial time.*

Proof (sketch). The statement follows from Lemma 2 and the fact that PerfectPartition can be solved in polynomial time for W-hedonic games. The latter is proved by a polynomial-time reduction of PerfectPartition to a polynomial-time solvable problem called *clique packing*.

We first introduce the more general notion of graph packing. Let $\mathscr{F}$ be a set of undirected graphs. An $\mathscr{F}$-packing of a graph G is a subgraph H such that each component of H is (isomorphic to) a member of $\mathscr{F}$. The size of $\mathscr{F}$-packing H is $|V(H)|$. We will informally say that vertex i is *matched* by $\mathscr{F}$-packing H if i is in a connected component in H. Then, a maximum $\mathscr{F}$-packing of a graph G is one that matches the maximum number of vertices. It is easy to see that computing a maximum $\{K_2\}$-packing of a graph is equivalent to maximum cardinality matching. Hell and Kirkpatrick [10] and Cornuéjols et al. [8] independently proved that there is a polynomial-time algorithm to compute a maximum $\{K_2, \ldots, K_n\}$-packing of a graph. Cornuéjols et al. [8] note that finding a $\{K_2, \ldots, K_n\}$-packing can be reduced to finding a $\{K_2, K_3\}$-packing.

We are now in a position to reduce `PerfectPartition` for W-hedonic games to computing a maximum $\{K_2, K_3\}$-packing. For a W-hedonic game (N, R), construct a graph $G = (N \cup (N \times \{0, 1\}), E)$ such that $\{(i, 0), (i, 1)\} \in E$ for all $i \in N$; $\{i, j\} \in E$ if and only if $i \in \max_{R_j}(N)$ and $j \in \max_{R_i}(N)$ for $i, j \in N$ such that $i \neq j$; and $\{i, (i, 0)\}, \{i, (i, 1)\} \in E$ if and only if $i \in \max_{R_i}(N)$ for all $i \in N$. Let H be a maximum $\{K_2, K_3\}$-packing of G.

It can then be proved that there exists a perfect partition of N according to R if and only if $|V(H)| = 3|N|$. We omit the technical details due to space restrictions.

Since `PerfectPartition` for W-hedonic games reduces to checking whether graph G can be packed perfectly by elements in $\mathscr{F} = \{K_2, K_3\}$, we have a polynomial-time algorithm to solve `PerfectPartition` for W-hedonic games. Denote by $CC(H)$ the set of connected components of graph H. If $|V(H)| = 3|N|$ and a perfect partition does exist, then $\{V(S) \cap N : S \in CC(H)\} \setminus \{\emptyset\}$ is a perfect partition. $\square$

Due to the second part of Lemma 2, the following is evident.

Theorem 8. *For W-hedonic games, it can be checked in polynomial time whether a given partition is Pareto optimal or weakly Pareto optimal.*

Our positive results for W-hedonic games also apply to hedonic games with $\mathscr{W}$-preferences.

5.4 B-Hedonic Games

We saw that for W-hedonic games, a Pareto optimal partition can be computed efficiently, even in the presence of unacceptable players. In the absence of unacceptable players, computing a Pareto optimal and individually rational partition is trivial in B-hedonic games, as the partition consisting of the grand coalition is a solution.

Interestingly, if preferences do allow for unacceptable players, the same problem becomes NP-hard. The statement is shown by a reduction from Sat.

Theorem 9. *For B-hedonic games, computing a Pareto optimal partition is NP-hard.*

By using similar techniques, the following can be proved.

Theorem 10. *For B-hedonic games, verifying whether a partition is weakly Pareto optimal is coNP-complete.*

We expect the previous result to also hold for Pareto optimality rather than weak Pareto optimality.

6 Conclusions

Pareto optimality and individual rationality are important requirements for desirable partitions in coalition formation. In this paper, we examined computational and structural issues related to Pareto optimality in various classes of hedonic games (see Table 1). We saw that unacceptability and ties are a major source of intractability when computing Pareto optimal outcomes. In some cases, checking whether a given partition is Pareto optimal can be significantly harder than finding one.

Table 1. Complexity of Pareto optimality in hedonic games: positive results hold for both Pareto optimality and individual rationality

Game	VERIFICATION	COMPUTATION
General	coNP-complete (Th. 4)	NP-hard (Th. 3)
General (strict)	coNP-complete (Th. 4)	in P (Prop. 1)
Roommate	in P (Th. 6)	in P (Th. 5)
B-hedonic	coNP-complete (Th. 10, weak PO)	NP-hard (Th. 9)
W-hedonic	in P (Th. 8)	in P (Th. 7)

It should be noted that most of our insights gained into Pareto optimality and the resulting algorithmic techniques—especially those presented in Section 3 and Section 4—do not only apply to coalition formation but to any discrete allocation setting.

References

1. Abdulkadiroğlu, A., Sönmez, T.: Random serial dictatorship and the core from random endowments in house allocation problems. Econometrica 66(3), 689–702 (1998)
2. Ballester, C.: NP-completeness in hedonic games. Games and Economic Behavior 49(1), 1–30 (2004)
3. Banerjee, S., Konishi, H., Sönmez, T.: Core in a simple coalition formation game. Social Choice and Welfare 18, 135–153 (2001)
4. Bogomolnaia, A., Jackson, M.O.: The stability of hedonic coalition structures. Games and Economic Behavior 38(2), 201–230 (2002)
5. Cechlárová, K.: On the complexity of exchange-stable roommates. Discrete Applied Mathematics 116, 279–287 (2002)
6. Cechlárová, K., Hajduková, J.: Stable partitions with $\mathcal{W}$-preferences. Discrete Applied Mathematics 138(3), 333–347 (2004)
7. Cechlárová, K., Hajduková, J.: Stability of partitions under $\mathcal{BW}$-preferences and $\mathcal{WB}$-preferences. International Journal of Information Technology and Decision Making 3(4), 605–618 (2004)
8. Cornuéjols, G., Hartvigsen, D., Pulleyblank, W.: Packing subgraphs in a graph. Operations Research Letters 1(4), 139–143 (1982)
9. Drèze, J.H., Greenberg, J.: Hedonic coalitions: Optimality and stability. Econometrica 48(4), 987–1003 (1980)
10. Hell, P., Kirkpatrick, D.G.: Packings by cliques and by finite families of graphs. Discrete Mathematics 49(1), 45–59 (1984)
11. Morrill, T.: The roommates problem revisited. Journal of Economic Theory 145(5), 1739–1756 (2010)
12. Ronn, E.: NP-complete stable matching problems. Journal of Algorithms 11, 285–304 (1990)

Externalities among Advertisers in Sponsored Search[*]

Dimitris Fotakis[1], Piotr Krysta[2], and Orestis Telelis[2]

[1] School of Electrical and Computer Engineering,
National Technical University of Athens, 157 80 Athens, Greece
fotakis@cs.ntua.gr
[2] Department of Computer Science,
University of Liverpool, L69 3BX Liverpool, UK
{p.krysta,o.telelis}@liverpool.ac.uk

Abstract. We introduce a novel computational model for single-keyword auctions in sponsored search, which models explicitly externalities among advertisers, an aspect that has not been fully reflected in the existing models, and is known to affect the behavior of real advertisers. Our model takes into account both positive and negative correlations between any pair of advertisers, so that the click-through rate of an ad depends on the identity, relative order and distance of other ads appearing in the advertisements list. In the proposed model we present several computational results concerning the Winner Determination problem for Social Welfare maximization. These include hardness of approximation and polynomial time exact and approximation algorithms. We conclude with an evaluation of the Generalized Second Price mechanism in presence of externalities.

1 Introduction

Sponsored search advertising is nowadays a predominant and arguably most successful paradigm for advertising products in a market, facilitated by the Internet. It constitutes a major source of income for popular search engines like Google, Yahoo! or MS Bing, who allocate up to $8 - 9$ advertisement slots in their sites, alongside the organic results of keyword searches performed by end users. Each time an end user makes a search for a keyword, slots are allocated to advertisers by means of an auction performed automatically; advertisers are ranked in non-increasing order of a *score*, defined as the product of their *bid* with a characteristic *relevance* quantity per advertiser. The relevance of each advertiser is interpreted as the probability that his ad will be clicked by an end user. The score corresponds then to the *declared expected revenue* of advertisers. Advertisers ranked higher receive higher slots. Each advertiser is charged per click an amount depending on the score (hence, the bid) of the one ranked below him.

The described auction is used in varying flavors by search engines. Apart from the mentioned *Rank-By-Revenue* rule, a plainer *Rank-By-Bid* rule has also been used (e.g. by Yahoo!). This auction, known as the *Generalized Second Price* (GSP) auction, constitutes a generalization of the well-known strategyproof Vickrey auction [25]. The

---

[*] Research partially supported by an NTUA Basic Research Grant (PEBE 2009), EPSRC grant EP/F069502/1, and DFG grant Kr 2332/1-3 within Emmy Noether Program.

G. Persiano (Ed.): SAGT 2011, LNCS 6982, pp. 105–116, 2011.

GSP auction is not strategyproof though; it encourages strategic bidding by the advertisers instead of eliciting their valuations truthfully. This induces a strategic game rich in Nash equilibria, and competitive behavior among advertisers that incurs significant revenue to the search engines. Pure Nash Equilibria of the GSP mechanism were first studied by Edelman, Ostrovsky, Schwartz [14] and Varian [24], under what came to be known as the *separable click-through rates* model. In this model every slot is associated with a *Click-Through Rate* (CTR), i.e. the probability that an ad displayed in this slot will be clicked. The joint probability that an ad is clicked is given by the product of the slot's CTR with the relevance of the ad. Since [14,24], a rich literature on keyword auctions has been published, concerning algorithmic and game theoretic issues such as bidding strategies, social efficiency, revenue, see, e.g. [22] (chapter 28).

Recent experiments [18] show that the probability of a displayed ad being clicked (hence, the utility of the advertiser) is affected by its relative order and distance to other ads on the list. E.g., competing ads displayed nearby each other may distract a user's attention from any of them, while related ads may profit from nearby display (e.g. a cars manufacturer and a spare parts supplier). We introduce a model for expressing such *externalities* in keyword auctions. Externalities result in complicated strategic competition among advertisers, precluding stable outcomes and social welfare optimization. A way of alleviating these effects is by solving optimally the *Winner Determination* problem [2,1,19]; i.e., the problem of selecting winners and their assignment to slots, to maximize the *Social Welfare* (the sum of the advertisers' utilities and the auctioneer's revenue – defined formally in Section 2). Such a solution, paired with payments of the Vickrey-Clarke-Groves mechanism [11] (chapter 1), yields a truthful mechanism. On the other hand, the study of the GSP auction's performance in presence of externalities [16] yields insights for the current practice of sponsored search. In our model we analyze exact and approximation algorithms for *Winner Determination*, and show how externalities can harm the GSP auction's stability and social efficiency.

Several recent works concern theoretical and experimental study of externalities in keyword auctions [1,6,15,16,19,17,18]. These works associate the occurrence of externalities with a model of how end users search through the list of ads and how this affects the probability of an ad being clicked. This search is commonly modeled by a top-down ordered scan of the list [6,1,19,16]. The popular *Cascade Model* [1,19], associates a *continuation probability* with every ad, i.e. the likelihood of the user continuing his ordered scan after viewing the ad. Using real data from keyword auctions, Jeziorski and Segal [18] found that previously proposed models fail to express externalities. The *Cascade Model* is contradicted by the fact that about half of the users *do not* click on the ads sequentially, i.e., they return to higher slots after clicking on lower slots. Jeziorski and Segal arrived at a structural model of end user behavior advocating that, after scanning all the advertisements, users focus on a subset of consecutive slots. Within this focus *window*, they observed externalities due to proximity and relative order of the displayed ads. They highlight that an ad's CTR on the list depends crucially on the ads displayed in the other slots, above and below it. We aim at quantifying rigorously these observations, while avoiding explicit modeling of end users' behavior.

Contribution and Techniques. By using a social context graph [5], we design a novel expressive model for describing positive and negative influences among advertisers' relevances in keyword auctions. Our model can describe influences depending on the advertisers': **(i)** IDs, **(ii)** relative order in the list of sponsored links and **(iii)** distance in the list. Motivated by [18], we assume that end users focus on any *window* of c consecutive slots, within which externalities take effect. The model expresses the practically relevant possibility of welfare maximization occurring under partial allocation of slots.

In our model we study the *Winner Determination* problem for social welfare maximization. We prove **APX**-hardness of the problem, even for window size 1 and *positive only* externalities. For positive only externalities we develop two approximation preserving reductions of the problem to the Weighted m-Set Packing problem with sets of size $m = 3$ and $m = 2c+1$ respectively. These reveal a tradeoff between approximability and computational efficiency; for k slots and n advertisers, we obtain algorithms with approximation factors: **(i)** $6c$ in time $O(kn^2 \log n)$, **(ii)** $4c$ in time $O(k^2 n^4)$ and **(iii)** $2(c + 1)$ in polynomial time for any $c = O(1)$. These results settle almost tightly the approximability of the problem for positive-only externalities and $c = O(1)$.

On the positive side, we build on the color coding technique [3] to obtain an exact algorithm for Winner Determination in the full generality of our model. A derandomization of our color coding is possible that yields a deterministic algorithm of running time $2^{O(k)} n^{2c+1} \log^2 n$, hence the class of practically interesting instances of the problem with $c = O(1)$ and even $k = O(\log n)$ is in **P**. This algorithm, paired with VCG payments [11] yields a truthful mechanism for social welfare maximization, when the advertisers' valuations are private information. Notice that the problem can be solved by exhaustive enumeration of $\binom{n}{k}$ tuples of ads, in $O(n^k)$ time. For any $c = O(1)$ our algorithm is significantly faster, even for non constant values of k. Thus the problem is *not* **NP**-hard for $k = O(\text{poly}(\log n))$, unless **NP** $\subseteq$ **DTIME**$(n^{\text{poly}(\log n)})$.

We conclude with an investigation of the GSP mechanism under externalities. We find that pure Nash equilibria do not exist in general, for *conservative bidders* that do not outbid their valuation [9,21]. Even when conservative equilibria exist, we show that their social welfare can be arbitrarily low compared to the social optimum. Due to space constraints several proofs are deferred to the full version of the paper.

2   A Model for Externalities in Keyword Auctions

We consider a set $N = [n] = \{1, 2, \ldots n\}$ of n advertisers (or players/bidders) and a set $K = \{1, 2, \ldots, k\}$ of $k \leq n$ advertisement slots. Each player $i \in N$ has valuation v_i per click and is associated with a probability q_i that her ad is clicked, independently of slot assignment. q_i is often termed *relevance* and measures the intrinsic "quality" of ad i. Each slot j is associated with a probability λ_j that an ad displayed in slot j will be clicked; this is called the *Click-Through Rate* (CTR) of the slot. The (overall) Click-Through Rate *of an ad* $i \in N$ occupying slot j is $\lambda_j \cdot q_i$ (separable CTRs). We assume $1 \geq \lambda_1 \geq \ldots \geq \lambda_k > 0$, i.e. that the top slot is slot 1, the second one is slot 2 and so on. Let $S \subseteq [n]$, $|S| \leq k$, be the set of winning ads and $\pi : S \to K$ their assignment to slots, i.e. $\pi(i) \in K$ is the slot of advertiser $i \in S$. Every advertiser $i \in S$ issues to the search engine (auctioneer) a payment p_i per received click, hence receives expected utility $u_i(S, \pi) = \lambda_{\pi(i)} \cdot q_i \cdot (v_i - p_i)$. The *Social Welfare* is then:

$$\text{sw}(S,\pi) = \sum_{i \in S} \lambda_{\pi(i)} \cdot q_i \cdot v_i = \sum_{i \in S} u_i(S,\pi) + \sum_{i \in S} \lambda_{\pi(i)} \cdot q_i \cdot p_i, \qquad (1)$$

where $\sum_{i \in S} \lambda_{\pi(i)} \cdot q_i \cdot p_i$ is the expected revenue of the auctioneer. Our modeling of externalities is built on top of the separable CTRs model and induces amplified or diminished actual relevance Q_i (compared to q_i) to the advertisers, depending on their relative position and distance of their ads on the list. We use a directed *social context graph* [5] $G(N, E^+, E^-)$, defined upon the set of advertisers N. Each edge $(i,j) \in E^+ \cup E^-$ is associated with a function $w_{ij} \in (0,1)$. An edge (j,i) in E^+ or in E^- denotes potential positive or negative influence respectively of i by j. Namely, edges in E^+ and E^- model respectively *positive* and *negative* externalities among advertisers. For any edge $(j,i) \in E^+ \cup E^-$, the potential influence of j to i is quantified by a function $w_{ji} : \{-k+1, -k+2, \ldots, -1, 1, \ldots, k-2, k-1\} \mapsto [0,1]$. If $d_\pi(j,i) = \pi(i) - \pi(j)$ is the distance of ad i from ad j in the list, $w_{ji}(d_\pi(j,i))$ is the probability that a user's interest in a displayed ad j may result in attraction/distraction of his attention to/from an ad i respectively, depending on whether $(j,i) \in E^+$ or $(j,i) \in E^-$ [1]. It is reasonable to assume that the closer i and j are in π, the stronger the influence of j on i is. Formally, for every ℓ, ℓ', with $|\ell|, |\ell'| \geq 1$, if $|\ell| < |\ell'|$, then $w_{ji}(\ell) \geq w_{ji}(\ell')$.

Let $S \subseteq N$ be the set of winners and π be the permutation assigning them to the slots. Define the subgraph $G_S(S, E_S^+, E_S^-)$ of the context graph, induced by S. The probability $Q_i(S,\pi)$ that a user is attracted by ad i is expressed as product $Q_i^+(S,\pi) \times Q_i^-(S,\pi)$. $Q_i^+(S,\pi)$ is the probability that i attracts the user's attention either by its intrinsic relevance q_i or by receiving positive influence. $Q_i^-(S,\pi)$ is the probability that the user's attention is *not* distracted from i due to negative influence of others. For each $i \in S$ define $N_i^+(S) = \{j \in S : (j,i) \in E_S^+\}$ and $N_i^-(S) = \{j \in S : (j,i) \in E_S^-\}$ to be the set of neighbors of i in G_S with positive and negative influence respectively. Let us derive $Q_i^+(S,\pi)$ first. A user's attention is *not* attracted by the ad of i with probability $(1 - q_i)$ and if, independently, i is not positively influenced by any $j \in N_i^+(S)$. The latter occurs either because j itself does not attract the user (with probability $1 - q_j$), or the positive influence of j to i does not occur (with probability $q_j(1 - w_{ji}(d_\pi(j,i)))$). Then j does not influence i with probability $(1 - q_j) + q_j(1 - w_{ji}(d_\pi(j,i))) = 1 - q_j w_{ji}(d_\pi(j,i))$ and:

$$Q_i^+(S,\pi) = 1 - (1 - q_i) \cdot \prod_{j \in N_i^+(S)} (1 - q_j w_{ji}(d_\pi(j,i))) \qquad (2)$$

For $Q_i^-(S,\pi)$ we use similar reasoning. A user's attention is not distracted from the ad of i due to negative influence of $j \in N_i^-(S)$ if either his attention is not captured by j or, if it is, j fails to influece i negatively. This event occurs with probability $(1 - q_j) + q_j(1 - w_{ji}(d_\pi(j,i))) = 1 - q_j w_{ji}(d_\pi(j,i))$. Assuming independence of the events for all $j \in N_i^-(S)$, we have:

$$Q_i^-(S,\pi) = \prod_{j \in N_i^-(S)} (1 - q_j w_{ji}(d_\pi(j,i)))$$

---

[1] We note that if $d_\pi(j,i) > 0$, i appears below j, and if $d_\pi(j,i) < 0$, j appears below i in π.

After deriving $Q_i(S, \pi) = Q_i^+(S, \pi) \times Q_i^-(S, \pi)$, we can restate the social welfare as:

$$\text{sw}(S, \pi, G) = \sum_{i \in S} \lambda_{\pi(i)} \cdot Q_i(S, \pi) \cdot v_i \,. \tag{3}$$

Arguably, users' attention and memory when they process the advertisements list have a bounded scope. Therefore, we assume that there is an integer constant $c > 0$, called *window size*, so that each ad j can only affect other ads at a distance at most c in π. Formally, if window size is c, for all $(j, i) \in E^+ \cup E^-$ and all integers ℓ with $|\ell| > c$, $w_{ji}(\ell) = 0$. Then the relevance Q_i only depends on the ads in $N_i^+(S) \cup N_i^-(S)$ assigned to slots at distance at most c from i.

Related Work. Edelman, Ostrovsky, Schwartz [14] and Varian [24] first modeled the game induced by the GSP auction mechanism under the assumption of separable CTRs, i.e. that the probability of a specific ad being clicked when displayed in a certain slot is given by the product of the slot's CTR and the ad's relevance. For this game model they identified socially optimal pure Nash equilibria. Prior to these works, Aggarwal, Goel and Motwani [2] had already designed a truthful mechanism for non-separable CTRs (in this case the VCG auction is not applicable). For separable CTRs they proved revenue equivalence of their mechanism to the VCG.

There has been a growing interest in modeling externalities in sponsored search and in how they affect the advertisers' bidding strategies and the properties of GSP equilibria. The unpublished work of Das *et al.* [13] is probably the closest in spirit to ours. Das et al. consider externalities among advertisers based on their relative quality (i.e., relevance). However, the model of [13] treats advertisers as anonymous, in the sense that externalities do not depend on their bids, but just on their relative quality. Our model makes a step further, since in addition to the advertisers' relative quality as measured by their relevance, we also take into account their IDs, their dependencies in the context graph, and their relative distance in the sponsored list.

Athey and Ellison [6] in one of the first models for externalities, they assumed an ordered top-down scan of slots by end users. By assuming a certain cost incurred to the end users for clicking on an ad, they derived the equilibria of the GSP auction for their game. Along similar lines, Aggarwal *et al.* [1] and Kempe and Mahdian [19] studied *Cascade Models* involving Markovian end-users. Every ad is associated with its individual CTR and a *continuation probability*, that an end-user will continue scanning the list of ads (in a top-down order) after viewing that particular ad. The authors studied the winner determination problem for social welfare maximization. Equilibria of the GSP auction mechanism in the cascade model were studied by Giotis and Karlin in [16]. Kuminov and Tennenholtz [20] study equilibria of the GSP and VCG (cf. [22, Ch. 9]) auctions in a similar model. The cascade model was assessed experimentally by Craswell *et al.* [12]. Gomes, Immorlica and Markakis [17] were the first to document externalities empirically under a cascade model, using real data.

Externalities among the bidders have also been considered in other similar settings. Ghosh and Mahdian [15] study the complexity of the Winner Determination problem and develop tractable incentive compatible mechanisms, under a model for externalities in online lead generation. Chen and Kempe [10] consider positive and negative social dependencies among bidders of single item auctions and study the properties of equilibria for the first and second price auctions.

Winner Determination Problems. We denote by MSW-E the problem of *Maximum Social Welfare with Externalities*, i.e. selecting a subset of winners S and a permutation π of S that maximize $\mathrm{sw}(S, \pi, G)$. The complexity and performance of our algorithms are parameterized by the window size c and we write MSW-E(c). An interesting special case of MSW-E(c) with *"positive-only"* externalities occurs for $E^- = \emptyset$; this is denoted by MSW-PE(c). Motivated by the Cascade Model, we also consider the case of *forward-only* positive-only externalities – denoted by MSW-FPE(c) – where an ad j may only influence an ad i iff $\pi(j) < \pi(i)$.

In presence of negative externalities, it may be profitable to select less than k ads and arrange them in a *broken list*, namely a list with empty slots appearing between slots occupied by negatively correlated ads. In practice however, this may not be feasible. Therefore, we restrict feasible solutions to so-called *unbroken lists*, namely lists where the selected ads occupy consecutive slots in the ad list starting from the first one[2]. Hence a feasible solution can select less than k adds, but it is not allowed to use empty slots and separate negatively correlated ads (i.e. the empty slots, if any, occupy the last $k - |S|$ slots in π). The analysis of our algorithms assumes the case of unbroken lists. However, it is not difficult to generalize our algorithmic results to the case of broken lists.

3   An Exact Algorithm Based on Color Coding

A PTAS reduction from the Traveling Salesman Problem with distances $1, 2$ [23] yields **APX**-hardness of MSW-E(c) even for $c = 1$, with uniform functions $w_{ji} = w$, position multipliers λ_j, valuations v_i and qualities q_i. Also, note that there is an elegant reduction from the Longest Path problem, that proves **NP**-hardness of MSW-FPE(1).

Theorem 1. MSW-FPE(1) *is* **APX**-*hard even in the special case of uniform position multipliers, valuations, and qualities.*

In this section we develop and analyze an exact algorithm for MSW-E(c). We employ *color coding* [3] and dynamic programming, to prove the following result:

Theorem 2. MSW-E(c) *can be solved optimally in* $2^{O(k)} n^{2c+1} \log^2 n$ *time.*

For simplicity, we assume that the optimal solution consists of k ads. We can remove this assumption by running the algorithm for every sponsored list size up to k, and keep the best solution. Since the running time is exponential in k, this does not change the asymptotics of the running time. To apply the technique of color coding, we consider a fixed coloring $h : N \mapsto [k]$ of the ads with k colors. A list (S, π) of k ads is *colorful* if all ads in S are assigned different colors by h. In the following, we formulate a dynamic programming algorithm that computes the best *colorful* list.

For each $2c$-tuple of ads $(i_1, \ldots, i_{2c}) \in N^{2c}$, with all ads assigned different colors by h, and each color set $C \subseteq [k], |C| \leq k - 2c$, that does not include any of the colors of

---

[2] A reasonable assumption that may justify the restriction above is that there exist an adequate number of "neutral" ads ($k - 1$ of them suffice) that do not negatively affect any other ad.

$i_1, \ldots, i_{2c}$, we compute $\mathrm{sw}(i_1, \ldots, i_{2c}, C)$, namely the maximum social welfare if the last $|C|$ positions in the list are colored according to C, and on top of them, there are ads $i_1, \ldots, i_{2c}$ in this order from top to bottom. More precisely, the solution corresponding to $\mathrm{sw}(i_1, \ldots, i_{2c}, C)$ assigns ad i_p to slot $k-(|C|+2c)+p$, $p = 1, \ldots, 2c$, and considers the best choice of ads colored according to C for the last $|C|$ slots. Clearly, there are at most $n^{2c} \, 2^{k-2c}$ different sw values to compute, and the maximum sw value for all colorful tuples $(i_1, \ldots, i_{2c}, C)$, with $|C| = k - 2c$, corresponds to the best colorful list of k ads. The proof of Theorem 2 follows:

Proof. For the basis of our dynamic programming, let $C = \emptyset$ and for all $2c$-tuples $(i_1, \ldots, i_{2c}) \in N^{2c}$ with all ads assigned different colors by h, we have:

$$\mathrm{sw}(i_1, \ldots, i_{2c}, \emptyset) = \sum_{p=1}^{c} \lambda_{k-2c+p} \cdot Q_{i_p}(i_1, \ldots, i_p, \ldots, i_{p+c}) \cdot v_{i_p}$$

$$+ \sum_{p=c+1}^{2c} \lambda_{k-2c+p} \cdot Q_{i_p}(i_{p-c}, \ldots, i_p, \ldots, i_{2c}) \cdot v_{i_p} ,$$

where $Q_{i_p}(i_1, \ldots, i_p, \ldots, i_{c+p})$ (resp. $Q_{i_p}(i_{p-c}, \ldots, i_p, \ldots, i_{2c})$) is the CTR of ad i_p given that the (only) ads in the list at distance at most c from i_p are $i_1, \ldots, i_{c+p}$ (resp. $i_{p-c}, \ldots, i_{2c}$) arranged in this order from top to bottom.

Given the values of sw for all $2c$-tuples of ads and all color sets of cardinality $s < k - 2c$, we compute the values of sw for all $2c$-tuples $(i_1, \ldots, i_{2c})$ and all color sets C of cardinality $s + 1$:

$$\mathrm{sw}(i_1, \ldots, i_{2c}, C) = \max_{i:h(i) \in C} \bigg\{$$

$$\mathrm{sw}(i_2, \ldots, i_{2c}, i, C - \{h(i)\}) + \lambda_{k-(|C|+2c)+1} \cdot Q_{i_1}(i_1, \ldots, i_{c+1}) \cdot v_{i_1} +$$

$$\sum_{p=2}^{c} \lambda_{k-(|C|+2c)+p} \cdot [Q_{i_p}(i_1, i_2, \ldots, i_p, \ldots, i_{p+c}) - Q_{i_p}(i_2, \ldots, i_p, \ldots, i_{p+c})] \cdot v_{i_p} +$$

$$\lambda_{k-|C|-c+1}[Q_{i_{c+1}}(i_1, i_2, \ldots, i_{c+1}, \ldots, i_{2c}, i) - Q_{i_p}(i_2, \ldots, i_{c+1}, \ldots, i_{2c}, i)]v_{i_{c+1}} \bigg\}$$

In the recursion above, the second term accounts for the additional social welfare due to i_1, and the third and the fourth term account for the difference in the social welfare due to ads $i_2, \ldots, i_{c+1}$, whose CTRs $Q_{i_2}, \ldots, Q_{i_{c+1}}$ are affected by i_1. Ads $i_{c+2}, \ldots, i_{2c}, i$ are used to calculate the difference in the CTRs $Q_{i_2}, \ldots, Q_{i_{c+1}}$. The CTRs of ads $i_{c+2}, \ldots, i_{2c}, i$ and of the ads at the bottom of the list with colors in $C - \{h(i)\}$ are not affected by i_1, since their distance to i_1 is greater than c.

Hence, for any fixed coloring h, the best colorful list of k ads can be computed in time $O(n^{2c+1} 2^k)$. If we choose a random coloring h, the probability that the optimal solution is colorful under h is $k!/k^k > e^{-k}$. If we run the algorithm for $e^k \ln n$ colorings chosen independently at random and keep the best solution, the probability of *not* finding the optimal solution is at most $1/n$. The approach can be derandomized using a k-perfect family of hash functions of size $2^{O(k)} \log^2 n$ ([3, Section 4] for details). □

Corollary 1. *For $k = O(\log n)$ and $c = O(1)$, MSW-E(c) is in* **P**.

Moreover, unless **NP** $\subseteq$ **DTIME**$(n^{\mathrm{poly}(\log n)})$, MSW-E$(c)$ is not **NP**-hard for $c = O(1)$ and $k = O(\mathrm{poly}(\log n))$. The exact algorithm can be paired with VCG payments to yield a truthful mechanism, when valuations are private to advertisers.

4 $O(c)$-Approximation Algorithms for Positive Externalities

In this section, we show how to use polynomial-time approximation algorithms for the Weighted m-Set Packing problem and approximate MSW-PE(c) within a factor of $O(c)$ in polynomial time. In Weighted m-Set Packing, we are given a collection of sets, each with at most m elements and a positive weight, and seek a collection of disjoint sets of maximum total weight. The greedy algorithm for Weighted m-Set Packing achieves an approximation ratio of m, the algorithm of [8] achieves an approximation ratio of $(2/3)m$ in time quadratic in the number of sets, and the algorithm of [7] achieves an approximation ratio of $(m + 1)/2$ in polynomial time for any constant m.

Theorem 3. *An α-approximation $T(\nu)$-time algorithm for Weighted 3-Set Packing with ν sets yields a $2\alpha c$-approximation $T(kn^2)$- time algorithm for* MSW-PE(c), *with n ads and k slots.*

Proof. We transform any instance of MSW-PE(c) to an instance of Weighted 3-Set Packing with $kn^2/4$ sets so that any α-approximation to the optimal set packing gives a $2\alpha c$-approximation to the optimal social welfare for the original MSW-PE(c) instance. To simplify the presentation, we assume that k is even. Our proof can be easily extended to the case where k is odd.

Given an instance of MSW-PE(c) with n ads and k slots, we partition the list into $k/2$ blocks of 2 consecutive slots each. The set packing instance has $\binom{n}{2}$ 3-element sets for each block. Namely, for every block $p = 1, 3, 5, \ldots, k - 1$ and every subset $\{i_1, i_2\}$ of 2 ads, there is a set $\{i_1, i_2, p\}$ in the set packing instance[3]. The weight $W(i_1, i_2, p)$ of each set $\{i_1, i_2, p\}$ is the maximum social welfare if i_1 and i_2 are assigned to slots p and $p + 1$, and i_1, i_2 are not influenced by any other ad in the list. Formally,

$$W(i_1, i_2, p) = \max\{\lambda_p Q_{i_1}(i_1, i_2)v_{i_1} + \lambda_{p+1} Q_{i_2}(i_1, i_2)v_{i_2},$$
$$\lambda_p Q_{i_2}(i_2, i_1)v_{i_2} + \lambda_{p+1} Q_{i_1}(i_2, i_1)v_{i_1}\}$$

where $Q_{i_1}(i_1, i_2) = 1 - (1 - q_{i_1})(1 - q_{i_2}w_{i_2 i_1}(1))$ (resp. $Q_{i_2}(i_1, i_2) = 1 - (1 - q_{i_2})(1 - q_{i_1}w_{i_1 i_2}(-1))$) denotes the relevance of ad i_1 (resp. i_2) given that the only ad in the list with an influence on i_1 (resp. i_2) is i_2 (resp. i_1) located just above (resp. below) i_1 (resp. i_2) in the list.

Given an instance of MSW-PE(c) with n advertisers and k slots, the corresponding instance of Weighted 3-Set Packing can be computed in $O(kn^2)$ time. To show that the transformation above is approximation preserving, we prove that **(i)** the optimal set packing has weight at least $1/(2c)$ of the maximum social welfare, and that **(ii)** given a set packing of weight W, we can efficiently compute a solution for the original instance of MSW-PE(c) with a social welfare of at least W.

To prove **(i)**, we assume (by renumbering the ads appropriately if needed) that the optimal list for the MSW-PE(c) instance is $(1, \ldots, k)$. We let Q_i^* be the relevance of ad

---

[3] Throughout the proof, we implicitly adopt the simplifying assumption that the range of block descriptors $1, 3, 5, \ldots, k - 1$ and the range of ad descriptors $1, \ldots, n$ are disjoint.

i in $(1, \ldots, k)$, and let $W^* = \sum_{i=1}^{k} \lambda_i Q_i^* v_i$ be the optimal social welfare. We construct a collection of $2c$ feasible set packings of total weight at least W^*. Thus, at least one of them has a weight of at least $W^*/(2c)$. The construction is based on the following claim, which can be proven by induction on c.

Claim 1. *Let c be any positive integer. Given a list $(1, \ldots, k)$, there is a collection of $2c$ feasible 3-set packings such that for each pair i_1, i_2 of ads in $(1, \ldots, k)$ with $|i_1 - i_2| \leq c$, the union of these packings contains a set $\{i_1, i_2, p\}$ with $p \leq \min\{i_1, i_2\}$.*

Intuitively, for each pair i_1, i_2 of ads located in $(1, \ldots, k)$ within a distance no more than the window size c, and thus possibly having a positive influence on each other, the collection of set packings constructed in the proof of Claim 1 includes a set $\{i_1, i_2, p\}$ whose weight accounts for the increase in i_1's and i_2's social welfare due to i_2's and i_1's positive influence, respectively. Summing up the weights of all those sets, we account for the positive influence between all pairs of ads in $(1, \ldots, k)$, and thus end up with a total weight of at least W^*.

Formally, let $W^{(j)}$ be the total weight of the j-th set packing constructed in the proof of Claim 1, and let i_1, i_2, with $i_1 < i_2$, be any pair of ads in $(1, \ldots, k)$ included in the same set $\{i_1, i_2, p\}$ of the j-th packing. Since each ad appears in each set packing at most once, we let $Q_{i_1}^{(j)} = Q_{i_1}(i_1, i_2)$ and $Q_{i_2}^{(j)} = Q_{i_2}(i_1, i_2)$ be the relevance of i_1 and i_2 in the calculation of $W(i_1, i_2, p)$. Since Claim 1 ensures that $p \leq i_1$, and since slot CTRs are non-increasing, λ_{i_1} (resp. λ_{i_2}) is no greater than λ_p (resp. λ_{p+1}). Therefore, $W(i_1, i_2, p) \geq \lambda_{i_1} Q_{i_1}^{(j)} v_{i_1} + \lambda_{i_2} Q_{i_2}^{(j)} v_{i_2}$. Setting $Q_i^{(j)} = 0$ for all ads i in $(1, \ldots, k)$ which do not appear in any set of the j-th packing, we obtain that $W^{(j)} \geq \sum_{i=1}^{k} \lambda_i Q_i^{(j)} v_i$. We show that

$$\sum_{j=1}^{2c} W^{(j)} \geq \sum_{i=1}^{k} \lambda_i v_i \sum_{j=1}^{2c} Q_i^{(j)} \geq \sum_{j=1}^{k} \lambda_i Q_i^* v_i = W^*$$

The first inequality follows from the discussion above and by changing the order of the summation. To establish the second inequality, we show that for every ad i in $(1, \ldots, k)$, $\sum_{j=1}^{2c} Q_i^{(j)} \geq Q_i^*$. To simplify the presentation, we focus on an ad i with $c < i \leq k - c$. Ads $1, \ldots, c$ and $k - c + 1, \ldots, k$ can be treated similarly.

We recall that $Q_i^* = 1 - (1 - q_i) \prod_{j=i-c, j \neq i}^{i+c} P_i(j)$, where for each ad j in the sublist $(i - c, \ldots, i - 1, i + 1, \ldots, i + c)$, $P_i(j) = (1 - q_j w_{ji}(j - i)) \in [0, 1]$ accounts for j's positive influence on i's relevance. Claim 1 ensures that for each j in $(i - c, \ldots, i - 1, i + 1, \ldots, i + c)$, ads i and j are included in the same set of some set packing. Since ad i appears in each set packing at most once, for simplicity, we can renumber the set packings of Claim 1, and say that i and j are included in the same set of the j-th set packing. Then, if $j < i$,

$$Q_i^{(j)} = 1 - (1 - q_i)(1 - q_j w_{ji}(-1)) \geq 1 - (1 - q_i) P_i(j),$$

because w_{ji} is non-increasing with the distance of j and i in the list, and thus $w_{ji}(-1) \geq w_{ji}(j-i)$. The same holds if $j > i$. Therefore, $Q_i^{(j)} \geq 1 - (1-q_i)P_i(j)$, for any j. To conclude the proof of (i), we observe that:

$$\sum_{j=i-c,j\neq i}^{i+c} (1 - (1-q_i)P_i(j)) \geq 1 - (1-q_i) \prod_{j=i-c,j\neq i}^{i+c} P_i(j) = Q_i^* \qquad (4)$$

To establish (4), we repeatedly apply that for every $x, y, z \in [0,1]$, $(1-xy)+(1-xz) \geq 1 - xyz$. We proceed to establish claim (ii), namely that given a set packing of weight W, we can efficiently construct a sponsored list of social welfare at least W for the original instance of MSW-PE(c). By construction, we can restrict our attention to set packings of the form $\{\{i_p, i_{p+1}, p\}\}_{p=1,3,\ldots,k-1}$, where the weight of the packing is $W = \sum_p W(i_p, i_{p+1}, p)$, and where ads i_p and i_{p+1} are indexed according to their best order, with respect to which $W(i_p, i_{p+1}, p)$ is calculated. Since we consider positive externalities, the sponsored list $(i_1, i_2, \ldots, i_{k-1}, i_k)$ has social welfare at least W. □

Combining Theorem 3 with the greedy 3-approximation algorithm for Weighted 3-Set Packing and with the algorithm of [8] we obtain respectively:

Corollary 2. *For n ads and k slots, the MSW-PE(c) problem can be approximated within factor $6c$ in $O(kn^2 \log n)$ time and within factor $4c$ in $O(k^2n^4)$ time.*

A similar reduction to the Weighted $(2c+1)$-Set Packing problem yields:

Theorem 4. *An $f(m)$–approximation $T(\nu, m)$-time algorithm for Weighted m–Set Packing with ν sets yields a $2f(2c+1)$-approximation $O(ckn^{2c} + T(kn^{2c}, 2c+1))$-time algorithm for MSW-PE(c) with n ads and k slots.*

5   On the GSP Mechanism with Externalities

In studying the GSP auction mechanism we make the reasonable assumption that only a snapshot of the players' intrinsic relevances q_i, $i \in N$ is available to the mechanism. In practice, estimates of the players' relevances are deduced by software of the sponsored search platform; therefore the mechanism will eventually extract information indicative of externalities. By that time however, associations among advertisers may have changed. This justifies the mechanism's unawareness of externalities. On the other hand, each advertiser is aware of associations that may harm him or boost his relevance. Given a social context graph $G(N, E^+, E^-)$, with functions w_{ji}, $(j,i) \in E^+ \cup E^-$ and window size c, we assume each $i \in N$ to know: $E_i^+ = \{(i', i) \in E^+ | i' \in N\}$, $E_i^- = \{(i', i) \in E^- | i' \in N\}$ and of c and w_{ji} for every $j \in E_i^+ \cup E_i^-$.

In a keyword auction each advertiser i bids b_i for receiving a slot in the list. The GSP mechanism in its most common flavors uses the *Rank-By-Revenue* (RBR) rule to assign advertisers to slots. Under RBR, advertisers are ranked in order of non-increasing score $q_i \cdot b_i$ and higher scores are assigned higher CTR slots. The score of a bidder is his *declared expected revenue* for a click. The plainer RBB rule is obtained by taking $q_i = 1$ for all $i \in N$. Given a bid vector b, let $\phi_b : K \to N$ denote the ranking of bidders in order of non-increasing expected revenue, i.e. $\phi_b(j)$ is the bidder assigned to slot j. According to the previously used definition of π, ϕ is an extension of π^{-1}. Every slot winning player $\phi_b(j)$, for $j = 1, \ldots, k$ pays *per click* a price $(q_{\phi_b(j+1)} \cdot b_{\phi_b(j+1)})/q_{\phi_b(j)}$; i.e., the score of the bidder occupying the next position under b, divided by the intrinsic relevance of $\phi_b(j)$. Using his knowledge of externalities that influence him, player $\phi_b(j)$ experiences a relevance $Q_{\phi_b(j)}(b)$ and estimates his expected profit (utility) as:

$$u_{\phi_b(j)}(b) = \lambda_j \cdot Q_{\phi_b(j)}(b) \times [v_{\phi_b(j)} - (q_{\phi_b(j+1)}/q_{\phi_b(j)}) \cdot b_{\phi_b(j+1)}] \tag{5}$$

We assume a complete information setting, as advertisers typically employ machine learning techniques to estimate how much they should outbid a competitor. Such techniques reveal the ranking information used by the GSP mechanism. Thus, in computing his best response under a bid vector b_{-i}, every advertiser $i \in N$ is assumed to know only $q_{i'}$ and $b_{i'}$ for each $i' \in N \setminus \{i\}$, and *not* the actual relevance $Q_{i'}$ perceived by i'. In studying pure Nash equilibria of the GSP mechanism we make the standard assumption of *conservative* bidders [21], i.e. that $b_i \leq v_i$ for all $i \in N$. Then:

Proposition 1. *The strategic game induced by the GSP mechanism under the RBR rule and deterministic tie-breaking does not generally have pure Nash equilibria in presence of forward positive externalities, even for 2 slots and 3 conservative players.*

The case is similar for bidirectional positive externalities (deferred to full version). Even when pure Nash equilibria exist, externalities may cause unbounded *Price of Stability* [4], in contrast to the favorable social efficiency shown in [9], for the GSP mechanism without externalities.

Proposition 2. *There is an infinite family of instances of the stategic game induced by the Generalized Second Price auction mechanism with unbounded Price of Stability, even with conservative bidders.*

Acknowledgments. Dimitris Fotakis thanks Vasileios Syrgkanis for many helpful discussions on the model of externalities presented in this work. The authors thank Moshe Tennenholtz for many interesting discussions and suggestions concerning this project.

References

1. Aggarwal, G., Feldman, J., Muthukrishnan, S., Pal, M.: Sponsored Search Auctions with Markovian Users. In: Papadimitriou, C., Zhang, S. (eds.) WINE 2008. LNCS, vol. 5385, pp. 621–628. Springer, Heidelberg (2008)
2. Aggarwal, G., Goel, A., Motwani, R.: Truthful auctions for pricing search keywords. In: Proceedings of 7th ACM Conference on Electronic Commerce (EC), pp. 1–7. ACM, New York (2006)
3. Alon, N., Yuster, R., Zwick, U.: Color Coding. Journal of the ACM 42, 844–856 (1995)
4. Anshelevich, E., Dasgupta, A., Kleinberg, J.M., Tardos, E., Wexler, T., Roughgarden, T.: The Price of Stability for Network Design with Fair Cost Allocation. SIAM Journal on Computing 38(4), 1602–1623 (2008)
5. Ashlagi, I., Krysta, P., Tennenholtz, M.: Social Context Games. In: Papadimitriou, C., Zhang, S. (eds.) WINE 2008. LNCS, vol. 5385, pp. 675–683. Springer, Heidelberg (2008)

6. Athey, S., Ellison, G.: Position Auctions with Consumer Search. Quartetly Journal of Economis (to appear, 2011)
7. Berman, P.: A $d/2$ Approximation for Maximum Weight Independent Set in d-Claw Free Graphs. In: Halldórsson, M.M. (ed.) SWAT 2000. LNCS, vol. 1851, pp. 214–219. Springer, Heidelberg (2000)
8. Berman, P., Krysta, P.: Optimizing misdirection. In: Proceedings of the ACM-SIAM Symposium on Discrete Algorithms (SODA), pp. 192–201. ACM, New York (2003)
9. Caragiannis, I., Kaklamanis, C., Kanellopoulos, P., Kyropoulou, M.: On the Efficiency of Equilibria in Generalized Second Price Auctions. In: Proceedings of the ACM Conference on Electronic Commerce (EC), pp. 81–90. ACM, New York (2011)
10. Chen, P., Kempe, D.: Bayesian Auctions with Friends and Foes. In: Mavronicolas, M., Papadopoulou, V.G. (eds.) SAGT 2009. LNCS, vol. 5814, pp. 335–346. Springer, Heidelberg (2009)
11. Cramton, P., Shoham, Y., Steinberg, R.: Combinatorial Auctions. MIT Press, Cambridge (2006)
12. Craswell, N., Zoeter, O., Taylor, M., Ramsey, B.: An experimental comparison of click position-bias models. In: WSDM 2008: Proceedings of the International Conference on Web Search and Web Data Mining, pp. 87–94. ACM, New York (2008)
13. Das, A., Giotis, I., Karlin, A., Mathieu, C.: On the Effects of Competing Advertisements in Keyword Auctions (2008) (unpublished manuscript)
14. Edelman, B., Ostrovsky, M., Schwarz, M.: Internet advertising and the generalized second price auction: Selling billions of dollars worth of keywords. American Economic Review 97(1), 242–259 (2007)
15. Ghosh, A., Mahdian, M.: Externalities in Online Advertising. In: Proceedings of the 17th International World Wide Web Conference (WWW), pp. 161–168. ACM, New York (2008)
16. Giotis, I., Karlin, A.R.: On the equilibria and efficiency of the GSP mechanism in keyword auctions with externalities. In: Papadimitriou, C., Zhang, S. (eds.) WINE 2008. LNCS, vol. 5385, pp. 629–638. Springer, Heidelberg (2008)
17. Gomes, R., Immorlica, N., Markakis, E.: Externalities in Keyword Auctions: an empirical and theoretical assessment. In: Leonardi, S. (ed.) WINE 2009. LNCS, vol. 5929, pp. 172–183. Springer, Heidelberg (2009)
18. Jeziorski, P., Segal, I.: What makes them click: Empirical analysis of consumer demand for search advertising. Working paper (Department of Economics, Stanford University (2009)
19. Kempe, D., Mahdian, M.: A Cascade Model for Externalities in Sponsored Search. In: Papadimitriou, C., Zhang, S. (eds.) WINE 2008. LNCS, vol. 5385, pp. 585–596. Springer, Heidelberg (2008)
20. Kuminov, D., Tennenholtz, M.: User Modeling in Position Auctions: Re-Considering the GSP and VCG Mechanisms. In: Proceedings of the 8th International Joint Conference on Autonomous Agents and Multiagent Systems (AAMAS), pp. 273–280. ACM, New York (2009)
21. Leme, R.P., Tardos, E.: Pure and Bayes-Nash Price of Anarchy for Generalized Second Price Auction. In: Proceedings of the 51th Annual IEEE Symposium on Foundations of Computer Science (FOCS), pp. 735–744. IEEE, Los Alamitos (2010)
22. Nisan, N., Roughgarden, T., Tardos, É., Vazirani, V.: Algorithmic Game Theory. Cambridge University Press, New York (2007)
23. Papadimitriou, C., Yannakakis, M.: The Travelling Salesman Problem with Distances One and Two. Mathematics of Operations Research 18(1), 1–11 (1993)
24. Varian, H.R.: Position auctions. International Journal of Industrial Organization 25(6), 1163–1178 (2007)
25. Vickrey, W.: Counterspeculation, Auctions and Competitive Sealed Tenders. Journal of Finance, 8–37 (1961)

Peer Effects and Stability in Matching Markets*

Elizabeth Bodine-Baron, Christina Lee, Anthony Chong,
Babak Hassibi, and Adam Wierman

California Institute of Technology,
Pasadena, CA 91125, USA
{eabodine,chlee,anthony,hassibi,adamw}@caltech.edu

Abstract. Many-to-one matching markets exist in numerous different
forms, such as college admissions, matching medical interns to hospitals
for residencies, assigning housing to college students, and the classic firms
and workers market. In all these markets, externalities such as complementarities and peer effects severely complicate the preference ordering
of each agent. Further, research has shown that externalities lead to serious problems for market stability and for developing efficient algorithms
to find stable matchings. In this paper we make the observation that
peer effects are often the result of underlying social connections, and we
explore a formulation of the many-to-one matching market where peer
effects are derived from an underlying social network. The key feature
of our model is that it captures peer effects and complementarities using utility functions, rather than traditional preference ordering. With
this model and considering a weaker notion of stability, namely two-
sided exchange stability, we prove that stable matchings always exist
and characterize the set of stable matchings in terms of social welfare.
To characterize the efficiency of matching markets with externalities, we
provide general bounds on how far the welfare of the worst-case stable
matching can be from the welfare of the optimal matching, and find that
the structure of the social network (e.g. how well clustered the network
is) plays a large role.

1 Introduction

Many-to-one matching markets exist in numerous forms, such as college admissions, the national medical residency program, freshman housing assignment, as
well as the classic firms-and-workers market. These markets are widely studied
in academia and also widely deployed in practice, and have been applied to other
areas, such as FCC spectrum allocation and supply chain networks [4,21].

In the conventional formulation, matching markets consist of two sets of
agents, such as medical interns and hospitals, each of which have preferences

* This work was supported in part by the National Science Foundation under grants
CCF-0729203, CNS-0932428 and CCF-1018927, by the Office of Naval Research
under the MURI grant N00014-08-1-0747, and by Caltech's Lee Center for Advanced
Networking.

G. Persiano (Ed.): SAGT 2011, LNCS 6982, pp. 117–129, 2011.

over the agents to which they are matched. In such settings it is important that matchings are 'stable' in the sense that agents do not have incentive to change assignments after being matched. The seminal paper on matching markets was by Gale and Shapley [13], and following this work an enormous literature has grown, e.g., [20,27,28,29] and the references therein. Further, variations on Gale and Shapley's original algorithm for finding a stable matching are in use by the National Resident Matching Program (NRMP), which matches medical school graduates to residency positions at hospitals [26].

However, there are problems with many of the applications of matching markets in practice. For example, couples participating in the NRMP often reject their matches and search outside the system. In housing assignment markets where college students are asked to list their preferences over housing options, there is often collusion among friends to list the same preference order for houses. These two examples highlight that 'peer effects', whether just couples or a more general set of friends, often play a significant role in many-to-one matchings. That is, agents care not only where they are matched, but also which other agents are matched to the same place. Similarly, 'complementarities' often play a role on the other side of the market. For example, hospitals and colleges care not only about which individual students are assigned to them, but also that the group has a certain diversity, e.g., of different specializations.

As a result of the issues highlighted above, there is a growing literature studying many-to-one matchings with externalities (i.e., peer effects and complementarities) [10,14,18,19,22,24,3,11,30] and the research has found that designing matching mechanisms is significantly more challenging when externalities are considered, e.g. incentive compatible mechanism design is no longer possible.

The reason for the difficulty is that there is no longer a guarantee that a stable many-to-one matching will exist when agents care about more than their own matching [26,28], and, if a stable matching does exist, it can be computationally difficult to find [25]. Consequently, most research has focused on identifying when stable matchings do and do not exist. Papers have proceeded by constraining the matching problem through restrictions of the possible preference orderings, [10,14,18,19,22,24], and by considering variations on the standard notion of stability [3,11,30].

The key idea of this paper is that *peer effects are often the result of an underlying social network*. That is, when agents care about where other agents are matched, it is often because they are friends. With this in mind, we construct a model in Section 2 that includes a weighted, undirected social network graph and allows agents to have utilities (which implicitly defines their preference ordering) that depend on where neighbors in the graph are assigned. The model is motivated by [3], which also considers peer effects defined by a social network but focuses on one-sided matching markets rather than two-sided matching markets.

We focus on *two-sided exchange-stable* matchings – see Section 2 for a detailed definition. We note that compared to the traditional notion of stability of [13], this is a distinct notion of stability, but one that is relevant to many situations where agents can compare notes with each other, such as the housing assignment or medical matching problem. For example, in [3,4,12], "pairwise-stability" is considered since they consider models where agents exchange offices or licenses in FCC spectrum auctions. Further, consider a situation where two hospital interns prefer to exchange the hospitals allocated to them by the NRMP. If this is a traditional stable matching, the hospitals would not allow the swap, even though the interns are highly unsatisfied with the match. Such a situation has been documented in [15], and has led to a similar type of stability, exchange stability, as defined in [1,8,9,15].

Given our model of peer effects, the focus of the paper is then on characterizing the set of two-sided exchange-stable matchings, as defined in Section 2. Our results concern (i) the existence of two-sided exchange-stable matchings and (ii) the efficiency of exchange-stable matchings (in terms of social welfare).

With respect to the existence of two-sided exchange-stable matchings (Section 3), it is not difficult to show that in our model stable matchings always exist. Given the contrast to the negative results that are common for many-to-one matchings, e.g., [11,25,26], these results are perhaps surprising.

With respect to the efficiency of exchange-stable matchings (Section 4), results are not as easy to obtain. In this context, we limit our focus to one-sided matching markets and simplify utility functions, but as a result we are able to attain bounds on the ratio of the welfare of the optimal matching to that of the worst stable matching, i.e., the 'price of anarchy'. We also demonstrate cases where our bounds are tight. When considering only one-sided markets, our model becomes similar to hedonic coalition formation, but with several key differences, as highlighted in Section 4. Our results (Theorems 3 and 4) show that the price of anarchy does not depend on the number of, say, interns, but does grow with the number of, say, hospitals – though the growth is typically sublinear. Further, we observe that the impact of the structure of the social network on the price of anarchy happens only through the clustering of the network, which is well understood in the context of social networks, e.g., [16,32]. Finally, it turns out that the price of anarchy has a dual interpretation in our context; in addition to providing a bound on the inefficiency caused by enforcing exchange-stability, it turns out to also provide a bound on the loss of efficiency due to peer effects.

2 Model and Notation

To begin, we define the model we use to study many-to-one matchings with peer effects and complementarities. There are four components to the model, which we describe in turn: (i) basic notation for discussing matchings; (ii) the model for agent utilities, which captures both peer effects and complementarities; (iii) the notion of stability we consider; and (iv) the notion of social welfare we consider.

To provide a consistent language for discussing many-to-one matchings, throughout this paper we use the setting of matching incoming students to residential houses. In this setting many students are matched to each house, and the students have preferences over the houses, but also have peer effects as a result of wanting to be matched to the same house as their friends. Similarly, the houses have preferences over the students, but there are additional complementarities due to goals such as maintaining diversity. It is clear that some form of stability is a key goal of this "housing assignment" problem.

Notation for Many-to-One Matchings. We define two finite and disjoint sets, $H = \{h_1, \ldots, h_m\}$ and $S = \{s_1, \ldots, s_n\}$ denoting the houses and students, respectively. For each house, there exists a positive integer *quota* q_h which indicates the number of positions a house has to offer. The quota for each house may be different.

Definition 1. *A matching is a subset $\mu \subseteq S \times H$ such that $|\mu(s)| = 1$ and $|\mu(h)| = q_h$, where $\mu(s) = \{h \in H : (s, h) \in \mu\}$ and $\mu(h) = \{s \in S : (s, h) \in \mu\}$.*[1]

Note that we use $\mu^2(s)$ to denote the set of student s's housemates (students also in house $\mu(s)$).

Friendship Network. The friendship network among the students is modeled by a weighted graph, $G = (V, E, w)$ where $V = S$ and the relationships between students are represented by the weights of the edges connecting nodes. The strength of a relationship between two students s and t is represented by the weight of that edge, denoted by $w(s, t) \in \mathbb{R}^+ \cup \{0\}$. We require that the graph is undirected, i.e., the adjacency matrix is symmetric so that $w(s, t) = w(t, s)$ for all s, t.

Additionally, we define a few metrics quantifying the graph structure and its role in the matching. Let the total weight of the graph be denoted by $|E| := \frac{1}{2} \sum_{s \in S} \sum_{t \in S} w(s, t)$. Further, let the weight of edges connecting houses h and g under matching μ be denoted by $E_{hg}(\mu) := \sum_{s \in \mu(h)} \sum_{t \in \mu(g)} w(s, t)$. Note that in the case of edges within the same house $E_{hh}(\mu) := \frac{1}{2} \sum_{s \in \mu(h)} \sum_{t \in \mu(h)} w(s, t)$. Finally, let the weight of edges that are within the houses of a particular matching μ be denoted by $E_{in}(\mu) := \sum_{h \in H} E_{hh}(\mu)$.

Agent utility functions. Each agent derives some utility from a particular matching and an agent (student or house) always strictly prefers matchings that give a strictly higher utility and is indifferent between matchings that give equal utility. This setup differs from the traditional notion of 'preference orderings' [13,28], but is not uncommon [2,3,4,7,12]. It is through the definitions of the utility functions that we model peer effects (for students) and complementarities (for houses).

Students derive benefit both from (i) the house they are assigned to and (ii) their peers that are assigned to the same house. We model each house h as having an desirability to student s of $D_h^s \in \mathbb{R}^+ \cup \{0\}$. If $D_h^s = D_h^t \; \forall s \neq t$ (objective desirability), this value can be seen as representing something like the U.S. News college rankings or hospital rankings – something that all students would agree on. This leads to a utility for student s under matching μ of

---

[1] If the number of students in $\mu(h)$, say r, is less than q_h, then $\mu(h)$ contains $q_h - r$ "holes" – represented as students with no friends and no preference over houses.

$$U_s(\mu) := D^s_{\mu(s)} + \sum_{t \in \mu^2(s)} w(s,t) \tag{1}$$

so that the total utility that a student derives from a match is a combination of how "good" a house is as well as how many friends they will have in that house.[2,3]

Similarly, the utility of a house h under matching μ is modeled by

$$U_h(\mu) := D^h_{\mu(h)}, \tag{2}$$

where D^h_σ denotes the desirability of a particular set of students σ for house h (the utility house h derives from being matched to the set of students σ). Note that this definition of utility allows general phenomena such as heterogeneous house preferences over groups of students.

Two-sided exchange stability. Under the traditional definition of stability, if a student and a house were to prefer each other to their current match (forming a blocking pair), the student is free to move to the preferred house and the house is free to evict (if necessary) another student to make space for the preferred student. In our model, however, we assume that students and houses cannot "go outside the system" (neither can students remain unmatched), like what medical students and hospitals do when they operate outside of the NRMP. As a result, we restrict ourselves to considering swaps of students between houses, similar to [3,4,12].

To define exchange stability, it is convenient to first define a *swap matching* μ^t_s in which students s and t switch places while keeping all other students' assignments the same.

Definition 2. *A swap matching* $\mu^t_s = \{\mu \setminus \{(s,h),(t,g)\}\} \cup \{(s,g),(t,h)\}$.

Note that the agents directly involved in the swap are the two students switching places and their respective houses – all other matchings remain the same. Further, one of the students involved in the swap can be a "hole" representing an open spot, thus allowing for single students moving to available vacancies. When two actual students are involved, this type of swap is a two-sided version of the "exchange" considered in [1,8,9,15] – *two-sided* exchange stability requires that houses approve the swap. As a result, while an exchange-stable matching may not exist in either the marriage or roommate problem, we show in Section 3 that a two-sided exchange-stable matching will always exist for the housing assignment problem.

Definition 3. *A matching μ is **two-sided exchange-stable (2ES)** if and only if there does not exist a pair of students (s,t) such that:*

---

[2] We note that the utility of any "holes" (such as what happens when a house's quota is not met), is simply $U_s(\mu) = 0$.

[3] Note also that if we remove D^s_h from the utility function and allow unlimited quotas, the matching problem becomes the coalitional affinity game from [7].

(i) $\forall\, i \in \{s, t, \mu(s), \mu(t)\}$, $U_i(\mu_s^t) \geq U_i(\mu)$ and
(ii) $\exists\, i \in \{s, t, \mu(s), \mu(t)\}$ such that $U_i(\mu_s^t) > U_i(\mu)$

This definition implies that a swap matching in which all agents involved are indifferent is two-sided exchange-stable. This avoids looping between equivalent matchings. Note that the above definition implies that if two students want to switch between two houses (or a single student wants to "switch" with a hole), the houses involved must "approve" the swap or if two houses want to switch two students, the students involved must agree to the swap (a hole will always be indifferent). This is natural for the house assignment problem and many other many-to-one matching markets, but would be less appropriate for some other settings, such as the college-admissions model.

Social welfare. One key focus of this paper is to develop an understanding of the "efficiency loss" that results from enforcing stability of assignments in matching markets. We measure the efficiency loss in terms of the "social welfare":

$$W(\mu) := \sum_{s \in S} U_s(\mu) + \sum_{h \in H} U_h(\mu)$$

Using this definition of social welfare, the efficiency loss can be quantified using the *Price of Anarchy* (PoA) and *Price of Stability* (PoS). Specifically, the PoA (PoS) is the ratio of the optimal social welfare over all matchings, not necessarily stable, to the minimum (maximum) social welfare over all stable matchings. Understanding the PoA and PoS is the focus of Section 4.

3   Existence of Stable Matchings

We begin by focusing on the existence of two-sided exchange-stable matchings. In most prior work, matching markets with externalities do not have guaranteed existence of a stable matching. In contrast, we prove that a 2ES matching always exists in the model considered in this paper. We begin by proposing a potential function $\Phi(\mu)$ for the matching game:

$$\Phi(\mu) = \sum_{h \in H} U_h(\mu) + \sum_{s \in S} D_{\mu(s)}^s + \frac{1}{2} \sum_{s \in S} \left(\sum_{x \in \mu^2(s)} w(s, x) \right) \tag{3}$$

Due to the symmetry of the social network, every approved swap will result in a strict increase of the potential function. As there is a finite set of matches, this results in the existence of a 2ES matching for every housing assignment market.

Theorem 1. *All local maxima of $\Phi(\mu)$ are two-sided exchange-stable.*

If we assume that there are no vacancies in any of the houses and students value houses according to the same rules (i.e., $D_h^s = D_h^t \,\forall\, s \neq t$), then each each approved swap will result in a strict increase in the *social welfare*. Note that this implies that the maximally efficient matching will be 2ES.

Theorem 2. *If house quotas are exactly met and $D_h^s = D_h^t \ \forall \ s \neq t$, all local maxima of $W(\mu)$ are two-sided exchange-stable.*

We omit the exact proofs here; see [5] for details. Note, however, that not all 2ES matchings are local maxima of $\Phi(\mu)$ or $W(\mu)$. Such a case arises when one student, for example, refuses a swap as her utility would decrease, but the other student involved stands to benefit a great deal from such a swap. If the swap were forced, the total potential function (or social welfare) could increase, but only at the expense of the first student.

The contrast between Theorem 1 and the results such as [26] and [28] can be explained by considering a few aspects of the model we study. In particular, we are using a distinct type of stability appropriate to our housing assignment market. Further, the assumption that the social network graph is symmetric is key to guaranteeing existence.

4 Efficiency of Stable Matchings

To this point, we have focused on the existence of two-sided exchange-stable matchings and how to find them. In this section our focus is on the "efficiency loss" due to stability in a matching market and the role peer effects play in this efficiency loss.

We measure the efficiency loss in a matching market using the price of stability (PoS) and the price of anarchy (PoA) as defined in Section 2. Interestingly, the price of anarchy has multiple interpretations in the context of this paper. First, as is standard, it measures the worst-case loss of social welfare that results due to enforcing two-sided exchange-stability. For example, the authors in [2] bound the loss in social welfare caused by individual rationality (by enforcing stable matchings) for matching markets without externalities. Second, it provides a competitive ratio for matching algorithms (like those described in [5]). Even a centralized mechanism with complete information may only find a stable matching, not necessarily the maximally efficient one. The price of anarchy gives us a bound on this worst-case. Third, we show later that the price of anarchy also has an interpretation as capturing the efficiency loss due to peer effects.

The results in this section all require one additional simplifying assumption to our model: *complementarities are ignored and only peer effects are considered*. Specifically, we assume, for all of our PoA results, $U_h(\mu) = 0$, and thus $W(\mu) = \sum_{s \in S} U_s(\mu)$. Under this assumption, the market is one-sided, with only students participating, and our notion of stability is simply exchange-stability. This assumption is limiting, but there are still many settings within which the model is appropriate. Two examples are the housing assignment problem in the case when students can swap positions without needing house approval, and the assignment of faculty to offices as discussed in [3], as clearly the offices have no preferences over which faculty occupy them. In order to simplify the analysis, we also make use of the assumptions in Theorem 2 : (i) $D_h^s = D_h^t \ \forall \ s \neq t$ and (ii) house quotas are exactly met.

4.1   Related Models

When the housing assignment problem is restricted to a one-sided market involving only students, we note that it becomes very similar to both (i) a hedonic coalition formation game with symmetric additively separable preferences, as described in [6], and (ii) a coalitional affinity game, as described in [7]. For a more detailed discussion of these types of games and their relation to the results in this paper, see [5].

While the one-sided housing assignment problem and hedonic coalition formation games appear to be very similar, there are a number of key differences. Most importantly, the housing assignment problem considers a *fixed* number of houses with a limited number of spots available; students cannot break away and form a new coalition/house, nor can a house have more students than its quota. In addition, our model considers exchange-stability, which is closest to the Nash stability of [6], but is still significantly different in that it involves a pair of students willing and able to swap. Finally, each student gains utility from the house they are matched with, in addition to the other members of that house, which is different from the original formulation of hedonic coalition games.

4.2   Discussion of Results

To begin the discussion of our results, note that under our simplifying assumptions the price of stability is 1 for our model because any social welfare optimizing matching is stable. However, the price of anarchy can be much larger than 1. In fact, depending on the social network, the price of anarchy can be unboundedly large, as illustrated in the following example.

Example 1 (Unbounded price of anarchy).
Consider a matching market with 4 students and 2 houses, each with a quota of 2, and two possible matchings illustrated by Figure 1. As shown in Figure 1 (a) and (b), respectively, in the optimal matching μ^*, $W(\mu^*) = k$; whereas there exists a exchange-stable matching with $W(\mu) = 2$. Thus, as k increases, the price of anarchy grows linearly in k.

Despite the fact that, in general, there is a large efficiency loss that results from enforcing exchange-stability, in many realistic cases the efficiency loss is actually quite small. The following two theorems provide insight into such cases.

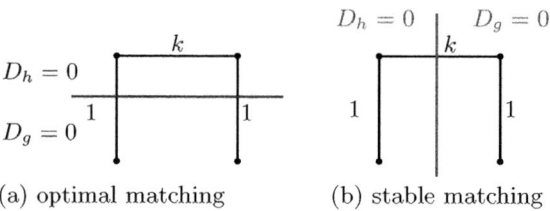

(a) optimal matching   (b) stable matching

Fig. 1. Arbitrarily bad exchange-stable matching

A key parameter in these theorems is γ_m^* which captures how well the social network can be "clustered" into a *fixed* number of m groups and is defined as follows.

$$\gamma_m(\mu) := \frac{E_{in}(\mu)}{|E|} \tag{4}$$

$$\gamma_m^* := \max_\mu \gamma_m(\mu) \tag{5}$$

Thus, γ_m^* represents the maximum edges that can be captured by a partition satisfying the house quotas. Note that γ_m^* is highly related to other clustering metrics, such as the conductance [17], [31] and expansion [23].

We begin by noting that due to the assumption that $\sum_{h \in H} U_h(\mu) = 0$, we can separate the social welfare function into two components:

$$W(\mu) = \sum_{s \in S} U_s(\mu) = \sum_{h \in H} \sum_{s \in \mu(h)} \left(D_h + \sum_{t \in \mu(h)} w(s,t) \right) = 2E_{in}(\mu) + \sum_{h \in H} q_h D_h.$$

Thus,

$$\frac{\max_\mu W(\mu)}{\min_{\mu \text{ is stable}} W(\mu)} = \frac{Q + \max_\mu \gamma_m(\mu)}{Q + \min_{\text{stable } \mu} \gamma_m(\mu)} \tag{6}$$

where

$$Q := \frac{\sum_{h \in H} q_h D_h}{2E}. \tag{7}$$

Note that the parameter Q is independent of the particular matching μ.

Our first theorem regarding efficiency is for the "simple" case of unweighted social networks with equal house quotas and/or equivalently valued houses.

Theorem 3. *Let $w(s,t) \in \{0,1\}$ for all students s,t and let $q_h \geq 2, D_h \in \mathbb{Z}^+ \cup \{0\}$ for all houses h. If $q_h = q$ for all h and/or $D_h = D$ for all h, then*

$$\min_{\text{stable } \mu} W(\mu) \geq \frac{\max_\mu W(\mu)}{1 + 2(m-1)\gamma_m^*}$$

The bound in Theorem 3 is tight, as illustrated by the example below.

Example 2 (Tightness of Theorem 3).
Consider a setting with m houses and $q_h = mk$ for all $h \in H$. Students are grouped into clusters of size $k > 2$, as shown for $m = 3$ in Figure 2. The houses have $D_h = k + 1$ and $D_g = D_i = 0$. Each student in the middle cluster in each row has k edges to the other students outside of their cluster (but none within), as shown.

The worst-case stable exchange-matching is represented by the vertical red lines. Note that since $D_h = k + 1$, this matching is stable, even though all edges are cut. Thus $\min_{\mu \text{ stable}} \gamma_m(\mu) = 0$. The optimal matching is represented by the

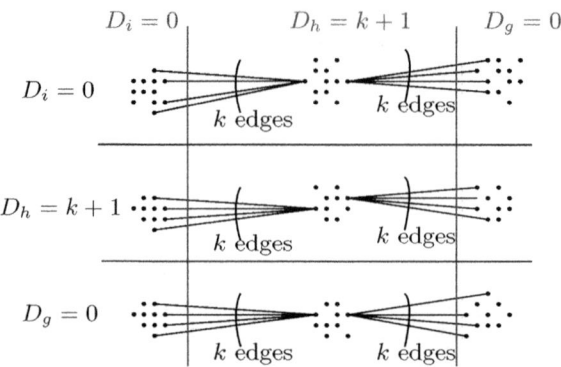

$D_i = 0$ $D_h = k + 1$ $D_g = 0$

Fig. 2. Network that achieves PoA bound

horizontal blue lines in the figure; note that $\gamma_m^* = 1$. To calculate the price of anarchy, we start from equations (6) and (7) and calculate

$$Q = \frac{\sum_{h \in H} q_h D_h}{2|E|} = \frac{mk(k+1)}{2mk(m-1)k} = \frac{k+1}{2(m-1)k},$$

which gives,

$$\frac{\max_\mu W(\mu)}{\min_{\text{stable } \mu} W(\mu)} = \frac{Q + \gamma_m^*}{Q + \min_\mu \text{ stable } \gamma_m(\mu)} = 1 + 2(m-1)\left(\frac{k}{k+1}\right).$$

Notice that as k becomes large, this approaches the bound of $1 + 2(m-1)\gamma_m^*$.

We note that the requirement $q_h = q$ for all h and/or $D_h = D$ for all h is key to the proof of Theorem 3 and in obtaining such a simple bound; otherwise, Theorem 4 applies. We omit the proofs of these theorems here for brevity; see [5] for the details.

Our second theorem removes the restrictions in the theorem above, at the expense of a slightly weaker bound. Define $q_{max} = \max_{h \in H} q_h$, $w_{max} = \max_{s,t \in S} w(s,t)$ and $D_\Delta = \min_{h,g \in H}(D_h - D_g)$, assuming that the houses are ordered in increasing values of D_h.

Theorem 4. *Let $w(s,t) \in \mathbb{R}^+ \cup \{0\}$ for all students s,t and $D_h \in \mathbb{R}^+ \cup \{0\}$, $q_h \in \mathbb{Z}^+$ for all houses h, then*

$$\min_{\text{stable } \mu} W(\mu) \geq \frac{\max_\mu W(\mu)}{1 + 2(m-1)\left(\gamma_m^* + \frac{q_{max} w_{max}}{D_\Delta}\right)}$$

Though Theorem 3 is tight, it is unclear at this point whether Theorem 4 is also tight. However, a slight modification of the above example does show that it has the correct asymptotics, i.e., there exists a family of examples that have price of anarchy $\Theta(m\gamma_m^* q_{max} w_{max} D_\Delta^{-1})$.

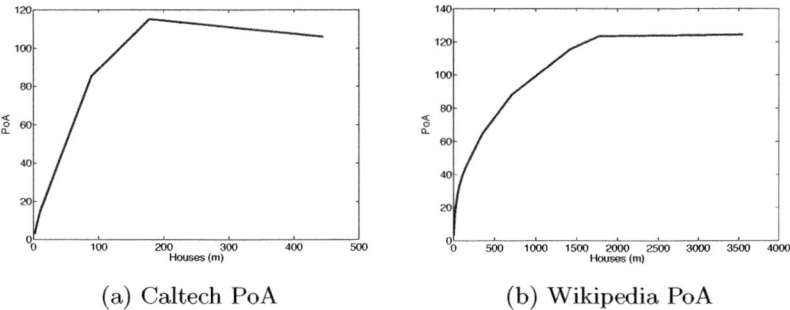

(a) Caltech PoA   (b) Wikipedia PoA

Fig. 3. Illustration of price of anarchy bounds in Theorem 3 for Caltech and Wikipedia networks

A first observation one can make about these theorems is that the price of anarchy has no direct dependence on the number of students. This is an important practical observation since the number of houses is typically small, while the number of students can be quite large (similar phenomena hold in many other many-to-one matching markets). In contrast, the theorems highlight that the degree of heterogeneity in quotas, network edge weights, and house valuations all significantly impact inefficiency.

A second remark about the theorems is that the only dependence on the social network is through γ_m^*, which measures how well the graph can be "clustered" into m groups. An important note about γ_m^* is that it is highly dependent on m, and tends to shrink quickly as m grows. A consequence of this behavior is that the price of anarchy is not actually linear in m in Theorems 3 and 4, as it may first appear, it turns out to be sublinear. This is illustrated in the context of real social network data in Figures 3a and 3b. Note that as we are increasing the number of houses, we are in fact creating finer allowable partitions of the network. The social networks used to generate the above plots are described in detail in [5].

Next, let us consider the impact of peer effects on the price of anarchy. Considering the simple setting of Theorem 3, we see that if there were no peer effects, this would be equivalent to setting $w(s,t) = 0$ for all s, t. This would imply that $\gamma_m^* = 0$, and so the price of anarchy is one. Thus, another interpretation of the price of anarchy in Theorem 3 is the efficiency lost as a result of peer effects.

5   Concluding Remarks

In this paper we have focused on many-to-one matchings with peer effects and complementarities. Typically, results on this topic tend to be negative, either proving that stable matchings may not exist, e.g., [26,28], or that stable matchings are computationally difficult to find, e.g., [25].

In this paper, our goal has been to provide positive results. To this end, we focus on the case when peer effects are the result of an underlying social network, and this restriction on the form of the peer effects allows us to prove that a two-sided exchange-stable matching always exists. Further, we provide bounds on the maximal inefficiency (price of anarchy) of any exchange-stable matching and show how this inefficiency depends on the clustering properties of the social network graph. Interestingly, in our context the price of anarchy has a dual interpretation as characterizing the degree of inefficiency caused by peer effects.

There are numerous examples of many-to-one matchings where the results in this paper can provide insight; one of particular interest to us is the matching of incoming undergraduates to residential houses which happens yearly at Caltech and other universities. Currently incoming students only report a preference order for houses, and so are incentivized to collude with friends and not reveal their true preferences. For such settings, the results in this paper highlight the importance of having students report not only their preference order on houses, but also a list of friends with whom they would like to be matched. Using a combination of these factors, the algorithms described in [5] and efficiency bounds presented in this paper provide a promising approach, for this specific market as well as any general market where peer effects change the space of stable matchings.

The results in the current paper represent only a starting point for research into the interaction of social networks and many-to-one matchings. There are a number of simplifying assumptions in this work which would be interesting to relax. For example, the efficiency bounds we have proven consider only a one-sided market, where students rate houses similarly and quotas are exactly met. These assumptions are key to providing simpler bounds, and they certainly are valid in some matching markets; however relaxing these assumptions would broaden the applicability of the work greatly.

References

1. Alcalde, J.: Exchange-proofness or divorce-proofness? stability in one-sided matching markets. Rev. Econ. Design 1, 275–287 (1994)
2. Anshelevich, E., Das, S., Naamad, Y.: Anarchy, stability, and utopia: Creating better matchings. In: Mavronicolas, M., Papadopoulou, V.G. (eds.) SAGT 2009. LNCS, vol. 5814, pp. 159–170. Springer, Heidelberg (2009)
3. Baccara, M., Imrohoroglu, A., Wilson, A., Yariv, L.: A field study on matching with network externalities, Working paper (2009)
4. Bajari, P., Fox, J.: Measuring the efficiency of an fcc spectrum auction. Working Paper 11671, National Bureau of Economic Research (2009)
5. Bodine-Baron, E., Lee, C., Chong, A., Hassibi, B.: A Wierman. Matching with friends: stability and peer effects in housing assignment, working paper, available on arxiv (2011)
6. Bogomolnaia, A., Jackson, M.: The stability of hedonic coalition structures. Games Econ. Behav. 38(2), 201–230 (2002)
7. Brânzei, S., Larson, K.: Coalitional affinity games and the stability gap. In: IJCAI, pp. 1319–1320 (2009)

8. Cechlarova, K., Manlove, D.: On the complexity of exchange-stable roommates. DAM 116(3), 279–287 (2002)
9. Cechlarova, K., Manlove, D.: The exchange-stable marriage problem. DAM 152(1-3), 109–122 (2005)
10. Dutta, B., Masso, J.: Stability of matchings when individuals have preferences over colleagues. J. Econ. Theory 75(2), 464–475 (1997)
11. Echenique, F., Yenmez, M.: A solution to matching with preferences over colleagues. Games Econ. Behav. 59(1), 46–71 (2007)
12. Fox, J.: Estimating matching games with transfers, Working paper (2010)
13. Gale, D., Shapley, L.S.: College admissions and the stability of marriage. AMM 69(1), 9–15 (1962)
14. Hafalir, I.E.: Stability of marriage with externalities. Int. J. Game Theory 37, 353–370 (2008)
15. Irving, R.: Stable matching problems with exchange restrictions. J. Comb. Opt. 16, 344–360 (2008)
16. Jackson, M.: Social and Economic Networks. Princeton University Press, Princeton (2008)
17. Kannan, R., Vempala, S., Vetta, A.: On clusterings: Good, bad and spectral. J. ACM 51(3), 497–515 (2004)
18. Klaus, B., Klijn, F.: Stable matchings and preferences of couples. J. Econ. Theory 121, 75–106 (2005)
19. Klaus, B., Klijn, F.: Paths to stability for matching markets with couples. Games Econ. Behav. 58, 154–171 (2007)
20. Kojima, F., Pathak, P.: Incentives and stability in large two-sided matching markets. Amer. Econ. Rev. 99(3), 608–627 (2009)
21. Ostrovsky, M.: Stability in supply chain networks. Amer. Econ. Rev. 98(3), 897–923 (2008)
22. Pycia, M.: Many-to-one matching with complementarities and peer effects, Working paper (2007)
23. Radicchi, F., Castellano, C., Cecconi, F., Loreto, V., Parisi, D.: Defining and identifying communities in networks. PNAS 101(9), 2658–2663 (2004)
24. Revilla, P.: Many-to-one matching when colleagues matter, Working paper (2004)
25. Ronn, E.: Np-complete stable matching problems. J. Alg. 11, 285–304 (1990)
26. Roth, A.E.: The evolution of the labor market for medical interns and residents: A case study in game theory. J. of Polit. Econ. 92, 991–1016 (1984)
27. Roth, A.E., Rothblum, U.G., Vande Vate, J.H.: Stable matchings, optimal assignments and linear programming. MOR 18(4), 803–828 (1993)
28. Roth, A.E., Sotomayor, M.: Two-sided matching: A study in game-theoretic modeling and analysis. Cambridge University Press, Cambridge (1990)
29. Roth, A.E., Vande Vate, J.H.: Random paths to stability in two-sided matching. Econometrica 58, 1475–1480 (1990)
30. Sasaki, H., Toda, M.: Two-sided matching problems with externalities. J. Econ. Theory 70(1), 93–108 (1996)
31. Shi, J., Malik, J.: Normalized cuts and image segmentation. IEEE Trans. of Pattern Analysis and Machine Intelligence 22(8), 888–905 (2000)
32. Wasserman, S., Faust, K.: Social Network Analysis: Methods and Applications. Cambridge University Press, Cambridge (1994)

Steady Marginality: A Uniform Approach to Shapley Value for Games with Externalities

Oskar Skibski*

Institute of Informatics, University of Warsaw,
Banacha 2, 02-097 Warsaw, Poland
oskar.skibski@mimuw.edu.pl

Abstract. The Shapley value is one of the most important solution concepts in cooperative game theory. In coalitional games without externalities, it allows to compute a unique payoff division that meets certain desirable fairness axioms. However, in many realistic applications where externalities are present, Shapley's axioms fail to indicate such a unique division. Consequently, there are many extensions of Shapley value to the environment with externalities proposed in the literature built upon additional axioms. Two important such extensions are "externality-free" value by Pham Do and Norde and value that "absorbed all externalities" by McQuillin. They are good reference points in a space of potential payoff divisions for coalitional games with externalities as they limit the space at two opposite extremes. In a recent, important publication, De Clippel and Serrano presented a marginality-based axiomatization of the value by Pham Do Norde. In this paper, we propose a dual approach to marginality which allows us to derive the value of McQuillin. Thus, we close the picture outlined by De Clippel and Serrano.

1   Introduction

The Shapley value is one of the most important and extensively studied solution concepts in coalitional game theory. In the environment where agents are allowed to cooperate, the Shapley value lays down a fair allocation of jointly achieved payoff. Here, fairness is built upon four axioms: (i) *Efficiency* (the whole payoff is distributed among agents); (ii) *Symmetry* (division of payoff does not depend on agents' names), (iii) *Additivity* (when two different games are combined, agent's share is equal to sum of shares in games considered independently); and (iv) *Null-player Axiom* (agent which has no impact on value of any coalition gets nothing). In his seminal work, Shapley showed that his division scheme is *unique*, i.e. no other division scheme meets all four axioms.

Since Shapley original work [1], his concept has been extended in a variety of directions. One of them are coalitional games with externalities, where a value of a coalition depends on formation of other coalitions in the system. Indeed,

---

* I would like to thank my supervisor Andrzej Szalas and Tomasz Michalak for their comments, suggestions and discussions.

G. Persiano (Ed.): SAGT 2011, LNCS 6982, pp. 130–142, 2011.

externalities occur in many real-life applications of coalitional games such as oligopolistic markets (where a merger is likely to affect other companies), a political scene (where the importance of a political party highly depends on the created coalitions), or a supply chain (where a large number of subcontractors increases the standardization costs). Unfortunately, they are substantially more challenging to the model than the conventional games. In particular, in the presence of externalities, the axioms proposed by Shapley are insufficient to determine a unique division of payoff. This problem was addressed in the literature by several authors who proposed more extended axiomatizations. Two important extensions of the Shapley value to games with externalities are "externality-free" value proposed by Pham Do and Norde [3] and McQuillin's value which "absorbed all externalities" [2]. Both can be considered as reference points for other extensions, as, under certain conditions, they limit the space of possible extensions at two opposite extremes.

The key role in the formula for the Shapley value is played by the marginality — an important economic concept in which evaluation of a player in a coalition is based on a difference between the coalition's values with and without the player. Specifically, the Shapley value is calculated as the weighted average of marginal contributions of players to all the coalitions in the game. This relationship of the Shapley value to the marginality was emphasized by the beautiful work of Young [8]. In Shapley's axiomatic characterization marginality comes from the Null-player Axiom which assigns zero value to every agent whose vector of marginal contributions to coalitions equals to zero. Unfortunately, as mentioned above, this axiom as well as the others do not determine the unique value in the case of games with externalities. Thus, it became a common practise in the literature, to add other, non-marginality based axioms. These axioms indeed allow for deriving the unique Shapley value extended to games with externalities, however, not in a way related to marginality as the original Shapley value did. In this context, the key additional axiom of McQuillin was recursion which required the value to be a fixed-point solution (i.e. if we consider a value to be a game by itself, then the value computed for such a game should not change). Nevertheless, marginality-based axiomatization of the McQuillin value that connects to the original Shapley value has remained unknown.

In this paper we close this gap by proposing an alternative approach to marginality. We present a new marginality axiom, which allows us to derive the extension of the Shapley value proposed by McQuillin. Our approach is dual to De Clippel and Serrano who in recent, important publication proposed a basic approach to marginality and derived "externality-free" value of Pham Do and Norde. In other words, we close the picture outlined by De Clippel and Serrano, so that the two opposite values for games with externalities that limit the space of many other extensions, are now based on the marginality principle.

Our new approach to marginality, which we call a steady marginality, differs from those proposed earlier in the literature ([5,7,6]). To compute an agent's marginal contribution to a coalition we compare the value of the coalition with the specific agent with the value of the coalition obtained by the transfer of the agent to another coalition, existing in the partition (so the number of coalitions is *steady*). We then do not include the value of a coalition in a partition when a specific agent forms its own singleton coalition.

The rest of the paper is organized as follows. In Section 2 we introduce the basic definitions and notation. In Section 3 we present our set of axioms, including our new marginality axiom which we relate to "externality-free" marginality. In Section 4 we define a new class of games and prove they form a basis of space of games with externalities. Finally, in Section 5 we prove that there exists only one value which satisfies all our axioms and that this value is equal to one proposed earlier by McQuillin. Section 6 presents different approaches to marginality in the literature. Section 7 concludes the paper and outlines future work.

2   Definitions and Notation

In this section we introduce the basic definitions and notation.

Let $N = \{1, 2, \ldots, n\}$ be the set of all agents. A *coalition* S is any subset of agents, $S \subseteq N$. A *partition* P *of* N is a set of disjoint coalitions which covers the whole set of agents, i.e., $P = \{S_1, S_2, \ldots, S_k\}$ and $\bigcup_{i \in N} S_i = N$, where $S_i \cap S_j = \emptyset$ for every $i, j \in \{1, \ldots, k\}$ with $i \neq j$. The set of all partitions is denoted by $\mathcal{P}$. A coalition S being a part of a partition P is called an *embedded coalition* and is denoted by (S, P). By $|P|$ we denote the number of coalitions in a partition P. The set of all embedded coalitions is denoted EC and formally defined as:

$$EC \stackrel{\text{def}}{=} \{(S, P) : P \in \mathcal{P}, S \in P\}$$

As common in the literature, for $S \subset N, i \notin S, j \in S$, we define $S_{+i} \stackrel{\text{def}}{=} S \cup \{i\}$ and $S_{-j} \stackrel{\text{def}}{=} S \setminus \{j\}$. If $S \in P$ then $P_{-S} \stackrel{\text{def}}{=} P \setminus \{S\}$.

The following notation will play an important part in our paper: let $S, T \in P$ and $i \in S$. A function $\tau_i^{S,T}$ will represent the transition of an agent i from S to T:

$$\tau_i^{S,T}(P) \stackrel{\text{def}}{=} P \setminus \{S, T\} \cup \{S_{-i}, T_{+i}\}$$

In the literature it is a common convention to assume that in every partition $P \in \mathcal{P}$ an empty, artificial coalition $\emptyset \in P$ exists. In our paper we accept a different assumption that only one partition (the one with the grand-coalition) contains a special empty coalition $\emptyset$.

Note 2.1. For technical convenience, in our paper we use the convention that in every partition there are at least two coalitions, so we assume that in partition with only one explicitly listed coalition there also exists an empty coalition. In such a case, the grand coalition takes the form $\{N, \emptyset\}$ (and $|\{N, \emptyset\}| = 2$). However, we will not consider $(\emptyset, \{N, \emptyset\})$ as a correct embedded coalition.

The *game* (in a partition-function form) is a function $v : EC \to \mathbb{R}$ which associates a real number with every embedded coalition. For convenience, we extend the domain of v and assign a zero value to every incorrect embedded coalition where S is empty: $v(\emptyset, P) = 0$.

Among a collection of games we distinguish a set of *games without externalities* (or, differently, *characteristic function games*), where the value of a coalition does not depend on a partition of other agents. Formally, for each coalition S and two partitions P_1, P_2 containing S we get $v(S, P_1) = v(S, P_2)$. In this case the definition of a game can be simplified to $\hat{v} : 2^N \to \mathbb{R}$, as the only argument is a coalition S. Conversely, we say that the game is *with externalities* when the value for at least one coalition depends on a structure of other agents: $v(S, P_1) \neq v(S, P_2)$ for some S and P_1, P_2 which contain S.

The *value* of the game is a function which assigns some payoff to every agent: $\varphi : v \to \mathbb{R}^N$. This payoff is meant to be the agent's share in the value achieved by all players united in the grand coalition: $v(N, \{N, \emptyset\})$. We are interested in a division of the payoff which is fair.

The Shapley value is defined as:

$$Sh_i(\hat{v}) = \sum_{S \subseteq N, i \in S} \frac{(|S| - 1)!(|N| - |S|)!}{|N|!}(\hat{v}(S) - \hat{v}(S_{-i}))$$

where by $\hat{v}$ we denote the game without externalities.

Shapley presented the following intuition behind his value. Assume that the agents enter the coalition in random order. Every agent i brings to the set S_{-i} of agents who already entered the coalition its marginal contribution $\hat{v}(S) - \hat{v}(S_{-i})$. Therefore, Shapley value of an agent i is the average of all its marginal contributions for every order of the agents' arrivals.

One of the most common approaches to the extension of Shapley value for games with externalities is the *average approach* proposed by Macho-Stadler et al [4]. In this approach, from the game v we create a simpler game $\hat{v}$ without externalities and define $\varphi_i(v) \overset{\text{def}}{=} Sh_i(\hat{v})$. The value of each coalition S in $\hat{v}$ is computed as the weighted average of values of a coalition S embedded in different partitions: $\hat{v}(S) = \sum_{P \in \mathcal{P}, S \in P} \alpha_{(S,P)} \cdot v(S, P)$. The different weights lead to the different values. Two extremes in those approaches are "externality-free" value and value which "absorbed all externalities".

The first one was proposed by Pham Do and Norde. It can be obtained using the average approach by defining $\hat{v}^{free}(S) \overset{\text{def}}{=} v(S, \{\{i\} : i \in N_{-S}\} \cup \{S\})$ and $\varphi_i^{free}(v) \overset{\text{def}}{=} Sh_i(\hat{v}^{free})$. Hence, the value of S is taken from the partition, in which no externalities from merging coalitions affect it.

The second one, proposed by McQuillin, is dual to Pham Do and Norde. Here, the value of S is taken from the partition, in which all other agents are in one coalition: $\hat{v}^{McQ}(S) \overset{\text{def}}{=} v(S, \{N_{-S}, S\})$. As we can see, this value of S is affected by all externalities from merging coalitions. McQuillin value takes the form $\varphi_i^{McQ}(v) \overset{\text{def}}{=} Sh_i(\hat{v}^{McQ})$.

3   Axiomatic Characterization

In this section we will present our axioms including a new definition of the marginal contribution. We will also briefly compare it to the definition proposed by De Clippel and Serrano.

Shapley value is based on four elementary axioms: Efficiency, Symmetry, Additivity and Null-player Axiom. The first three are easily translated to games with externalities.

Definition 3.1. *(Efficiency) Function φ satisfies Efficiency if the whole payoff is distributed among agents, i.e. $\sum_{i \in N} \varphi_i(v) = v(N, \{N, \emptyset\})$ for every game v.*

Let $\sigma : N \to N$ be a permutation of the set of agents. Then
- for every coalition $S \subseteq N$, $\sigma(S) \stackrel{\text{def}}{=} \{\sigma(i) : i \in S\}$
- for every partition $P \in \mathcal{P}$, $\sigma(P) \stackrel{\text{def}}{=} \{\sigma(S) : S \in P\}$.

The *permutation of game* $\sigma(v)$ is a game defined on every embedded coalition by $\sigma(v)(S, P) \stackrel{\text{def}}{=} v(\sigma(S), \sigma(P))$ and the *permutation of* $\sigma(\varphi(v))$ is the vector $(\varphi_{\sigma(i)}(v))_{i \in N}$.

Definition 3.2. *(Symmetry) Function φ satisfies Symmetry if agents' values do not depend on their names, i.e. $\varphi(\sigma(v)) = \sigma(\varphi(v))$ for every game v and every permutation σ.*

It is widely accepted ([5,4,2]) to translate Additivity as the Linearity in the context of externalities.[1]

Definition 3.3. *(Linearity) Function φ satisfies Linearity if:*

(a) for every two games v_1, v_2, we have $\varphi(v_1 + v_2) = \varphi(v_1) + \varphi(v_2)$, where $v_1 + v_2$ is a game defined by $(v_1 + v_2)(S, P) = v_1(S, P) + v_2(S, P)$
(b) for every game v and constant $\lambda \in \mathbb{R}$, we have $\varphi(\lambda v) = \lambda \cdot \varphi(v)$, where λv is a game defined by $(\lambda v)(S, P) = \lambda \cdot v(S, P)$.

Our key axiom will be based on the marginality principle. When there are no externalities, the marginal contribution of an agent i to a coalition S can be easily calculated as a difference between the coalition value with and without an agent i: $\hat{v}(S) - \hat{v}(S_{-i})$. But when the externalities exist, the value of a coalition S (embedded in P) without an agent i depends on where the agent i is. Let us define the *elementary marginal contribution* $mc_{(i,S,P,T)}$ of an agent i to (S, P) in comparison to i being in $T \in P_{-S} \cup \{\emptyset\}$ as a difference between the value of (S, P) and $(S_{-i}, \tau_i^{S,T}(P))$. Then the marginal contribution is the (weighted) average of the elementary marginal contributions:

$$mc_{(i,S,P)}(v) = \sum_{T \in P_{-S} \cup \{\emptyset\}} \alpha_{(i,S,P,T)}(v(S, P) - v(S_{-i}, \tau_i^{S,T}(P)))$$

---

[1] Shapley based his value on Additivity – part (a) of our axiom – as it (combined with his three other axioms) implies part (b) – a very intuitive assumption that when we multiply every value in the game by some scalar, agents' share will increase respectively (i.e. the ratio of agents' share will not change). As shown in [4], the standard Shapley's axioms translated to the games with externalities are too weak to imply full Linearity. Thus, in the presence of externalities, Additivity is usually strengthened to the Linearity.

The empty set in the sum corresponds to the partition in which i is in a singleton coalition $\{i\}$.[2] For our later discussion it would be convenient to consider the elementary marginal contribution $v(S, P) - v(S_{-i}, \tau_i^{S,T}(P))$ as a cost of the agent's i transfer to a coalition T.

De Clippel and Serrano used only one non-zero weight for the transfer of i to the empty coalition: $\alpha_{(i,S,P,\emptyset)} = 1$ and $\alpha_{(i,S,P,T)} = 0$ for $T \in P_{-S}$. Their definition of the marginal contribution takes the form:

$$mc_{(i,S,P)}^{free}(v) \overset{\text{def}}{=} v(S, P) - v(S_{-i}, P_{-S} \cup \{S_{-i}, \{i\}\})$$

This approach is justified by treating the transfer as a two-step process. In the first step, agent i leaves the coalition S and for a moment remains alone (i.e., creates a singleton coalition). An optional second step consists of agent i joining some coalition from P_{-S} (in coalition terms, $\{i\}$ merges with some other coalition). Although both steps may change the value of S_{-i}, the authors argue that only the first one corresponds to the *intrinsic* marginal contribution – the influence from the second step comes rather from the external effect of merging coalitions, not from i leaving S. Discarding the impact of merging coalitions in marginal contribution allowed them to derive an "externality-free" value.

We will consider the transfer of i in a different way. Our first step will consist of leaving coalition S and joining one of the other coalitions in partition. In the second step, agent i can exit his new coalition and create his own. Thus, we look at *creating new coalition* as an extra action, which should not be included in the effect of i leaving coalition S. According to this, the natural way to define the *steady marginal contribution* of an agent i to (S, P), is to take into account only the transfer to the other existing coalition.

Definition 3.4. *The* steady marginal contribution *of an agent* $i \in S$ *to the embedded coalition* $(S, P) \in EC$ *is defined as:*

$$mc_{(i,S,P)}^{full}(v) \overset{\text{def}}{=} \sum_{T \in P_{-S}} (v(S, P) - v(S_{-i}, \tau_i^{S,T}(P)))$$

Then, $mc_i^{full}(v) \overset{\text{def}}{=} (mc_{(i,S,P)}^{full}(v))_{(S,P) \in EC, i \in S}$ *is a vector of steady marginal contributions.*

Our approach can be justified by these real life examples in which creation of a new coalition is rare and not likely. These include political parties or million-dollar industries (such as oil oligopoly). In all such situations, our approach is likely to lead to more proper results.

Based on the definition of steady marginal contribution we can introduce the last axiom, which is our version of the standard Null-player Axiom. In the literature on games with externalities, it is common to assume ([4,5]) that agent i is a null-player when all of his elementary marginal contributions are equal to zero $(v(S, P) - v(S_{-i}, \tau_i^{S,T}(P)) = 0$ for each $(S, P) \in EC$ such that $i \in S$ and

---

[2] Note that when $(S, P) = (N, \{N, \emptyset\})$ then $P_{-S} \cup \{\emptyset\} = \{\emptyset\}$, so $mc_{(i,N,\{N,\emptyset\})}(v) = \alpha_{(i,N,\{N,\emptyset\},\emptyset)}(v(N, \{N, \emptyset\}) - v(N_{-i}, \{N_{-i}, \{i\}\}))$.

$T \in P_{-S} \cup \{\emptyset\}$). Our definition of null-player will differ – we will consider an agent as a null-player when all of his steady marginal contributions are equal to zero.[3]

Definition 3.5. *(Null-player Axiom in a steady marginal contribution sense)*
Function φ satisfies Null-player Axiom if for every agent i such that vector of steady marginal contributions $mc_i^{full}(v)$ is a zero vector occurs $\varphi_i(v) = 0$.

Thus, an agent is a null-player when for every embedded coalition his *expected value* of transfer to the other existing coalition equals zero.[4]

4   Constant-Coalition Games

In this section we will introduce a new class of simple games – *constant-coalition games*. This games will play a key role in a proof of the uniqueness of the value in the next section. The name comes from the fact that, in a given game, every partition in which a coalition with non-zero value is embedded, has exactly the same number of coalitions. We show that the collection of such games is a basis of partition function games.

First, we will need some additional notation.

Definition 4.1. *($R_1 \preceq R_2$) Let R_1, R_2 be two proper, non-empty subsets of two partitions. We say that R_2 can be reduced to R_1 (denoted $R_1 \preceq R_2$) if three conditions are met:*
(a) all agents which appear in R_1, appear in R_2 (i.e. $\bigcup_{T_1 \in R_1} T_1 \subseteq \bigcup_{T_2 \in R_2} T_2$)
(b) two agents which are in the same coalition in R_1, are in the same coalition in R_2
(c) two agents which are not in the same coalition in R_1 are not in the same coalition in R_2.

Assume $R_1 \preceq R_2$. Based on the presented conditions, as we delete agents from R_2 which are not in R_1, we get exactly the R_1 configuration. This observation can be expressed in an alternative definition of the $\preceq$-operator.

Proposition 4.2. *Let R_1, R_2 be two proper, non-empty subsets of two partitions. Then:[5]*

$$R_1 \preceq R_2 \Leftrightarrow \exists_{S \subseteq N} R_1 = \{T_2 \setminus S : T_2 \in R_2 \text{ and } T_2 \not\subseteq S\} \vee R_1 = \{\emptyset\}$$

---

[3] It is easy to prove that our axiom strengthens the standard one, as the fact that all of the agent's i elementary marginal contributions equal zero implies that all his steady marginal contributions are equal to zero as well.

[4] I thank the anonymous reviewer for this interpretation.

[5] The equivalence of the definitions when R_1 and R_2 contain only non-empty coalitions is easy to see. As the only partition which contains an empty set is $\{N, \emptyset\}$ then the only proper, non-empty subset of the partition which contains an empty set is $\{\emptyset\}$. Based on the first definition $\{\emptyset\} \preceq R_2$ for every R_2 and $R_1 \preceq \{\emptyset\}$ implies $R_1 = \{\emptyset\}$ (as we don't allow empty R_1).

For example $\{\{1,2\},\{3\}\} \preceq \{\{1,2,4\},\{3\},\{5\}\}$ but $\{\{1,2\},\{3\}\} \not\preceq \{\{1,2,3\}\}$ and $\{\{1,2\},\{3\}\} \not\preceq \{\{1\},\{3,4\}\}$.

Now we can introduce our new basis for games with externalities.

Definition 4.3. *For every embedded coalition (S,P), the* constant-coalition *game $e^{(S,P)}$ is defined by*

$$e^{(S,P)}(\tilde{S},\tilde{P}) \stackrel{\mathrm{def}}{=} \begin{cases} (|P|-1)^{-(|\tilde{S}\setminus S|)} & \text{if } (\tilde{P}_{-\tilde{S}} \preceq P_{-S}) \text{ and } (|P| = |\tilde{P}|), \\ 0 & \text{otherwise,} \end{cases}$$

for every $(\tilde{S},\tilde{P}) \in EC$.

Note that $(\tilde{P}_{-\tilde{S}} \preceq P_{-S})$ implies $S \subseteq \tilde{S}$ as we get $\tilde{N} \setminus \tilde{S} \subseteq N \setminus S$ from the (a) condition in $\preceq$-operator definition. So, in our game, $e^{(S,P)}$ non-zero values have only embedded coalitions formed from (S,P) by some transition of agents from $P \setminus \{S\}$ to S which does not change the number of the coalitions.

Lemma 4.4. *The collection of constant-coalition games is a basis of the partition function games.*

Proof. Let $e = (e^{(S,P)})_{(S,P) \in EC}$ be the vector of all games.

First, we will show that the constant-coalition games are linearly independent. Suppose the contrary. Then there exists a vector of weights $\alpha = (\alpha_{(S,P)})_{(S,P) \in EC}$ with at least one non-zero value such that $\alpha \times e = \sum_{(S,P) \in EC} \alpha_{(S,P)} e^{(S,P)}$ is a zero vector. Let (S^*, P^*) be the embedded coalition with a non-zero weight $\alpha_{(S^*,P^*)} \neq 0$ and minimal S^* (i.e. (S^*, P^*) is the minimal element of the embedded-coalition relation r: $(S_1, P_1) r (S_2, P_2) \Leftrightarrow S_1 \subseteq S_2$). So, for any other game $e^{(S,P)}$ either $\alpha_{(S,P)} = 0$ or $S \not\subseteq S^* \Rightarrow e^{(S,P)}(S^*, P^*) = 0$ (the implication follows from the remarks after Definition 4.3). Then

$$\sum_{(S,P) \in EC} \alpha_{(S,P)} e^{(S,P)}(S^*, P^*) = \alpha_{(S^*,P^*)} e^{(S^*,P^*)}(S^*, P^*) = \alpha_{(S^*,P^*)} \neq 0,$$

contrary to the previous assumption.

The size of a collection of all the constant-coalition games is equal to the dimension of the partition function games space, hence the collection must be a basis. □

5 Uniqueness of the Value

In this section we show that there exists only one value that satisfies all the introduced axioms and that it is equivalent to the value proposed by McQuillin.

Theorem 5.1. *There is a unique value φ^{full} satisfying Efficiency, Symmetry, Linearity and Null-player Axiom (in a steady marginality sense).*

Proof. We will show that in every coalition-constant game there exists only one value which satisfies those axioms. Based on Linearity and Lemma 4.4 this will imply our theorem.

Let $e^{(S,P)}$ be one of the coalition-constant games. We will show that any player i from the coalition other than S is a null-player (in a steady marginality sense). Based on the Definition 3.5 we have to prove that $mc_i(e^{(S,P)})$ is a zero vector. So, for every $(\tilde{S}, \tilde{P}) \in EC$ such that $i \in \tilde{S}$:

$$mc_{(i,\tilde{S},\tilde{P})}(e^{(S,P)}) = \sum_{\tilde{T} \in \tilde{P} \setminus \{\tilde{S}\}} e^{(S,P)}(\tilde{S}, \tilde{P}) - e^{(S,P)}(\tilde{S}_{-i}, \tau_i^{\tilde{S},\tilde{T}}(\tilde{P})) = 0$$

We divide the proof into two cases with zero and non-zero value of $e^{(S,P)}(\tilde{S}, \tilde{P})$.

Lemma 5.2. *If $e^{(S,P)}(\tilde{S}, \tilde{P}) = 0$ then, for every $T \in \tilde{P} \setminus \{\tilde{S}\}$,*

$$e^{(S,P)}(\tilde{S}_{-i}, \tau_i^{\tilde{S},T}(\tilde{P})) = 0.$$

Proof. Based on the definition of the coalition-constant games we can deduce that at least one of the following conditions occurs:

- $\tilde{P}_{-\tilde{S}} \not\preceq P_{-S}$ - from the definition of $\preceq$-operator we know that there is an agent in $\tilde{P}_{-\tilde{S}}$ which is not in P_{-S}, or there is a pair of agents which are together in one and not together in the other structure; it is easy to see, that adding player i to some coalition in $\tilde{P}_{-\tilde{S}}$ will not fix any of these anomalies;
- $|P| \neq |\tilde{P}|$ - if i is alone ($\tilde{S} = \{i\}$) then, for every $T \in \tilde{P} \setminus \{\tilde{S}\}$,

$$e^{(S,P)}(\tilde{S}_{-i}, \tau_i^{\tilde{S},T}(\tilde{P})) = e^{(S,P)}(\emptyset, \tau_i^{\tilde{S},T}(\tilde{P})) = 0;$$

otherwise, as we only consider the transfer of an agent i to the other existing coalition, the number of the coalitions remains intact: $|P| \neq |\tau_i^{\tilde{S},T}(\tilde{P})|$. □

Lemma 5.3. *If $e^{(S,P)}(\tilde{S}, \tilde{P}) = x$ for $x > 0$ then there exists only one $T^* \in \tilde{P} \setminus \{\tilde{S}\}$ such that $e^{(S,P)}(\tilde{S}_{-i}, \tau_i^{\tilde{S},T^*}(\tilde{P}))$ has non-zero value. Moreover, this value is equal to $x(|P| - 1)$.*

Proof. If $e^{(S,P)}(\tilde{S}, \tilde{P}) > 0$ and $e^{(S,P)}(\tilde{S}_{-i}, \tau_i^{\tilde{S},T}(\tilde{P})) > 0$ then from the definition:

$$e^{(S,P)}(\tilde{S}_{-i}, \tau_i^{\tilde{S},T}(\tilde{P})) = (|P| - 1)^{-|\tilde{S}_{-i} \setminus S|} = (|P| - 1) \cdot (|P| - 1)^{-|\tilde{S} \setminus S|}$$
$$= (|P| - 1) \cdot e^{(S,P)}(\tilde{S}, \tilde{P})$$

which proves the values equality part.

First we will consider a special case when $\tilde{P}$ contains the empty coalition. Then $\tilde{P} = \{N, \emptyset\}$ (as it is the only partition with the empty coalition) and the only transition of i is allowed to the empty coalition: $(\tilde{S}, \tau_i^{\tilde{S},T}(\tilde{P}))$ is equal $(N_{-i}, \{N_{-i}, \{i\}\})$. As we assumed $i \notin S$, we know that $\{i\} \preceq P_{-S}$ and as we're not changing the partition size ($|\{N, \emptyset\}| = |\{N_{-i}, \{i\}\}|$) this follows that $e^{(S,P)}(N_{-i}, \{N_{-i}, \{i\}\}) > 0$ and finishes this case.

Now let's assume that all the coalitions in $\tilde{P}$ are non-empty. Let $T_i \in P \setminus S$ be the agent's i coalition. From the constant-games definition we know that P_{-S} can be reduced to $\tilde{P}_{-\tilde{S}}$ and that in both there is the same number of the coalitions: $|P_{-S}| = |\tilde{P}_{-\tilde{S}}|$. As the agents from one coalition cannot be separated, there must be some non-empty coalition $\tilde{T}_i$ in $\tilde{P}_{-\tilde{S}}$ which can be reduced from T_i by deleting agents from $\tilde{S}$. It must contain at least one agent denoted by j (and $j \neq i$, because $i \in \tilde{S}$). So when we consider a transition to any other coalition than $\tilde{T}_i$ we will separate i and j agents which will violate (c) condition in $\preceq$-operator definition and imply zero value in $e^{(S,P)}$ game. But in $\tau_i^{\tilde{S},\tilde{T}_i}(P)$ all the conditions will be satisfied – (a) is obviously satisfied as $i \notin S$ and $\tilde{P}_{-\tilde{S}}$ was already a subset of P_{-S}; (b) and (c) are satisfied because the relations between an additional i agent are equal to the relations of j who is already in the structure.

Again, we do not change the size of the partition. We have to check only one special case when $\tilde{S} = \{i\}$. But from $e^{(S,P)}(\tilde{S}, \tilde{P}) > 0$ we get $S \subseteq \tilde{S}$ and as we know that $i \notin S$ we get $S = \emptyset$ which means that the game $e^{(S,P)}$ is incorrect.

So, finally: $e^{(S,P)}(\tilde{S}_{-i}, \tau_i^{\tilde{S},\tilde{T}_i}(\tilde{P})) > 0$. $\qquad\qquad\square$

From Lemma 5.2 and Lemma 5.3 we have that every agent $i \notin S$ is a null-player (in a steady marginality sense). Based on our Null-player Axiom (Definition 3.5), $\varphi_i^{full}(e^{(S,P)}) = 0$ and based on Symmetry (Definition 3.2) and Efficiency (Definition 3.1) we get:

$$\varphi_j^{full}(e^{(S,P)}) = \frac{1}{|S|} \cdot \sum_{j \in S} \varphi_j^{full}(e^{(S,P)}) = \frac{1}{|S|} \cdot e^{(S,P)}(N, \{N, \emptyset\}).$$

As our value φ^{full} clearly satisfies Efficiency, Linearity and Symmetry, the only observation we need to add is that it also satisfies Null-player Axiom. As agents not from S are null-players and get nothing it would be sufficient to show that no agent from S is a null-player. But every agent j from S has a non-zero marginal contribution to (S,P): $e^{(S,P)}(S,P) = 1$ and $e^{(S,P)}(S_{-j}, \tau_j^{S,T}(P)) = 0$ for every $T \in P \setminus S$. That finishes the proof of Theorem 5.1. $\qquad\qquad\square$

Now let us show that our unique value is indeed equal to the value proposed by McQuillin.

Theorem 5.4. *Let v be a game with externalities. Then $\varphi^{McQ}(v) = \varphi^{full}(v)$.*

Proof. Again, based on Linearity, we will show the adequacy on the constant-coalition games. In the proof of Theorem 5.1 we have showed that $\varphi_i^{full}(e^{(S,P)}) = 0$ for every $i \notin S$ and $\varphi_j^{full}(e^{(S,P)}) = \frac{1}{|S|} \cdot e^{(S,P)}(N, \{N, \emptyset\})$ for every $j \in S$.

Let (S,P) be an embedded coalition. Assume that $|P| > 2$. As $|\{N, \emptyset\}| = 2 \neq |P|$, based on the definition of the constant-coalition games (Definition 4.3) we get $e^{(S,P)}(N, \{N, \emptyset\}) = 0$ and $\varphi_i^{full}(e^{(S,P)}) = 0$ for every agent $i \in N$. Also $\varphi_i^{McQ}(e^{(S,P)}) = 0$ for every agent $i \in N$, because $\hat{v}^{McQ}(S) = 0$ for every $S \subseteq N$ as no embedded coalition of form $(\tilde{S}, \{\tilde{S}, N \setminus \tilde{S}\})$ has a non-zero value in $e^{(S,P)}$ (the reason here is the same – the partitions sizes do not match).

If $|P| = 2$ then embedded coalition has the form $(S, \{S, N\setminus S\})$ and $e^{(S,\{S,N\setminus S\})}$ assigns a non-zero value (equal 1) only to an embedded coalition $(\tilde{S}, \{\tilde{S}, N \setminus \tilde{S})$ such that $S \subseteq \tilde{S}$. Hence, $\hat{v}^{McQ}(\tilde{S}) = 1$ when $S \subseteq \tilde{S}$ and $\hat{v}(\tilde{S}) = 0$ otherwise. Based on the basic Shapley's axioms for $\hat{v}^{McQ}$ we get that $\varphi_i^{McQ}(e^{(S,P)}) = 0$ for $i \notin S$ and $\varphi_j^{McQ}(e^{(S,P)}) = Sh_j(\hat{v}) = \frac{1}{|S|}$ for $j \in S$.

Let's check if our value has the same results. As mentioned at the beginning of the proof, for $i \notin S$, $\varphi_i^{full}(e^{(S,P)}) = 0$ and for $j \in S$, $\varphi_j^{full}(e^{(S,P)}) = \frac{1}{|S|}$. $e^{(S,P)}(N, \{N, \emptyset\}) = \frac{1}{|S|}$ which completes the proof. $\qquad\square$

6 A Comparison of Various Marginality Definitions

In this section we examine various approaches to marginality more broadly and we compare them with steady marginality.

In Section 3 we have presented the universal definition of marginality:

$$mc_{(i,S,P)}(v) = \sum_{T \in P_{-S} \cup \{\emptyset\}} \alpha_{(i,S,P,T)}(v(S,P) - v(S_{-i}, \tau_i^{S,T}(P)))$$

Based on the accepted definition of marginal contribution we define a vector of marginal contributions $mc_i(v) = (mc_{(i,S,P)}(v))_{(S,P) \in EC, i \in S}$.

It seems reasonable to normalize the weights by assuming that the sum of them is equal 1: $\sum_{T \in P_{-S} \cup \{\emptyset\}} \alpha_{(i,S,P,T)} = 1$ for each $(S,P) \in EC$ such that $i \in S$. As we have considered only 0-1 weights we have omitted this step to increase the clarity of the presentation. But it is important to notice, that with respect to the conventional axioms based on the marginality this normalization is not significant as we compare $mc_{(i,S,P)}$ only to the same marginal contribution in other game (in Bolger [5] and De Clipper and Serrano [7] from $mc_i(v_1) = mc_i(v_2)$ we conclude $\varphi_i(v_1) = \varphi_i(v_2)$) or to zero (in Hu and Yang [6] and our paper from $mc_i(v_1) = 0$ we conclude that $\varphi_i(v) = 0$). Thus, the only important aspect is the weight ratio.

In all the definitions of marginal contribution proposed in the literature $\alpha_{(i,N,\{N,\emptyset\},\emptyset)} = 1.$[6] Thus, in the rest of this section we assume, that $P \neq \{N, \emptyset\}$.

Chronologically, the first definition of marginality proposed is also the most intuitive one. Bolger [5] defined the marginality as a simple average of all the elementary marginal contributions:

$$\alpha_{(i,S,P,T)}^B = 1 \text{ for } T \in P_{-S} \cup \{\emptyset\}$$

In his paper, Bolger studied the *basic games* – games with only 0-1 values. Thus, his marginal contribution is the number of partitions from which agent's transfer turns the S_{-i} coalition value (negative if value of S is zero). Unfortunately, there is no closed form expression for Bolger's value.

Steady marginality is quite similar to Bolger's one:

$$\alpha_{(i,S,P,T)}^{full} = 1 \text{ for } T \in P_{-S} \quad \text{and} \quad \alpha_{(i,S,P,T)}^{full} = 0 \text{ otherwise}$$

We have also already introduced the marginality proposed by De Clipper and Serrano:

$$\alpha_{(i,S,P,T)}^{free} = 1 \text{ for } T = \emptyset \quad \text{and} \quad \alpha_{(i,S,P,T)}^{free} = 0 \text{ otherwise}$$

---

[6] Assume that $\alpha_{(i,N,\{N,\emptyset\},\emptyset)} = 0$, then $mc_{(i,N,\{N,\emptyset\})} = 0$ regardless of $v(N, \{N, \emptyset\})$ and $v(N_{-i}, \{N, \{i\}\})$. So, in game with only one non-zero value for grand-coalition $v(N, \{N, \emptyset\}) = 1$ all the marginal contributions are equal zero which is unintuitive and with every axiom based on marginality results in a contradiction.

It is easy to see, that our marginality complements De Clipper and Serrano's marginality to Bolger's one.

Another approach was proposed by Hu and Yang [6]. Their marginality assigns the same weight to the transfer to every existing coalition and *higher* value for the partition where i is alone:

$$\alpha^{HY}_{(i,S,P,T)} = 1 \text{ for } T \in P_{-S} \quad \text{and} \quad \alpha^{HY}_{(i,S,P,T)} = 1 + r \text{ otherwise}$$

where $1 + r = \frac{|\{P^* \in \mathcal{P} : (\tau_i^{S,\emptyset}(P)\setminus\{S_{-i}\}) \preceq P^*\}|}{|\{P^* \in \mathcal{P} : (\tau_i^{S,T}(P)\setminus\{S_{-i}\}) \preceq P^*\}|}$ for any $T \in P_{-S}$.[7] It can be shown, that $r \geq 0$.

Let us consider a simple environment in which there exists m political parties – $S^{(j)}$ for every $j \in \{1, 2, \ldots, m\}$ – and one independent agent i (i.e. $N = \{i\} \cup \bigcup_j S^{(j)}$). Let $P = \{S^{(j)} : j \in \{1, 2, \ldots, m\}\} \cup \{\{i\}\}$ and define $P^{(j)} = \tau_i^{\{i\},S^{(j)}}(P)$. Assume that an agent i does not have any political power on his own, but by joining one of the parties he increases the coalition value by $(m-1)$ and decreases the values of other parties by 1, thus: $v(S^{(j)}_{+i}, P^{(j)}) = v(S^{(j)}, P) + (m-1)$ and $v(S^{(j)}, P^{(k)}) = v(S^{(j)}, P) - 1$ for every $S^{(j)}, S^{(k)} \in P$ and $S^{(j)} \neq S^{(k)}$.

Consider the marginal contribution of an agent i to the coalition $S^{(j)}_{+i}$. As we normalize weights (in a way mentioned before) based on non-existing marginality we get $mc^{full}_{(i,S^{(j)}_{+i},P^{(j)})} = m$ as we exclude a non-realistic situation from the consideration – when agent i decides to waste his potential. De Clippel and Serrano consider only this one situation and get $mc^{free}_{(i,S^{(j)}_{+i},P^{(j)})} = m - 1$. Bolger reaches some compromise, as he does not differentiate partitions: $mc^B_{(i,S^{(j)}_{+i},P^{(j)})} = m - \frac{1}{m}$. The last marginal contribution $mc^{HY}_{(i,S^{(j)}_{+i},P^{(j)})}$ is slightly smaller than Bolger's and depends on the size of $S^{(j)}$ and the number of parties.

7 Conclusions

In this paper we have studied the problem of finding a fair division of jointly gained payoff in coalitional games with externalities. We have presented an innovative approach to marginality in which the contribution of an agent to the coalition is evaluated only in reference to the partitions where the agent is not alone. This allowed us to define a new version of Null-player Axiom which, together with Efficiency, Symmetry and Linearity, uniquely determines a division scheme. We have proved that this value is equal to the one proposed earlier by McQuillin.

Our work can be extended in various directions. It is not clear if any of the adopted axioms can be dropped (just as De Clippel and Serrano based their value only on Symmetry, Efficiency and their version of Marginality axiom). Another

---

[7] That definition may seem wrong, as it might not be obvious, why $|\{P^* \in \mathcal{P} : (\tau_i^{S,T}(P) \setminus \{S_{-i}\}) \preceq P^*\}|$ is equal for every $T \in P_{-S}$. It appears, that $|\{P^* \in \mathcal{P} : \tilde{P}_{-\tilde{S}} \preceq P^*\}|$ depends only on $|\tilde{P}|$ and $|\tilde{S}|$ and grows with increasing $|\tilde{P}|$. That also explains why the numerator is larger than the denominator in $1 + r$ fractional definition.

question is whether there exist any other definitions of marginality which lead to McQuillin's value. Looking a bit further, it would be interesting to find any universal link between the definition of marginality and the formula for the value derived from it. Finally, other approaches to marginality axioms can be studied.

References

1. Shapley, L.S.: A value for n-person games. In: Kuhn, H.W., Tucker, A.W. (eds.) Contributions to the Theory of Games, volume II, vol. II, pp. 307–317. Princeton University Press, Princeton (1953)
2. McQuillin, B.: The extended and generalized Shapley value: Simultaneous consideration of coalitional externalities and coalitional structure. Journal of Economic Theory (144), 696–721 (2009)
3. Pham Do, K.H., Norde, H.: The Shapley value for partition function form games. International Game Theory Review 9(2), 353–360 (2007)
4. Macho-Stadler, I., Perez-Castrillo, D., Wettstein, D.: Sharing the surplus: An extension of the shapley value for environments with externalities. Journal of Economic Theory (135), 339–356 (2007)
5. Bolger, E.M.: A set of axioms for a value for partition function games. International Journal of Game Theory 18(1), 37–44 (1989)
6. Hu, C.-C., Yang, Y.-Y.: An axiomatic characterization of a value for games in partition function form. SERIEs: Journal of the Spanish Economic Association, 475–487 (2010)
7. de Clippel, G., Serrano, R.: Marginal contributions and externalities in the value. Econometrica 76(6), 1413–1436 (2008)
8. Young, H.P.: Monotonic Solutions of Cooperative Games. International Journal of Game Theory 14, 65–72 (1985)

Scheduling without Payments

Elias Koutsoupias

Department of Informatics, University of Athens
elias@di.uoa.gr

Abstract. We consider mechanisms without payments for the problem of scheduling unrelated machines. Specifically, we consider truthful in expectation randomized mechanisms under the assumption that a machine (player) is bound by its reports: when a machine lies and reports value $\tilde{t}_{ij}$ for a task instead of the actual one t_{ij}, it will execute for time $\tilde{t}_{ij}$ if it gets the task—unless the declared value $\tilde{t}_{ij}$ is less than the actual value t_{ij}, in which case, it will execute for time t_{ij}. Our main technical result is an optimal mechanism for one task and n players which has approximation ratio $(n + 1)/2$. We also provide a matching lower bound, showing that no other truthful mechanism can achieve a better approximation ratio. This immediately gives an approximation ratio of $(n + 1)/2$ and $n(n + 1)/2$ for social cost and makespan minimization, respectively, for any number of tasks.

1 Introduction

A major challenge today is to design algorithms that work well even when the input is reported by selfish agents or when the algorithm runs on a system with selfish components. The classical approach is to use mechanism design [13], that is, to design algorithms that use payments to convince the selfish agents to reveal the truth and then compute the outcome using the reported values. Central to the mechanism design approach is the use of dominant strategies as the equilibrium concept. Mechanism design is a very important framework with many unexpected results and it remains a very active research area trying to address some beautiful and challenging problems. Nevertheless, one major problem with mechanism design with payments is that in many situations, the use of payments may not be feasible.

Partly for this reason, there is a lot of recent interest in mechanisms that use no payments (mechanism design without payments) [13]. Given that in many problems we have obtained very poor results using mechanisms with payments, it will be really surprising if the substantially more restricted class of mechanisms without payments can achieve any positive results. For the general unrestricted domain with at least 3 outcomes and truthful mechanisms without payments, the Gibbard-Satterthwaite theorem [6,17] states that only dictatorial mechanisms are truthful; dictatorial mechanisms are those in which a particular player determines the outcome. Contrast this to Roberts theorem [16], which states that if we allow payments, the truthful mechanisms are the affine maximizers,

G. Persiano (Ed.): SAGT 2011, LNCS 6982, pp. 143–153, 2011.

a much richer class than the dictatorial mechanisms (yet a very poor class in comparison to the set of all possible algorithms).

When we restrict the domain to scheduling unrelated machines, perhaps the most influential problem in algorithmic game theory, the results so far have also been disappointing. The best approximation ratio for the makespan that we know by truthful in expectation mechanisms is $(n + 1)/2$ [3], where n is the number of players. The best known lower bound of 2 [3,11] leaves the possibility open for improved mechanisms. The situation for deterministic mechanisms is similar: the upper bound is easily n [12], and the best known lower bound is 2.61 [8]. It may seem surprising that in this work we can achieve comparable results using mechanisms without payments (our main result, Theorem 2, is a truthful in expectation mechanism with approximation ratio $n(n + 1)/2$). But a moments thought will reveal that we can get a slightly weaker bound (approximation ratio n for one task and hence approximation ratio n^2 for many tasks) with the natural mechanism which allocates each task independently and with probability inversely proportional to the execution times (Proposition 1). The assumption that the players are bound by their declarations plays also a crucial role; without it, no positive result is possible. The value of our main result is that it gives a definite answer (tight upper and lower bounds, albeit only for one task) for this fundamental problem.

Mechanism design without payments is a major topic in game theory [13], although it has not been studied as intensively as the variant with payments in algorithmic game theory. There are however many recent publications which study such mechanisms. Procaccia and Tennenholtz [15] proposed to study approximate mechanism design without payments for combinatorial problems and they studied facility location problems. This work was substantially extended (see for example [10,9,1]). Such mechanisms studied also by Dughmi and Gosh [4] who consider mechanisms assignment problems and by Guo and Conitzer [7] who consider selling items without payments to 2 buyers. Conceptually, closer to our approach of assuming that the players pay their declared values is the notion of impositions of Nissim, Smorodinsky, and Tennenholtz [14] which was further pursued by Fotakis and Tzamos [5].

2   Model

We study the problem of scheduling tasks when the machines are selfish. We formulate the problem in its more general form, the unrelated machines version: there are n selfish machines and m tasks; the machines are lazy and prefer not to execute any tasks. Machine i needs time $t_{i,j}$ to execute task j. The tasks are allocated to machines with the objective of minimizing the makespan (or the social welfare which is the negation of the sum of executing times of all machines). Let a be an (optimal) allocation for input t, where $a_{i,j}$ is an indicator variable about the event of allocating task j to machine i. The execution time of machine

i is $\sum_{j=1}^{m} a_{i,j} t_{i,j}$ and the makespan is $\max_{i=1,...,n} \sum_{j=1}^{m} a_{i,j} t_{i,j}$. For randomized algorithms, the allocation variables $a_{i,j}$ are not integral but real values in $[0,1]$ which is the probability to allocate task j to machine i[1].

In this work, we consider direct revelation mechanisms; a mechanism is simply a randomized scheduling algorithm S which computes an allocation based on the declarations of the machines. There are no payments. More precisely, every machine i reports its private values $\tilde{t}_{i,j}$, one for each task and we apply the algorithm S on this input. The notation $\tilde{t}_{i,j}$ instead of $t_{i,j}$ is used because the machine may lie and not declare its true values $t_{i,j}$. There is however a very important difference with the standard Nisan-Ronen framework [12] for this mechanism design problem. *We assume that the machines are bound by their reports.* More precisely, if a machine i declares a value $\tilde{t}_{i,j} \geq t_{i,j}$ for task i and is allocated the task, *its actual cost is the declared value $\tilde{t}_{i,j}$ and not $t_{i,j}$.* One justification for this assumption is that in some environments the machines can be observed during the execution of the tasks and cannot afford to be caught lying about the execution times. Similar assumptions have been used for other problems. Our assumption is similar to the notion of imposition [14,5]; for example, in the facility location problem, the players may be forced to use the facility which is closer to their declared position instead of letting them freely choose between the facilities. To complete our assumptions, we need to specify what happens when a machine declares a smaller value, i.e., $\tilde{t}_{i,j} < t_{i,j}$. In this case, we make the simple assumption that the actual cost is the true value $t_{i,j}$; it would be simpler to assume that the machines are not allowed to lie in this direction, but we prefer the weakest assumption since it does not affect our results. To summarize, the cost of machine i for task j is $\max(t_{i,j}, \tilde{t}_{i,j})$.

Our framework now is simple. We design a randomized scheduling algorithm S. The selfish machines report values $\tilde{t}_{i,j}$ and we apply algorithm S to the reported values. This induces a game among the machines: the pure strategies of machine i are the vectors $\left(\tilde{t}_{i,j}\right)_{j=1}^{m}$ with $\tilde{t}_{i,j} \geq 0$. The cost is the execution time computed by algorithm S on input $\tilde{t}$. To be more precise, if $a = S(\tilde{t})$ is the allocation computed by algorithm S on input $\tilde{t}$, the cost of machine i is $c_i = \sum_{j=1}^{m} a_{i,j} \max(t_{i,j}, \tilde{t}_{i,j})$. We seek mechanisms which minimize the makespan $\max_{i=1,...,n} c_i$ or the social cost $\sum_{i=1}^{n} c_i$.

Our assumption is related to mechanisms with verification [12], in which the mechanism learns the actual execution time of the machines and pays after the execution. Because of the delayed payments, these mechanisms are much more powerful than the mechanisms of our framework; for example, they can enforce that the machines are bound by their declarations by imposing a very high penalty for lying. A similar framework was proposed in [2], but in this case the power of the mechanism is limited: it can only deny payment at the end when a machine is caught lying by not being able to finish the tasks within the declared time.

---

[1] We may also interpret the probabilities as fractional allocations. Naturally, our results apply to the fractional allocations as well.

2.1 The Case of a Single Task

We focus on the simple case of one task and n machines. Let $t_1, \ldots, t_n$ be the true values of the machines for the task and let $\tilde{t}_1, \ldots, \tilde{t}_n$ be the declared values; we drop the second subscript since there is only one task. Let $p_i(\tilde{t})$ be the probability of allocating the task to player i. The expected cost of player i is $c_i = c_i(t_i, \tilde{t}) = p_i(\tilde{t}) \max(t_i, \tilde{t}_i)$, while the social cost of the algorithm is $\sum_{i=1}^{n} c_i$ (in the case of the single task, the makespan and social cost are identical).

The mechanism is called truthful if for every t_{-i}, the expected cost c_i of player i is minimized when $\tilde{t}_i = t_i$. This notion of truthfulness, truthful in expectation, is the weakest notion of truthfulness which contains a richer class of mechanisms than the standard notion on truthfulness or universal truthfulness. It is not ex post truthful, meaning that after the players see the outcome of the coins, they may want to change their declarations. It is trivial to see that the stronger notion of truthfulness (universally truthful) cannot achieve any positive result. Notice however, that for the fractional version of the scheduling problem, in which we can allocate parts of the same tasks to different machines, we can consider deterministic algorithms and consequently the strongest notion of truthfulness.

Claim. An algorithm is truthful if and only if for every i and t_{-i}, $t_i p_i(t)$ is non-decreasing in t_i and $p_i(t)$ is non-increasing in t_i.

Proof. Indeed, if $t_i p_i(t)$ is non-decreasing in t_i, the player prefers t_i to higher values, i.e., when $\tilde{t}_i \geq t_i$, the cost $c_i = p_i(\tilde{t}_i, t_{-i})\tilde{t}_i$ is minimized at $\tilde{t}_i = t_i$. Similarly, if $p_i(t_i)$ is non-increasing, the player prefers t_i to smaller values, i.e., when $\tilde{t}_i \leq t_i$, the cost $c_i = p_i(\tilde{t}_i, t_{-i})t_i$ is again minimized at $\tilde{t}_i = t_i$.

Conversely, if there exist $x < y$ with $p_i(t_{-i}, y)y < p_i(t_{-i}, x)x$, then player i gains by lying: when $t_i = x$ he prefers to declare y. Similarly, if there exist $x > y$ with $p_i(t_{-i}, x) > p_i(t_{-i}, y)$, the player would again prefer to declare y when $t_i = x$.

A mechanism is defined simply by the probability functions $p_i(t)$. The expected makespan is $\sum_i c_i$ and its approximating ratio is $\sum_i c_i / \min_i t_i$. We seek truthful mechanisms with small approximation ratio.

3 Truthful Mechanisms for One Task

In this section, we study the case of a single task. We first consider a natural mechanism, the *proportional* algorithm: It allocates the task to machine i with probability inversely proportional to the declared value t_i, i.e., $p_i(t) = t_i^{-1} / \sum_k t_k^{-1}$.

Proposition 1. *The proportional algorithm is truthful and achieves approximation ratio n.*

Proof. Indeed, to verify that the mechanism is truthful, it suffices to observe that $p_i(t)$ is non-increasing in t_i and that $t_i p_i(t)$ is non-decreasing it t_i. The expected makespan of this mechanism is $n/\sum_i t_i^{-1}$ while the optimal makespan is $\min_i t_i$. It follows that the approximation ratio is at most n and that it can be arbitrarily close to n (when for example one value is 1 and the other $n-1$ values are arbitrarily high).

It is natural to ask whether there are better mechanisms than the proportional mechanism. In the next subsection, we give a positive answer by designing an optimal mechanism, albeit with not substantially better approximation ratio.

3.1   An Optimal Truthful Mechanism

In this subsection, we study truthful algorithms that have optimal approximation ratio.

To find an optimal truthful mechanism, we want to find functions $p_i(t)$ such that for every t:

- for every i: $t_i p_i(t)$ is non-decreasing in t_i,
- for every i: $p_i(t)$ is non-increasing in t_i,
- $\sum_i p_i(t) = 1$

and which minimize $\max_t \sum_i t_i p_i(t)/\min_i t_i$. The first two conditions capture truthfulness and the third condition the natural property that the probabilities add to 1.

We will show the following theorem:

Theorem 1. *There is a truthful in expectation mechanism without payments with approximation ratio $(n+1)/2$. Conversely, no truthful in expectation mechanism without payments can have approximation ratio better than $(n+1)/2$.*

Before proceeding with the proof of the theorem, it is instructive to consider first the case of $n=2$ players. We will consider a *symmetric* mechanism, so it suffices to give the probabilities $p_i(t)$ of assigning the task to player i when $t_1 \leq t_2$. We claim that the mechanism with probabilities

$$p_1(t) = 1 - \frac{t_1}{2t_2} \qquad\qquad p_2(t) = \frac{t_1}{2t_2}$$

is truthful and has approximation ratio $3/2$. To clarify: these are the probabilities when $t_1 \leq t_2$; by symmetry, we can compute the probabilities when the declared value of the second player is smaller than the declared value of the first player.

Let us verify that this mechanism is truthful. Specifically we want to show that the expected cost of player i is minimized when he declares his true value; this must hold for every value t_{-i} of the other player. Consider first player 1 (the one with true value less than the declared value of the other player).

- He has no reason to declare something less than t_1, because $p_1(t)$ is non-increasing in t_1, consequently $t_i p_i(t) \leq t_i p_i(\tilde{t}_i, t_{-i})$.
- He has no reason to declare something in (t_1, t_2), because $t_1 p_1(t) = t_1 - \frac{t_1^2}{2t_2} = \frac{t_2^2 - (t_2 - t_1)^2}{2t_2}$ is increasing in t_1 for $t_1 < t_2$.

– Finally, he has no reason to declare something in $[t_2, \infty)$. In this case, his lie changes the order of the values, and, by the definition of the mechanism, the probability of getting the task will be $p_2(t_2, \tilde{t}_1)$. Nevertheless, we still have

$$t_1 p_1(t) = \frac{t_2^2 - (t_2 - t_1)^2}{2t_2} \leq \frac{t_2}{2} = \tilde{t}_1 \frac{t_2}{2\tilde{t}_1} = \tilde{t}_1 p_2(t_2, \tilde{t}_1).$$

We work similarly for the second player (the one with true value greater than the declared value of the other player). If he declares his true value t_2, his expected cost is $t_2 p_2(t) = t_1/2$.

– He has no reason to declare something less than t_1, because in this case the probability $p_1(t_1, \tilde{t}_2)$ of getting the task is at least $1/2$ and his cost will be at least $t_2/2 \geq t_1/2$.
– He has no reason to declare any other value greater than t_1 because his cost is going to be $\tilde{t}_2 p_2(t_1, \tilde{t}_2) = t_1/2$, anyway.

The above mechanism has approximation ratio $3/2$, because the cost is

$$t_1 p_1(t) + t_2 p_2(t) = t_1 - \frac{t_1^2}{2t_2} + \frac{t_1}{2} = \frac{3}{2} t_1 - \frac{t_1^2}{2t_2} \leq \frac{3}{2} t_1.$$

Trivially, the approximation ratio tends to $3/2$ as t_2 tends to ∞.

We proceed to generalize the above to more than two players. Again, we define a symmetric mechanism, so it suffices to describe it when $t_1 \leq \cdots \leq t_n$:

$$p_1 = \frac{1}{t_1} \int_0^{t_1} \prod_{i=2}^n \left(1 - \frac{y}{t_i} \right) dy \tag{1}$$

$$p_k = \frac{1}{t_k t_1} \int_0^{t_1} \int_0^y \prod_{i=2..n, i \neq k} \left(1 - \frac{x}{t_i} \right) dx\, dy \qquad \text{for } k \neq 1$$

For example, for $n = 2$ we get the mechanism discussed above, and for $n = 3$ the probabilities are

$$p_1 = 1 - \frac{t_1(t_2 + t_3)}{2t_2 t_3} + \frac{t_1^2}{3t_2 t_3} \qquad p_2 = \frac{t_1}{2t_2} - \frac{t_1^2}{6t_2 t_3} \qquad p_3 = \frac{t_1}{2t_3} - \frac{t_1^2}{6t_2 t_3}.$$

This definition is not arbitrary, but it is the natural solution to the requirements at the beginning of this subsection. This will become apparent as we proceed to show that this is an optimal algorithm for our problem.

First we verify that the mechanism is well-defined: We need to show that these probabilities are nonnegative and add up to 1. Indeed, consider the quantities q_i inside the integrals

$$q_1(y) = \prod_{i=2}^n \left(1 - \frac{y}{t_i} \right)$$

$$q_k(y) = \frac{1}{t_k} \int_0^y \prod_{i=2..n, i \neq k} \left(1 - \frac{x}{t_i} \right) dx$$

for which

$$p_i = \frac{1}{t_1} \int_0^{t_1} q_i(y)\, dy.$$

Since the integral is for $y \leq t_1$ and t_1 is the minimum among the t_k's, all factors in these expressions are nonnegative. This shows that $q_i(y) \geq 0$ for every $i = 1, \ldots, n$. We also observe that $q_1'(y) = -\sum_{i=2}^n q_k'(y)$ which shows that $\sum_{i=1}^n q_i(y)$ is constant; taking $y = 0$, we see that this constant is 1. In summary the p_i's are nonnegative and their sum is $\frac{1}{t_1}\int_0^{t_1}\sum_{i=1}^n q_i(y)\, dy = \frac{1}{t_1}\int_0^{t_1} 1\, dy = 1$.

We also need to verify that this indeed a symmetric mechanism; otherwise the above definition is not complete. Specifically, we need to show that if $t_1 = t_2$ then $p_1 = p_2$. Indeed, it is straightforward to verify it by employing the following easy identity for every function g:

$$\int_0^a (a-y)g(y)\, dy \leq \int_0^a \int_0^y g(x)\, dx\, dy^2$$

In this case, $g(y) = \prod_{i=3}^n \left(1 - \frac{y}{t_i}\right)$.

We now proceed to establish that the mechanism is truthful.

Lemma 1. *The symmetric mechanism defined by the probabilities in (1) is truthful.*

Proof. To show that the algorithm is truthful, we observe that

- The probabilities p_i are non-increasing in t_i. This is trivially true for $i \neq 1$ and it can be easily verified for $i = 1$. The fact that the probabilities are non-increasing in t_i shows that no player i has a reason to lie and declare a value $\tilde{t}_i < t_i$. To see this, fix the values of the remaining players and assume that player i changes his value to $\tilde{t}_i < t_i$. We will argue that the probability of getting the task with the new value is greater than or equal to the original probability; this suffices because the cost of the player is this probability times t_i.

  If, after the change, the order of the players remains the same, the probability does not decrease by this change because p_i is non-increasing in t_i. We need to show the same when the change affects the order of the players. This turns out to be easy but we note that we need to be careful, because the expressions that define the probabilities of the mechanism depend on the order of the values. However, we don't need the expressions of the probabilities but a

---

[2] Proof: Let $G(y) = \int_0^y g(x)\, dx$. Then

$$\int_0^a (a-y)g(y) - \int_0^a \int_0^y g(x)\, dx\, dy = \int_0^a (a-y)g(y) - G(y)\, dy$$
$$= \int_0^a ag(y) - (yG(y))'\, dy$$
$$= aG(a) - aG(a) = 0.$$

simpler argument: imagine that we lower the value from t_i to $\tilde{t}_i$ in stages: from t_i to t_{i-1} to t_{i-2}, and so on, to t_{i-k}, and finally to $\tilde{t}_i$. In each stage, the order of the values remains the same and therefore the probability can only increase. It follows that it does not decrease from the total change. The fact that the mechanism is symmetric is crucial in this argument because the algebraic expressions of the probability of the player may change from stage to stage, but the values at the boundaries $t_{i-1}, \dots, t_{i-k}$ of successive stages are the same.

- every player k for $k \neq 1$ is truthful since $t_k p_k$ is independent of his value t_k; this holds even when the player reports a higher value which may change the order of the players
- player 1 has no reason to lie and report a value in $(t_1, t_2]$ because $t_1 p_1$ is increasing in t_1; furthermore, player 1 has no reason to report a higher value than t_2 which will change the order of the players because the cost will change from $t_1 p_1$ to $t'_k p'_k = \frac{1}{t_2} \int_0^{t_2} \int_0^y \prod_{i=3}^n \left(1 - \frac{x}{t_i}\right) dx\, dy$ (because now the minimum value is t_2). It suffices therefore to show

$$t_1 p_1 \leq t'_k p'_k \qquad \Leftrightarrow$$

$$\int_0^{t_1} \prod_{i=2}^n \left(1 - \frac{y}{t_i}\right) dy \leq \frac{1}{t_2} \int_0^{t_2} \int_0^y \prod_{i=3}^n \left(1 - \frac{x}{t_i}\right) dx\, dy \qquad \Leftarrow$$

$$\int_0^{t_2} \prod_{i=2}^n \left(1 - \frac{y}{t_i}\right) dy \leq \frac{1}{t_2} \int_0^{t_2} \int_0^y \prod_{i=3}^n \left(1 - \frac{x}{t_i}\right) dx\, dy \qquad \Leftrightarrow$$

$$\int_0^{t_2} (t_2 - y) \prod_{i=3}^n \left(1 - \frac{y}{t_i}\right) dy \leq \int_0^{t_2} \int_0^y \prod_{i=3}^n \left(1 - \frac{x}{t_i}\right) dx\, dy$$

The last holds because of Equation (2).

Putting everything together, we see that the mechanism is indeed truthful.

Lemma 2. *The symmetric mechanism defined by the probabilities in* (1) *has approximation ratio* $(n+1)/2$.

Proof. Now that we have established that the algorithm is truthful, we proceed to bound its approximation ratio. The approximation ratio is $\sum_{i=1}^n t_i p_i / t_1$. Clearly, $t_1 p_1 \leq t_1$ and for $k > 1$

$$t_k p_k = \frac{1}{t_1} \int_0^{t_1} \int_0^y \prod_{i=2..n, i \neq k} \left(1 - \frac{x}{t_i}\right) dx\, dy$$
$$\leq \frac{1}{t_1} \int_0^{t_1} \int_0^y 1\, dx\, dy$$
$$= t_1/2.$$

Therefore, $\sum_{i=1}^n t_i p_i \leq t_1 + (n-1)t_1/2 = t_1(n+1)/2$, which shows that the approximation ratio is at most $(n+1)/2$. It is trivial that if we fix the other values and let t_1 tend to 0, the above inequalities are almost tight, and therefore the approximation ratio can be arbitrarily close to $(n+1)/2$.

We will now show that no other truthful algorithm has a better approximation ratio.

Lemma 3. *No truthful in expectation mechanism without payments has approximation ratio smaller than $(n+1)/2$.*

Proof. We will employ instances with values of the form $(1, 1, m, \ldots, m)$ and $(1, m, \ldots, m)$ where m is some large value (which we will allow to tend to infinity to obtain the lower bound). Without loss of generality we can assume that the mechanism is symmetric: if not, consider a non-symmetric mechanism M and create a new mechanism which first permutes the values and then applies mechanism M; the new mechanism has approximation ratio smaller or equal to the approximation ratio of M. Let p and p' be the probabilities assigned by an algorithm to the above instances. For the instance $(1, 1, m, \ldots, m)$, we have $2p_2 + (n-2)p_3 = 1$ and the approximation ratio is at least $r \geq 2p_2 + m(n-2)p_3 = m - 2p_2(m-1)$. Similarly for the other instance $(1, m, \ldots, m)$ we have $p'_1 + (n-1)p'_2 = 1$ and $r \geq p'_1 + (n-1)mp'_2 = 1 + (m-1)(n-1)p'_2$.

The crucial step is to use the truthfulness of player 2 to connect the two instances. Specifically, the second player is truthful only when $1 \cdot p_2 \leq m \cdot p'_2$. Substituting the value of p'_2 in the bound of the second instance, we get that the approximation is at least $1 + p_2(m-1)(n-1)/m$.

In summary, the approximation ratio is at least $\max\{m - 2p_2(m-1), 1 + p_2(m-1)(n-1)/m\}$. The first expression is decreasing in p_2 while the second one is increasing in p_2; the minimum approximation ratio is achieved when the expressions are equal, that is, when $p_2 = m/(2m+n-1)$ which gives ratio at least $(n+1)/2 - (n^2-1)/(4m+2n-2)$. As m tends to infinity, the approximation ratio tends to $(n+1)/2$.

4   Extension to Many Tasks and Discussion

In the previous section, we gave an optimal mechanism for one task. We can use it to get a mechanism for many tasks by running it independently for every task. The resulting mechanism is again truthful. If the objective is the social cost (i.e., to minimize the sum of the cost of all players), the mechanism clearly retains its approximation ratio. If the objective however is the makespan, then the approximation ratio is at most $n(n+1)/2$, for the simple reason that $\max_i c_i \leq \sum_i c_i \leq n \max_i c_i$. So we get

Theorem 2. *There is a truthful in expectation mechanism without payments for the problem of scheduling unrelated machines with approximation ratio $(n+1)/2$ when the objective is the social cost and $n(n+1)/2$ when the objective is the makespan.*

It follows from the case of one task that the approximation ratio $(n+1)/2$ is tight for the social cost. It is not clear that the pessimistic way we used to bound the approximation ratio of the makespan is tight. It remains open to estimate the approximation ratio of the given mechanism for many tasks. It is also open whether there are (task-independent or not) mechanisms without payments for many tasks with better approximation ratio.

References

1. Alon, N., Feldman, M., Procaccia, A., Tennenholtz, M.: Strategyproof approximation of the minimax on networks. Mathematics of Operations Research 35(3), 513–526 (2010)
2. Auletta, V., De Prisco, R., Penna, P., Persiano, G.: The power of verification for one-parameter agents. In: Díaz, J., Karhumäki, J., Lepistö, A., Sannella, D. (eds.) ICALP 2004. LNCS, vol. 3142, pp. 171–182. Springer, Heidelberg (2004)
3. Christodoulou, G., Koutsoupias, E., Kovács, A.: Mechanism design for fractional scheduling on unrelated machines. In: Arge, L., Cachin, C., Jurdziński, T., Tarlecki, A. (eds.) ICALP 2007. LNCS, vol. 4596, pp. 40–52. Springer, Heidelberg (2007)
4. Dughmi, S., Ghosh, A.: Truthful assignment without money. In: Proceedings of the 11th ACM Conference on Electronic Commerce, pp. 325–334. ACM, New York (2010)
5. Fotakis, D., Tzamos, C.: Winner-imposing strategyproof mechanisms for multiple facility location games. In: Saberi, A. (ed.) WINE 2010. LNCS, vol. 6484, pp. 234–245. Springer, Heidelberg (2010)
6. Gibbard, A.: Manipulation of voting schemes: a general result. Econometrica: Journal of the Econometric Society, 587–601 (1973)
7. Guo, M., Conitzer, V.: Strategy-proof allocation of multiple items between two agents without payments or priors. In: Proceedings of the 9th International Conference on Autonomous Agents and Multiagent Systems, vol. 1, pp. 881–888. International Foundation for Autonomous Agents and Multiagent Systems (2010)
8. Koutsoupias, E., Vidali, A.: A lower bound of $1+\varphi$ for truthful scheduling mechanisms. In: Kučera, L., Kučera, A. (eds.) MFCS 2007. LNCS, vol. 4708, pp. 454–464. Springer, Heidelberg (2007)
9. Lu, P., Sun, X., Wang, Y., Zhu, Z.: Asymptotically optimal strategy-proof mechanisms for two-facility games. In: Proceedings of the 11th ACM Conference on Electronic Commerce, pp. 315–324. ACM, New York (2010)
10. Lu, P., Wang, Y., Zhou, Y.: Tighter bounds for facility games. In: Leonardi, S. (ed.) WINE 2009. LNCS, vol. 5929, pp. 137–148. Springer, Heidelberg (2009)
11. Mu'alem, A., Schapira, M.: Setting lower bounds on truthfulness: extended abstract. In: Proceedings of the Eighteenth Annual ACM-SIAM Symposium on Discrete Algorithms, pp. 1143–1152. Society for Industrial and Applied Mathematics, Philadelphia (2007)
12. Nisan, N., Ronen, A.: Algorithmic mechanism design (extended abstract). In: Proceedings of the Thirty-First Annual ACM Symposium on Theory of Computing, pp. 129–140. ACM, New York (1999)
13. Nisan, N., Roughgarden, T., Tardos, E., Vazirani, V.: Algorithmic game theory. Cambridge Univ. Pr., Cambridge (2007)

14. Nissim, K., Smorodinsky, R., Tennenholtz, M.: Approximately optimal mechanism design via differential privacy. Arxiv preprint arXiv:1004.2888 (2010)
15. Procaccia, A., Tennenholtz, M.: Approximate mechanism design without money. In: Proceedings of the Tenth ACM Conference on Electronic Commerce, pp. 177–186. ACM, New York (2009)
16. Roberts, K.: The characterization of implementable choice rules. Aggregation and Revelation of Preferences, 321–348 (1979)
17. Satterthwaite, M.: Strategy-proofness and Arrow's conditions: Existence and correspondence theorems for voting procedures and social welfare functions. Journal of Economic Theory 10(2), 187–217 (1975)

Combinatorial Agency of Threshold Functions

Shaili Jain[1] and David C. Parkes[2]

[1] Yale University, New Haven, CT
shaili.jain@yale.edu
[2] Harvard University, Cambridge, MA
parkes@eecs.harvard.edu

Abstract. We study the combinatorial agency problem introduced by Babaioff, Feldman and Nisan [5] and resolve some open questions posed in their original paper. Our results include a characterization of the transition behavior for the class of *threshold* functions. This result confirms a conjecture of [5], and generalizes their results for the transition behavior for the OR technology and the AND technology. In addition to establishing a (tight) bound of 2 on the Price of Unaccountability (POU) for the OR technology for the general case of $n > 2$ agents (the initial paper established this for $n = 2$, an extended version establishes a bound of 2.5 for the general case), we establish that the POU is unbounded for all other threshold functions (the initial paper established this only for the case of the AND technology). We also obtain characterization results for certain compositions of anonymous technologies and establish an unbounded POU for these cases.

1 Introduction

The classic principal-agent model of microeconomics considers an agent with unobservable, costly actions, each with a corresponding distribution on outcomes, and a principal with preferences over outcomes [9,15]. The principal cannot contract on the action directly (e.g. the amount of effort exerted), but only on the final outcome of the project. The main goal is to design contracts, with a payment from the principal to the agent conditioned upon the outcome, in order to maximize the payoff to the principal in equilibrium with a rational, self-interested agent.

The principal-agent model is a classic problem of moral hazard, with agents with potentially misaligned incentives and private actions. A related theory has considered the problem of moral hazard on teams of agents [4,14,13]. Much of this work involves a continuous action choice by the agent (e.g., effort) and a continuous outcome function, typically linear or concave in the effort of the agents. Moreover, rather than considering the design of an optimal contract that maximizes the welfare of a principal, considering the loss to the principal due to transfers to agents, it is more typical to design contracts that maximize the total value from the outcome net the cost of effort, and without consideration of the transfers other than requiring some form of budget balance.

G. Persiano (Ed.): SAGT 2011, LNCS 6982, pp. 154–165, 2011.

Babaioff et al. [5] introduce the *combinatorial agency* problem. This a version of the moral hazard in teams problem in which the agents have binary actions and the outcome is binary, but where the outcome technology is a *complex combination* of the inputs of a team of agents. Each agent is able to exert high or low effort in its own hidden action, with the success or failure of an overall project depending on the specific technology function. In particular, these authors consider a number of natural technology functions such as the AND technology, the OR technology, the majority technology, and nested models such as AND-of-ORs and OR-of-ANDs. This can be conceptualized as a problem of moral hazard in teams where agents are situated on a graph, each controlling the effort at a particular vertex.

The combinatorial agency framework considers the social welfare, in terms of the cost to agents and the value to the principal, that can be achieved in equilibrium under an optimal contract where the principal seeks a contract that maximizes payoff, i.e. value net of transfers to the agents, in equilibrium. Thus the focus is on contracts that would be selected by a principal, not be a designer interested in finding an equilibrium that maximizes social welfare. In particular, Babaioff et al. consider the (social) *Price of Unaccountability* (POU), which is the worst case ratio between the *optimal social welfare* when actions are observable as compared to when they are not observable. The worst-case is taken over different probabilities of success for an individual agent's actions (and thus different, uncertain technology functions), and over the principal's value for a successful outcome. The optimal social welfare is obtained by requesting a particular set of agents to exert effort, in order to maximize the total expected value to the principal minus the cost incurred by these agents. In the agency case, the social welfare is again this value net cost, but optimized under the contract that maximizes the expected payoff of the principal.

The main contribution of this work is to characterize the *transition behavior* for the k-out-of-n (or *threshold*) technology, for n agents and $k \in \{1, \ldots, n\}$. The threshold technology is anonymous, meaning that the probability of a successful outcome only depends on the number of agents exerting high effort, not the specific set of agents. Because of this, the transition behavior — a characterization of the optimal contract, which specifies which agents to contract with, as a function of the principal's valuation — can be explained in terms of the number of agents with whom the principal contracts. We establish that the transition behavior (in both the non-strategic and agency cases) includes a transition from contracting between 0 and l agents for some $1 \leq l \leq n$, followed by all $n - l$ remaining transitions, for any $0 < \alpha < \beta < 1$, where α (resp. β) is the probability that the action of a low effort (resp. high effort) action by an agent results in a successful local outcome. This generalizes the prior result of Babaioff et al. [5] for the AND gate (a single transition from 0 agents to all n agents contracted) and the OR gate (all n transitions), and closes an important open question.

Considering the POU, we establish a tight bound of 2 for the OR technology, for all values of n, α and $\beta = 1 - \alpha$. The initial paper established this POU for the case of $n = 2$ agents only, while an extended version of the paper provides a bound of $n = 2.5$ for the general $n > 2$ case [6]. In addition, we establish that the POU is unbounded for the threshold technology for the general case of $k \geq 2, n \geq 2$. The initial paper established this result only for AND technology, and so our result closes this for the more general threshold case for any $0 < \alpha < \beta < 1$. In addition, we consider non-anonymous technology functions such as Majority-of-AND, Majority-of-OR, and AND-of-Majority, and study their transition behavior.

We believe that this work is an interesting step in extending the combinatorial agency model in a direction of interest for crowdsourcing [16,3,1,2]. Combinatorial agency is relevant to applications where neither the effort nor the individual outcome of each worker is observable. All that is observable is the ultimate success or failure. Perhaps the boundaries between individual contributions are hard to define, or workers prefer to hide individual contributions in some way (e.g., to protect their privacy.) Perhaps it is extremely costly, or even impossible, to determine the quality of the work performed by an individual worker when studied in isolation. The threshold technology seems natural in modeling crowdsourcing problems in which success requires getting enough suitable contributions.

Related Work. A characterization of the transition behavior was first conjectured for the Majority technology in Babaioff et al. [5], but almost all of the subsequent literature is restricted to read-once networks [7,8,11,12]. The combinatorial agency problem has also been studied under mixed Nash equilibria [7]. Babaioff et al. [8] study "free labor" and whether the principal can benefit from having certain agents reduce their effort level, even when this effort is free. The principal is hurt by free labor under the OR technology, because free labor can lead to free riding. Another variation allows the principal to audit some fraction of the agents, and discover their individual private action [10]. Some computational complexity results for identifying optimal contracts have also been developed. This problem is NP-hard for OR technology [11], and the difficulty is later shown to be a property of unobservable actions [12]. This is in contrast to the AND technology, which admits a polynomial time algorithm for computing the optimal contract. An FPTAS is developed for OR technology, and extended to almost all series-parallel technologies [11].

2   Model

In the combinatorial agency model, a principal employs a set of n self-interested agents. Each agent i has an action space A_i and a cost (of effort) associated with each action $c_i(a_i) \geq 0$ for every $a_i \in A_i$. We let $\boldsymbol{a}_{-i} = (a_1, \ldots, a_{i-1}, a_{i+1}, \ldots, a_n)$ denote the action profile of all other agents besides agent i. Similar to Babaioff et al. [5], we focus on a binary-action model. That is, agents either exert effort ($a_i = 1$) or do not exert effort ($a_i = 0$), and the cost function becomes c_i if $a_i = 1$ and 0 if $a_i = 0$. If agent i exerts effort, she succeeds with probability β_i. If agent i does not exert effort, she succeeds with probability α_i, where $0 < \alpha_i < \beta_i < 1$. We deal with the case of homogenous agents (e.g. $\beta_i = \beta$, $\alpha_i = \alpha$ and $c_i = c$ for all i), though some of the prior work deals with the case of heterogenous agents. Sometimes we use the additional assumption of [5], that $\beta = 1 - \alpha$, where $0 < \alpha < \frac{1}{2}$.

Completing the description of the technology is the *outcome function* f, which determines the success or failure of the overall project as a function of the success or failure of each agent. Let $\boldsymbol{x} = (x_1, \ldots, x_n)$, with $x_i \in \{0, 1\}$ to denote the success or failure of the action of agent i given its selected effort level. Following Babaioff et al. [5] we focus on a binary outcome setting, so that the outcome is 1 (= success) or 0 (= failure.) Given this, we study the following outcome functions:

1. AND technology: $f(x_1, x_2, \ldots, x_n) = \wedge_{i \in N} x_i$. In other words, the project succeeds if and only if all agents succeed in their tasks.
2. OR technology: $f(x_1, x_2, \ldots, x_n) = \vee_{i \in N} x_i$. In other words, the project succeeds if and only if at least one agent succeeds in her task.
3. Majority technology: $f(x) = 1$ if a majority of the x_i are 1. In other words, the project succeeds if and only if a majority of the agents succeed at their tasks.
4. Threshold technology: We can generalize the majority technology into a threshold technology, where $f(x) = 1$ if and only if at least k of the x_i are 1, e.g. at least k of the n agents succeed in their tasks.

In fact, the threshold technology is a generalization of the OR, AND and majority technologies, since the $k = 1$ case is equivalent to the OR technology, the $k = n$ case is equivalent to the AND technology, and the $k = \lceil \frac{n}{2} \rceil$ case is equivalent to the majority technology. It should be noted that the set of threshold technologies is exactly the set of threshold functions. It is easy to see that each of these outcome functions is *anonymous*, meaning that the outcome is invariant to a permutation on the agent identities.

Given outcome function f, and success probabilities α and β, then action profile $\boldsymbol{a}$ induces a probability $p(\boldsymbol{a}) \in [0, 1]$ with which the project will succeed. This is just

$$p(\boldsymbol{a}) = E_{\boldsymbol{x}}[f(\boldsymbol{x}) \mid \boldsymbol{x} \sim \boldsymbol{a}] \qquad (1)$$

where the local outcomes $\boldsymbol{x}$ are distributed according to α, β and as a result of the effort $\boldsymbol{a}$ by agents. Since p considers the combined effect of technology f, α and β, then we refer to p as the *technology function*.

The principal has a value v for a successful outcome and 0 for an unsuccessful outcome. Like [5], we assume that the principal is risk-neutral and seeks to maximize expected value minus expected payments to agents. The principal is unable to observe either the actions $\boldsymbol{a}$ or the (local) outcomes $\boldsymbol{x}$. The only thing the principal can observe is the success or failure of the overall project. Based on this, a *contract* specifies a payment $t_i \geq 0$ to each agent i when the project succeeds, with a payment of zero otherwise. The principal can pay the agents, but not fine them. It is convenient to include in a contract the set of agents that the principal intends to exert high effort; this is the set of agents that *will* exert high effort when the principal selects an appropriate payment function.

The utility to agent i under action profile $\boldsymbol{a}$ is $u_i(\boldsymbol{a}) = t_i \cdot p(\boldsymbol{a}) - c_i$ if the agent exerts effort, and $u_i(\boldsymbol{a}) = t_i \cdot p(\boldsymbol{a})$ otherwise. The principal's expected utility is $u(\boldsymbol{a}) = v \cdot p(\boldsymbol{a}) - \sum_{i \in N} t_i \cdot p(\boldsymbol{a})$. The principal's task is to design a contract so that its utility is maximized under an action profile $\boldsymbol{a}$ that is a Nash equilibrium. We make the same assumption as Babaioff et al. [5], that if there are multiple Nash equilibria (NE), the principal can contract for the best NE. The *social welfare* for an action profile $\boldsymbol{a}$ is given by $u(\boldsymbol{a}) + \sum_{i \in N} u_i(\boldsymbol{a}) = v \cdot p(\boldsymbol{a}) - \sum_{i \in N} c_i \cdot a_i$, with payments from the principal to the agents canceling out.

Throughout, we focus on outcome functions that are *monotonic*, so that $f(\boldsymbol{x}) = 1 \Rightarrow f(x'_1, \boldsymbol{x}_{-1}) = 1$ for $x'_1 \geq x_1$. Based on this, then the technology function p is also *monotonic* in the amount of effort exerted, that is for all i and all $\boldsymbol{a}_{-i} \in \{0,1\}^{n-1}$, $p(1, \boldsymbol{a}_{-i}) \geq p(0, \boldsymbol{a}_{-i})$. Similarly, a technology function p is *anonymous* if it symmetric with respect to the players. That is, it is anonymous if it only depends on the number of agents that exert effort and is indifferent to permutations of the joint action profile $\boldsymbol{a}$. This is true whenever the underlying outcome function is anonymous.

In the *non-strategic* variant of the problem, the principal can choose which agents exert effort and these agents need not be "motivated", the principal can simply bear their cost of exerting effort. Let S_a^* and S_{ns}^* denote the optimal set of agents to contract with in the agency case and the non-strategic case respectively. That is, these sets of agents are those that maximize the expected value to the principal net cost, first where the sets must be induced in a Nash equilibrium and second when they can be simply selected.

Definition 1. *[5] The Price of Unaccountability (POU) for an outcome function f is defined as the worst case ratio (over v, α and β) of the social welfare in the non-strategic case and the social welfare of the agency case:*

$$POU(f) = \sup_{v>0, \alpha, \beta} \frac{p(S_{ns}^*(v)) \cdot v - \sum_{i \in S_{ns}^*(v)} c_i}{p(S_a^*(v)) \cdot v - \sum_{i \in S_a^*(v)} c_i}, \tag{2}$$

where p is the technology function induced by f, α and β, with $0 < \alpha < \beta < 1$.

In studying the POU, it becomes useful to characterize the transition behavior for a technology. The transition behavior is, for a fixed technology function p, the optimal set of contracted agents as a function of the principal's value v. We know that when $v = 0$ it is optimal to contract with 0 agents and likewise, as $v \to \infty$, it is optimal to contract with all agents. However, we would like to understand what are the optimal sets of agents contracted between these two extreme cases. There are, in fact, two sets of transitions, for both the agency and the non-strategic case. For anonymous technologies, there can be at most n transitions in either case, since the number of agents in the optimal contract is (weakly) monotonically increasing in the principal's value. We seek to understand how many transitions occur, and the nature of each "jump" (i.e. the change in number of agents contracted with at a transition.)

We also consider compositions of these technologies such as majority-of-AND, Majority-of-OR, and AND-of-Majority. These technologies are no longer anonymous. With non-anonymous technologies, one needs to specify the contracted set of agents, in addition to the number of agents contracted. In considering composition of anonymous technologies, we assume we are composing identical technology functions, e.g. each AND gate in the majority-of-AND technology consists of the same number of agents.

3   Transition Behavior of the Optimal Contract

Below we will characterize the transition behavior of the threshold technology. We show that there exists an $l \in \{1, ..., n\}$ such that the first transition is from 0 to l agents followed by all remaining transitions. This result holds for any value of α, β such that $0 < \alpha < \beta < 1$.

Our proof builds on the framework of Babaioff et al. [5]. In Babaioff et al., it was shown that the AND technology always exhibits "increasing returns to scale" (IRS) and the OR technology always exhibits "decreasing returns to scale" (DRS). It was also shown that any anonymous technology that exhibits IRS has a single transition from 0 to n agents for the optimal contract in the non-strategic case and that any anonymous technology that exhibits DRS exhibits all n transitions in the non-strategic case. Similar to the non-strategic case, it was shown in Babaioff et al. that the AND technology always exhibits overpayment (OP), in the agency case, where the OP condition guarantees a single transition from 0 to n, and the OR technology always exhibits increasing relative marginal payment (IRMP), in the agency case, where the IRMP condition guarantees all n transitions.

We show that the threshold technology exhibits IRS up to a certain number of agents contracted and DRS thereafter, which gives the transition characterization for the non-strategic case. Likewise, we show that the threshold function exhibits OP to a point and IRMP in the agency case, which is sufficient to give the transition characterization for the agency case. Our analysis is new, in the sense that we consider the possibility that a single technology can exhibit IRS up to a certain number of agents contracted, followed by DRS and likewise, that it can exhibit OP up to a certain number of agents contracted, followed by IRMP. Babaioff et al. only considered the possibility a function exhibits either IRS or DRS, and likewise, either OP or IRMP. In addition to this insight, we use properties of (log) convex functions to establish this result. We state our main theorems below:

Theorem 1. For any threshold technology (any k, n, c, α and β) in the non-strategic case, there exists an $1 \leq l_{ns} \leq n$ where, such that the first transition is from 0 to l_{ns} agents, followed by all remaining $n - l_{ns}$ transitions.

Theorem 2. For any threshold technology (any k, n, c, α and β) in the agency case, there exists an $1 \leq l_a \leq l_{ns}$ such that the first transition is from 0 to l_a agents, followed by all remaining $n - l_a$ transitions.

The following observations give us the optimal payment rule for any technology and establish a monotonic property for the optimal contract as a function of v.

Definition 2. *[5] The marginal contribution of agent i for a given a_{-i} is denoted by $\Delta_i(\boldsymbol{a}_{-i}) = p(1, \boldsymbol{a}_{-i}) - p(0, \boldsymbol{a}_{-i})$, and is the difference in the probability of success of the technology function when agent i exerts effort and when she does not.*

For anonymous technologies, if exactly j entries in $\boldsymbol{a}_{-i}$ are 1, then $\Delta_i = p_{j+1} - p_j$, where p_j is the probability of success when exactly j agents exert effort. Since p is strictly monotone, we have $\Delta_i > 0$ for all i.

Remark 1. [5] The best contracts (from the principal's point of view) that induce the action profile $\boldsymbol{a} \in \{0, 1\}^n$ as a Nash equilibrium are $t_i = 0$ when the project is unsuccessful and $t_i = \frac{c_i}{\Delta_i(\boldsymbol{a}_{-i})}$ when the project succeeds and the principal requests effort $a_i = 1$ from agent i.

The following lemma gives a set of sufficient conditions for an anonymous technology to have a first transition from 0 to l, for some $l \in \{1, ..., n\}$, followed by all remaining $n - l$ transitions. This lemma holds for both the non-strategic case (where $Q_i = i \cdot c$) and the agency case (where $Q_i = \frac{i \cdot c}{\Delta_i}$). We view this lemma as a generalization of Theorem 9 from [5] and it follows a similar proof structure in that it uses Lemmas 12 and 13 from [5] that relate the principal's utility of contracting with a fixed number of agents to the Q_i values. This lemma states that as long as a technology function exhibits OP up to a certain number of agents contracted followed by IRMP, then the transition behavior involves a first transition from 0 to l, for some $l \in \{1, ..., n\}$, followed by all remaining $n - l$ transitions.

Lemma 1. *Any anonymous technology function that satisfies:*

1. $\frac{Q_i}{Q_l} > \frac{p_i - p_0}{p_l - p_0}$ *for all $i \neq l$*
2. $\frac{Q_{l+1} - Q_l}{p_{l+1} - p_l} > \frac{Q_l}{p_l - p_0}$
3. $\frac{Q_{i+1} - Q_i}{p_{i+1} - p_i} > \frac{Q_i - Q_{i-1}}{p_i - p_{i-1}}$ *for all $i > l$*

for some $l \in \{1, ..., n\}$ has a first transition from 0 to l and then all $n - l$ subsequent transitions, where Q_i is defined appropriate for the non-strategic case or the agency case.

Now that we have established a set of sufficient conditions for an anonymous technology to exhibit a first transition from 0 to l, followed by all remaining transitions (for either the non-strategic case or the agency case), we interpret what these conditions are for the non-strategic case.

Lemma 2. *Any anonymous technology that has a probability of success function that satisfies:*

1. $\frac{p_i - p_0}{i} > \frac{p_{i-1} - p_0}{i-1}$ *for all $2 \leq i \leq l$ and $\frac{p_i - p_0}{i} < \frac{p_{i-1} - p_0}{i-1}$ for all $i > l$*
2. $\frac{1}{p_{i+1} - p_i} > \frac{1}{p_i - p_{i-1}}$ *for all $i > l$*

for some $l \in \{1, ..., n\}$ has a first transition from 0 to l and then all $n - l$ subsequent transitions for the nonstrategic version of the problem.

In establishing that the threshold technology satisfies the conditions outlined in Lemma 2, it becomes useful to define a property of the probability of success function.

Definition 3. *We say that a probability of success p for a particular technology is unimodal if it satisfies one of three alternatives:*

1. *$p_i - p_{i-1} > p_{i-1} - p_{i-2}$ for all $2 \leq i \leq j$ and $p_i - p_{i-1} < p_{i-1} - p_{i-2}$ for all $i > j$*
2. *$p_i - p_{i-1} > p_{i-1} - p_{i-2}$ for all $2 \leq i \leq n$*
3. *$p_i - p_{i-1} < p_{i-1} - p_{i-2}$ for all $2 \leq i \leq n$*

Let $f(i) = \frac{p_i - p_0}{i}$. This function is useful to consider, because in order to establish the first condition of Lemma 2, we need to show that $f(i)$ is unimodal.

Lemma 3. *If the probability of success function is unimodal over the set $\{1, ..., n\}$, then we know that $f(i)$ is also unimodal.*

Corollary 1. *For any anonymous technology function (p, c) that has a unimodal probability of success, there exists an $1 \leq l \leq n$ such that the first transition in the non-strategic case is from 0 to l agents (where l is the smallest value that satisfies $\frac{p_l - p_0}{l} > \frac{p_{l+1} - p_0}{l+1}$) followed by all remaining $n - l$ transitions.*

Therefore, it suffices to show that p is unimodal in order to establish that the technology (p, c) exhibits a first transition from 0 to l, for some $l \in \{1, ..., n\}$, followed by all remaining $n - l$ transitions, in the non-strategic case.

Lemma 4. *The probability of success function for any threshold technology is unimodal.*

The characterization of the transition behavior of the threshold technology in the non-strategic case follows from Lemmas 2, 3, and 4.

Theorem 1. *For any threshold technology (any k, n, c, α and β) in the non-strategic case, there exists an $1 \leq l_{ns} \leq n$ where, such that the first transition is from 0 to l_{ns} agents, followed by all remaining $n - l_{ns}$ transitions.*

Now that we have characterized the transition behavior of the threshold technology, for any k, in the non-strategic case, we focus on establishing the conditions of Lemma 1, for the agency case. The following lemma is used to show that the first condition in Lemma 1 is satisfied by the threshold technology.

Lemma 5. *The discrete valued function, $\frac{Q_i}{p_i - p_0}$, is convex.*

Lemma 6. *There exists a value of $1 \leq l_a \leq n$ such that $\frac{Q_i}{Q_{l_a}} > \frac{p_i - p_0}{p_{l_a} - p_0}$ for all $i \neq l_a$.*

Since there exists an l_a such that $\frac{Q_i}{p_i - p_0} > \frac{Q_{i+1}}{p_{i+1} - p_0}$ for all $1 \leq i < l_a$ and $\frac{Q_i}{p_i - p_0} < \frac{Q_{i+1}}{p_{i+1} - p_0}$ for all $l_a \leq i < n$, we have the following corollary.

Corollary 2. *We have* $\frac{Q_{l_a+1}-Q_{l_a}}{p_{l_a+1}-p_{l_a}} > \frac{Q_{l_a}}{p_{l_a}-p_0}$, *where* $1 \le l_a \le n$ *satisfies* $\frac{Q_i}{Q_{l_a}} > \frac{p_i-p_0}{p_{l_a}-p_0}$ *for all* $i \ne l_a$.

Lemma 7. *We have* $\frac{Q_{i+1}-Q_i}{p_{i+1}-p_i} > \frac{Q_i-Q_{i-1}}{p_i-p_{i-1}}$ *for all* $i > l_a$ *where* l_a *is the smallest value such that* $\frac{Q_{l_a}}{p_{l_a}-p_0} < \frac{Q_{l_a+1}}{p_{l_a+1}}$.

Lemmas 1, 6, and 7 and Corollary 2 establish the following result.

Theorem 2. *For any threshold technology (any k, n, c, α and β) in the agency case, there exists an $1 \le l_a \le l_{ns}$ such that the first transition is from 0 to l_a agents, followed by all remaining $n - l_a$ transitions.*

These results beg the question, how do the values of l_a and l_{ns} relate to k? Below we give the trend in transition behavior as a function of β, when $\alpha = 0$. This remark holds for both the non-strategic and agency cases. We also provide a technical lemma regarding the value of l_a and l_{ns} as $\alpha \to 0$. This lemma is used in the next section to establish an unbounded POU for the threshold function.

Remark 2. For any threshold technology with fixed $k \ge 2$, n, c and $\alpha = 0$, we have that $l = k$ for β close enough to 1 and $l = n$ for β close enough to 0.

Lemma 8. *As $\alpha \to 0$, we know that $k \le l_a \le l_{ns}$, where l_a is the first transition in the agency case and l_{ns} is the first transition in the non-strategic case.*

We note that it is not always the case that $l_a \ge k$. For example, when $\alpha = \frac{1}{2} - \epsilon$, $\beta = \frac{1}{2} + \epsilon$ and ϵ is sufficiently small, we have all n transitions, regardless of the value of k.

4   Price of Unaccountability

Lemma 9. *[5] For any technology function, the price of unaccountability is obtained at some value v which is a transition point, of either the agency or the non-strategic cases.*

We are able to improve upon this result, for the OR technology, which is needed to establish Theorem 4.

Lemma 10. *For the OR technology, the price of unaccountability occurs at a transition in the agency case, as opposed to a transition in the non-strategic case.*

The following theorem is a result of Babaioff et al. [5], where they derive the price of unaccountability for AND technology where $\beta = 1 - \alpha$.

Theorem 3. *[5] For the AND technology with $\alpha = 1 - \beta$, the price of unaccountability occurs at the transition point of the agency case and is POU $= (\frac{1}{\alpha} - 1)^{n-1} + (1 - \frac{\alpha}{1-\alpha})$.*

Remark 3. [5] The price of unaccountability for the AND technology is not bounded. More specifically, $POU \to \infty$ as $\alpha \to 0$ and $POU \to \infty$ as $\beta \to 0$.

Babaioff et al. [5] show that the Price of Unaccountability for the OR technology is bounded by 2 for exactly 2 agents and give an upper bound of 2.5 for the general case [6], when $\beta = 1 - \alpha$. We extend these results for the $\beta = 1 - \alpha$ case and show that the Price of Unaccountability is bounded above by 2 for any OR technology (i.e. for all n). This result is tight, namely, as $\alpha \to 0$, $POU \to 2$. We suspect that these results hold for the more general $0 < \alpha < \beta < 1$ case, but we have been unable to prove it for all values of α, β.

Theorem 4. *The POU for the OR technology is bounded by 2 for all $\alpha, \beta = 1 - \alpha$ and n.*

In contrast to the OR technology, we show that the POU for the threshold technology with $k \geq 2$ is unbounded. This result holds for any $0 < \alpha < \beta < 1$.

Theorem 5. *The Price of Unaccountability for the threshold technology is not bounded for all values of $k \geq 2$ and n. More specifically, when $\alpha \to 0$, $POU \to \infty$.*

5   Composition of Anonymous Technologies

5.1   Majority-of-ANDs

We prove the transition behavior for the majority-of-AND technology in the non-strategic case. These results hold for the more general threshold-of-ORs case. For the following assume that in the majority-of-AND technology, the majority gate contains q AND gates, each with m agents. This builds on a conjecture of Babaioff et al. who conjecture the following behavior for both the non-strategic and agency cases. We are unable to prove the transition behavior for the agency case.

Lemma 11. *If the principal decides to contract with $j \cdot m + a$ agents for some $j \in Z^+$ and some $0 \leq a < m$, the probability of success is maximized by fully contracting j AND gates and contracting with a remaining agents on the same AND gate.*

Lemma 12. *For any principal's value v, the optimal contract involves a set of fully contracted AND gate.*

Theorem 6. *The transition behavior for the majority-of-AND technology in the non-strategic case has a first transition to l fully contracted AND gates, where $1 \leq l \leq n$, followed by each subsequent transition of fully contracted AND gates.*

While we are unable to characterize the transition behavior for the majority-of-AND technology in the agency case, we know that the first transition in the agency case must involve contracting at most $l \cdot m$ agents. This allows us to prove that the Price of Unaccountability is unbounded.

Theorem 7. *The Price of Unaccountability is unbounded for the majority-of-AND technology.*

5.2 Majority of ORs

We will characterize the transition behavior for the non-strategic case of the majority of ORs below. In what follows, we assume that each OR gate has j agents and there are m of them comprising a majority function (i.e. $n = j \cdot m$). We also assume that $k = \lceil \frac{m}{2} \rceil$. In considering the majority-of-OR case, we further assume that $\beta = 1 - \alpha$ and $0 < \alpha < \frac{1}{2}$.

Lemma 13. *Consider an integer i such that $i = a \cdot m + b$, where $0 \le b < m$. Fixing i, the probability of success for a majority-of-ORs function is maximized when $a + 1$ agents are contracted on each of b OR gates and a agents are contracted on each of $n - b$ OR gates.*

The following lemma gives the complete transition behavior in the majority-of-OR technology in the nonstrategic case.

Lemma 14. *The first transition for the non-strategic case of the majority-of-OR technology jumps from contracting with 0 agents to l agents, where $1 \le l \le k$, followed by all remaining transitions, where the transitions proceed in such a way so that no OR gate has more than 1 more agent contracted as compared to any other OR gate.*

We conjecture that a similar transition behavior holds in the agency case, but we have thus far been unable to prove it. However, we do know that as $\alpha \to 0$, the first transition jumps to k. This is enough to determine that the POU is unbounded.

Theorem 8. *The Price of Unaccountability is unbounded for the majority-of-OR technology.*

5.3 AND of Majority

In what follows, we will also characterize the transition behavior of AND-of-majorities. These results hold for the more general AND-of-threshold's. We give a result from [5] that allows for the characterization of the transition behavior of AND-of-majority. Let g and h be two Boolean functions on disjoint inputs with any cost vectors, and let $f = g \wedge h$. An optimal contract S for f for some v is composed of some agents from the g-part (denoted by the set R) and some agents from the h-part (denoted by the set T).

Lemma 15. *[5] Let S be an optimal contract for $f = g \wedge h$ on v. Then, T is an optimal contract for h on $v \cdot t_g(R)$, and R is an optimal contract for g on $v \cdot t_h(T)$.*

The previous lemma gives us a characterization of the transition behavior in the AND-of-majorities technology. The statement of this result is analogous to the result given in [5] for the AND-of-ORs technology. Since the previous lemma holds for both the non-strategic and agency variations of the problem, the following theorem holds for both the non-strategic and agency variations of the problem.

Theorem 9. *Let h be an anonymous majority technology and let $f = \bigwedge_{j=1}^{n_c}$ be the AND of majority technology that is obtained by a conjunction of n_c of these majority technology functions on disjoint inputs. Then for any value v, an optimal contract contracts with the same number of agents in each majority component.*

Theorem 9 gives us a complete characterization of the transition behavior in the AND-of-majorities technology for both the non-strategic and the agency cases. Since we know that the first transition in both the agency and non-strategic cases for the AND-of-majority technology occurs to a value greater than 1, we have the following result.

Theorem 10. *The Price of Unaccountability is unbounded for the AND-of-majority technology.*

References

1. http://answers.yahoo.com/
2. https://networkchallenge.darpa.mil/
3. http://www.netflixprize.com/
4. Alchian, A.A., Demsetz, H.: Production, information costs, and economic organization. The American Economic Review 62(5), 777–795 (1972)
5. Babaioff, M., Feldman, M., Nisan, N.: Combinatorial agency. In: ACM Conference on Electronic Commerce, pp. 18–28 (2006)
6. Babaioff, M., Feldman, M., Nisan, N.: Combinatorial agency. Full Version (2006)
7. Babaioff, M., Feldman, M., Nisan, N.: Mixed strategies in combinatorial agency. In: Spirakis, P.G., Mavronicolas, M., Kontogiannis, S.C. (eds.) WINE 2006. LNCS, vol. 4286, pp. 353–364. Springer, Heidelberg (2006)
8. Babaioff, M., Feldman, M., Nisan, N.: Free-riding and free-labor in combinatorial agency. In: Mavronicolas, M., Papadopoulou, V.G. (eds.) SAGT 2009. LNCS, vol. 5814, pp. 109–121. Springer, Heidelberg (2009)
9. Bolton, P., Dewatripont, M.: Contract Theory. MIT Press, Cambridge (2005)
10. Eidenbenz, R., Schmid, S.: Combinatorial agency with audits. In: IEEE International Conference on Game Theory for Networks, GameNets (2009)
11. Emek, Y., Feldman, M.: Computing optimal contracts in series-parallel heterogeneous combinatorial agencies. In: Leonardi, S. (ed.) WINE 2009. LNCS, vol. 5929, pp. 268–279. Springer, Heidelberg (2009)
12. Emek, Y., Haitner, I.: Combinatorial agency: The observable action model (2006) (manuscript)
13. Hermalin, B.E.: Toward an economic theory of leadership: Leading by example. The American Economic Review 88(5), 1188–1206 (1998)
14. Holmstrom, B.: Moral hazard in teams. The Bell Journal of Economics 13(2), 324–340 (1982)
15. Laffont, J.-J., Martimort, D.: The Theory of Incentives: The Principal-Agent Model. Princeton University Press, Princeton (2001)
16. von Ahn, L., Dabbish, L.: Labeling images with a computer game. In: Proceedings of the 2004 Conference on Human Factors in Computing Systems (CHI), pp. 319–326 (2004)

Lower Bound for Envy-Free and Truthful Makespan Approximation on Related Machines[*]

Lisa Fleischer and Zhenghui Wang

Dept. of Computer Science, Dartmouth College, Hanover, NH 03755, USA
{lkf,zhenghui}@cs.dartmouth.edu

Abstract. We study problems of scheduling jobs on related machines so as to minimize the makespan in the setting where machines are strategic agents. In this problem, each job j has a length l_j and each machine i has a private speed t_i. The running time of job j on machine i is $t_i l_j$. We seek a mechanism that obtains speed bids of machines and then assign jobs and payments to machines so that the machines have incentive to report true speeds and the allocation and payments are also envy-free. We show that

1. A deterministic envy-free, truthful, individually rational, and anonymous mechanism cannot approximate the makespan strictly better than $2 - 1/m$, where m is the number of machines. This result contrasts with prior work giving a deterministic PTAS for envy-free anonymous assignment and a distinct deterministic PTAS for truthful anonymous mechanism.
2. For two machines of different speeds, the unique deterministic scalable allocation of any envy-free, truthful, individually rational, and anonymous mechanism is to allocate all jobs to the quickest machine. This allocation is the same as that of the VCG mechanism, yielding a 2-approximation to the minimum makespan.
3. No payments can make any of the prior published monotone and locally efficient allocations that yield better than an m-approximation for $Q||C_{max}$ [1,3,5,9,13] a truthful, envy-free, individually rational, and anonymous mechanism.

Keywords: Mechanism Design, Incentive Compatible, Envy-Free, Makespan Approximation.

1 Introduction

We study problems of scheduling jobs on related machines so as to minimize the makespan (i.e. $Q||C_{max}$) in a strategic environment. Each job j has a length l_j and each machine i has a *private* speed t_i, which is only known by that machine. The speed t_i is the time it takes machine i to process one unit length of a job — t_i is the inverse of the usual sense of speed. The running time of job j on machine i is $t_i l_j$. A single job cannot be performed by more than one machine

---

[*] This work was supported in part by NSF grants CCF-0728869 and CCF-1016778.

G. Persiano (Ed.): SAGT 2011, LNCS 6982, pp. 166–177, 2011.
© Springer-Verlag Berlin Heidelberg 2011

(indivisible), but multiple jobs can be assigned to a single machine. The *workload* of a machine is the total length of jobs assigned to that machine and the *cost* is the running time of its workload. The scheduler would like to schedule jobs to complete in minimum time, but has to pay machines to run jobs. The *utility* of a machine is the difference between the payment to the machine and its cost. The mechanism used by the scheduler asks the machines for their speeds and then determines an allocation of jobs to machines and payments to machines. Ideally, the mechanism should be fair and efficient. To accomplish this, the following features of mechanism are desirable.

Individually rational. A mechanism is *individually rational* (IR), if no agent gets negative utility when reporting his true private information, since a rational agent will refuse the allocation and payment if his utility is negative. In order that each machine accepts its allocation and payment, the payment to a machine should exceed its cost of executing the jobs.

Truthful. A mechanism is *truthful* or *incentive compatible* (IC), if each agent maximizes his utility by reporting his true private information. Under truth-telling, it is easier for the designer to design and analyze mechanisms, since agents' dominant strategies are known by the designer. In a truthful mechanism, an agent does not need to compute the strategy maximizing his utility, since it is simpler to report his true information.

Envy-free. A mechanism is *envy-free* (EF), if no agent can improve his utility by switching his allocation and payment with that of another. Envy-freeness is a strong concept of fairness [10,11]: each agent is happiest with his allocation and payment.

Prior work on envy-free mechanisms for makespan approximation problems assumes that all machine speeds are public knowledge [6,15]. We assume that the speed of a machine is private information of that machine. This assumption makes it harder to achieve envy-freeness. Only if the mechanism is also truthful, can the mechanism designer ensure that the allocation is truly envy-free.

In this paper, we prove results about *anonymous* mechanisms. A mechanism is anonymous, roughly speaking, if when two agents switch their bids, their allocated jobs and payments also switch. This means the allocation and payments depend only on the agents' bids, not on their names. Anonymous mechanisms are of interest in this problem for two reasons. On the one hand, to the best of our knowledge, all polynomial-time mechanisms for $Q||C_{max}$ are anonymous [1,3,5,9,15]. On the other hand, in addition to envy-freeness, anonymity can be viewed as an additional characteristic of fairness [4].

We also study *scalable* allocations. Scalability means that multiplying the speeds by the same positive constant does not change the allocation. Intuitively, the allocation function should not depend on the "units" in which the speed are measured, and hence scalability is a natural notion. But allocations based on rounded speeds of machines are typically not scalable [1,5,13].

The truthful mechanisms and envy-free mechanisms for $Q||C_{max}$ are both well-understood. There is a payment scheme to make an allocation truthful if and only if the allocation is *monotone decreasing* [3]. For $Q||C_{max}$, an allocation is monotone decreasing if no machine gets more workload by bidding a slower speed than its true speed. On the other hand, a mechanism for $Q||C_{max}$ can be envy-free if and only if its allocation is *locally efficient* [15]. An allocation is locally efficient if a machine never gets less workload than a slower one.

The complexity of truthful mechanisms and, separately, envy-free mechanisms have been completely settled. $Q||C_{max}$ is strongly NP-hard, so there is no FPTAS for this problem, assuming P $\neq$ NP. On the other hand, there is a deterministic monotone PTAS [5] and a distinct deterministic locally efficient PTAS [15]. This implies the existence of truthful mechanisms and distinct envy-free mechanisms that approximate the makespan arbitrarily closely. However, neither of these payment functions make the mechanisms both truthful and envy-free.

The VCG mechanism for $Q||C_{max}$ is truthful, envy-free, individually rational, and anonymous [8]. However, since the VCG mechanism maximizes the social welfare (i.e. minimizing the total running time), it always allocates all jobs to the quickest machines, yielding a m-approximation of makespan for m machines in the worst case. So a question is whether there is a truthful, envy-free, individually rational and anonymous mechanism that approximates the makespan better than the VCG mechanism. Since there already exists many allocation functions that are both monotone and locally efficient, one natural step to answer this question could be checking whether some of these allocation functions admit truthful and envy-free payments.

Our Results. We show that

1. A deterministic envy-free, truthful, individually rational, and anonymous mechanism cannot approximate the makespan strictly better than $2 - 1/m$, where m is the number of machines. (Section 3). This result contrasts with prior results [5,15] discussed above.
2. For two machines of different speeds, the unique deterministic scalable allocation of any envy-free, truthful, individually rational, and anonymous mechanism is to allocate all jobs to the quickest machine. (Section 5). This allocation is the same as that of the VCG mechanism, yielding a 2-approximation of makespan for this case.
3. No payments can make any of the prior published monotone and locally efficient allocations that yield better than an m-approximation for $Q||C_{max}$ [1,3,5,9,13] a truthful, envy-free, individually rational, and anonymous mechanism.

Related Work. Hochbaum and Shmoys [12] give a PTAS for $Q||C_{max}$. Andelman, Azar, and Sorani [1] give a 5-approximation deterministic truthful mechanism. Kovács improves the approximation ratio to 3 [13] and then to 2.8 [14]. Randomization has been successfully applied to this problem. Archer and Tardos [3] give a 3-approximate randomized mechanism, which is improved to 2 in [2]. Dhangwatnotai [9] et. al. give a monotone randomized PTAS. All these randomized mechanisms are truthful-in-expectation. However, we can show that no payment function can form a truthful, envy-free, individually rational and anonymous mechanism with any allocation function of these mechanisms. We give a proof for a deterministic allocation [13] in Section 5 and proofs for other allocations are similar to it.

When players have different finite valuation spaces, it is known that a monotone and locally efficient allocation function may not admit prices to form a simultaneously truthful and envy-free mechanism for allocating goods among players [7]. In this paper, we consider mechanisms where all players have identical infinite valuation spaces.

Cohen et. al. [8] study the truthful and envy-free mechanisms on combinatorial auctions with additive valuations where agents have a upper capacity on the number of items they can receive. They seek truthful and envy-free mechanisms that maximize social welfare and show that VCG with Clarke Pivot payments is envy-free if agents' capacities are all equal. Their result can be interpreted in our setting by viewing that each agent has the same capacity n and the valuation of each agent is the reverse of its cost. So their result implies that the VCG mechanism for $Q||C_{max}$ is truthful and envy-free; but the VCG mechanism does not give a good approximation guarantee for makespan.

2   Preliminaries

There are m machines and n jobs. Each agent will report a bid $b_i \in \mathbb{R}$ to the mechanism. Let t denote the vector of true speeds and b the vector of bids.

A mechanism consists of a pair of functions (w, p). An *allocation* w maps a vector of bids to a vector of allocated workload, where $w_i(b)$ is the workload of agent i. For all bid vectors b, $w(b)$ must correspond to a valid job assignment. An allocation w is called *scalable* if $w_i(b) = w_i(c \cdot b)$ for all bid vectors b, all $i \in \{1 \ldots m\}$ and all scalars $c > 0$. A *payment* p maps a vector of bids to a vector of payments, i.e. $p_i(b)$ is the payment to agent i.

The *cost* machine i incurs by the assigned jobs is $t_i w_i(b)$. Machine i's private value t_i measures its cost per unit work. Each machine i attempts to maximize its *utility*, $u_i(t_i, b) := p_i(b) - t_i w_i(b)$.

The *makespan* of allocation $w(b)$ is defined as $\max_i w_i(b) \cdot t_i$. A mechanism (w, p) is *c-approximate* if for all bids b and values t, the makespan of the allocation given by w is within c times the makespan of the optimal allocation, i.e.,

$$\max_i w_i(b) \cdot t_i \leq c \cdot OPT(t),$$

where $OPT(t)$ is the minimum makespan for machines with speeds t .

Vector b is sometimes written as (b_i, b_{-i}), where b_{-i} is the vector of bids, not including agent i. A mechanism (w, p) is *truthful* or *incentive compatible*, if each agent i maximizes his utility by bidding his true value t_i, i.e., for all agent i, all possible t_i, b_i and b_{-i},

$$p_i(t_i, b_{-i}) - t_i w_i(t_i, b_{-i}) \geq p_i(b_i, b_{-i}) - t_i w_i(b_i, b_{-i})$$

A mechanism (w, p) is *individually rational*, if agents who bid truthfully never incur a negative utility, i.e. $u_i(t_i, (t_i, b_{-i})) \geq 0$ for all agents i, true value t_i and other agents' bids b_{-i}.

A mechanism (w, p) is *envy-free* if no agent wishes to switch his allocation and payment with another. For all $i, j \in \{1, \ldots, m\}$ and all bids b,

$$p_i(b) - b_i w_i(b) \geq p_j(b) - b_i w_j(b).$$

Notice that we use bids b instead of the true speeds t in this definition, because a mechanism can determine the envy-free allocation only based on the bids. However, a mechanism can ensure the outcome is envy-free, only if it is also truthful.

A mechanism (w, p) is anonymous if for every bid vector $b = (b_1, \ldots, b_m)$, every k such that b_k is unique and every $l \neq k$,

$$w_l(\ldots, b_{k-1}, b_l, b_{k+1}, \ldots, b_{l-1}, b_k, b_{l+1}, \ldots) = w_k(b)$$

and

$$p_l(\ldots, b_{k-1}, b_l, b_{k+1}, \ldots, b_{l-1}, b_k, b_{l+1}, \ldots) = p_k(b).$$

The condition that b_k is unique is important, because in some case the mechanism may have to allocate jobs of different lengths to agents with the same bids. If mechanism (w, p) is anonymous and the bid of an agent is unique, the workload of that agent stays the same no matter how that agent is indexed. So we can write $w_i(b_i, b_{-i})$ simply as $w(b_i, b_{-i})$ for unique b_i to represent the workload of agent i. Similarly, we can write $p_i(b_i, b_{-i})$ simply as $p(b_i, b_{-i})$ for unique b_i.

Characterization of Truthful Mechanisms

Lemma 1 ([3]). *The allocation $w(b)$ admits a truthful payment scheme if and only if w is monotone decreasing, i.e., $w_i(b_i', b_{-i}) \leq w_i(b_i, b_{-i})$ for all $i, b_{-i}, b_i' \geq b_i$. In this case, the mechanism is truthful if and only if the payments satisfy*

$$p_i(b_i, b_{-i}) = h_i(b_{-i}) + b_i w_i(b_i, b_{-i}) - \int_0^{b_i} w_i(u, b_{-i})\, du, \qquad \forall i \qquad (1)$$

where the h_is can be arbitrary functions.

By Lemma 1, the only flexibility in designing the truthful payments for allocation w is to choose the terms $h_i(b_{-i})$. The utility of truth-telling agent i is $h_i(b_{-i}) - \int_0^{b_i} w_i(u, b_{-i})du$, because his cost is $t_i w_i(t_i, b_{-i})$, which cancels out the second term in the payment formula. Thus, in order to make the mechanism individually rational, the term $h_i(b_{-i})$ should be at least $\int_0^{b_i} w_i(u, b_{-i})\, du$ for any b_i. Since b_i can be arbitrarily large, h_i should satisfy

$$h_i(b_{-i}) \geq \int_0^{\infty} w_i(u, b_{-i})\, du, \qquad \forall i, b_{-i}. \qquad (2)$$

Characterization of envy-free mechanisms. An allocation function w is *envy-free implementable* if there exists a payment function p such that the mechanism $M = (w, p)$ is envy-free. An allocation function w is *locally efficient* if for all bids b, and all permutations π of $\{1, \cdots, m\}$,

$$\sum_{i=1}^{m} b_i \cdot w_i(b) \leq \sum_{i=1}^{m} b_i \cdot w_{\pi(i)}(b).$$

Lemma 2 ([15]). *Allocation w is envy-free implementable if and only if w is locally efficient.*

The proof of sufficiency constructs a payment scheme that ensures the envy-freeness for any locally efficient allocation w. Specifically, assuming $b_1 \geq b_2 \geq \ldots \geq b_m$, the payments for related machines are the following:

$$p_i(b) = \begin{cases} b_1 \cdot w_1(b) & \text{for } i = 1 \\ p_{i-1}(b) + b_i \cdot (w_i(b) - w_{i-1}(b)) & \text{for } i \in \{2, \ldots, m\} \end{cases}$$

These payments are not truthful payments, since $p_1(b)$ is clearly not in the form of (1). But the set of envy-free payments is a convex polytope for fixed w, since payments satisfying linear constraints $\forall i, j \quad p_i(b) - b_i w_i(b) \geq p_j(b) - b_i w_j(b)$ are envy-free. So there could be other payments that are both envy-free and truthful.

3   Lower Bound on Anonymous Mechanisms

In this section, we will prove an approximation lower bound for truthful, envy-free, individually rational, and anonymous mechanisms.

Theorem 1. *Let $M = (w, p)$ be a deterministic, truthful, envy-free, individually rational, and anonymous mechanism. Then M is not c-approximate for $c < 2 - \frac{1}{m}$.*

Since the only flexibility when designing payments in a truthful mechanism is to choose the h_is, we need to know what kind of h_is are required for envy-free anonymous mechanisms. The following two lemmas give necessary conditions on h_is.

Lemma 3. *If a mechanism (w, p) is both truthful and anonymous, then there is a function h such that $h_i(v) = h(v)$ in (1) for all bid vector $v \in \mathbb{R}_+^{m-1}$ and machines i.*

Proof. Let β be a real number such that $\beta < \min_j v_j$. For all $i \in \{1, \ldots, m-1\}$, define vector $b^{(i)} = (v_1, \ldots, v_{i-1}, \beta, v_i, \ldots, v_{m-1})$, $b'^{(i)} = (v_1, \ldots, v_i, \beta, v_{i+1}, \ldots, v_{m-1})$. Since $v = b^{(i)}_{-i} = b'^{(i)}_{-(i+1)}$ and M is anonymous, it must be that $p_i(b^{(i)}) = p_{i+1}(b'^{(i)})$ and $w_i(b^{(i)}) = w_{i+1}(b'^{(i)})$. Since $\alpha < v_j$ for any $0 < \alpha < \beta, j \in$

$\{1 \ldots m-1\}$, we also have $w_i(\alpha, v) = w_{i+1}(\alpha, v)$ by anonymity. Thus, for truthful payments, we have

$$h_i(v) = p_i(b^{(i)}) - \beta w_i(b^{(i)}) + \int_0^\beta w_i(\alpha, v) \, d\alpha$$

$$= p_{i+1}(b'^{(i)}) - \beta w_{i+1}(b'^{(i)}) + \int_0^\beta w_{i+1}(\alpha, v) \, d\alpha = h_{i+1}(v). \qquad \square$$

Lemma 4. *Let $L = \sum_k l_k$. If mechanism $M = (w, p)$ is truthful, envy-free, and anonymous, then*

$$h(t_{-i}) - h(t_{-j}) \leq L \cdot t_i + (t_j - t_i) w_i(t), \qquad (3)$$

for all $t \in \mathbb{R}_+^m$ and $i, j \in \{1, \ldots, m\}$.

Proof. If machine j does not envy machine i, then $p_j(t) - t_j w_j(t) \geq p_i(t) - t_j w_i(t)$. Using (1) to substitute in for p_i and p_j, and Lemma 3, this yields

$$\left(h(t_{-j}) + t_j w_j(t) - \int_0^{t_j} w(x, t_{-j}) \, dx \right) - t_j w_j(t)$$
$$\geq \left(h(t_{-i}) + t_i w_i(t) - \int_0^{t_i} w(x, t_{-i}) \, dx \right) - t_j w_i(t).$$

Rearranging terms gives

$$h(t_{-i}) - h(t_{-j}) \leq \int_0^{t_i} w(x, t_{-i}) \, dx - \int_0^{t_j} w(x, t_{-j}) dx + (t_j - t_i) w_i(t)$$
$$\leq \int_0^{t_i} L \, dx - \int_0^{t_j} 0 \, dx + (t_j - t_i) w_i(t)$$
$$= L \cdot t_i + (t_j - t_i) w_i(t). \qquad \square$$

Proof of Theorem 1. Consider $n = m$ jobs of length $l = (1, \ldots, 1, m)$. Let $L := 2m - 1$ denote the total length of the jobs. Define speed vector $t = (m\alpha, \ldots, m\alpha, \alpha)$, where α is a real number that only depends on m and c and will be determined at the end of this section. We will show that if M is deterministic, truthful, envy-free, individually rational and anonymous, it should allocate all jobs to the quickest machine in this instance.

Claim 5. *For speed vector $t = (m\alpha, \ldots, m\alpha, \alpha)$ and jobs $l = (1, \ldots, 1, m)$, if M is c-approximate and $w_i(t) \geq 1$ for some $i \in \{1, \ldots, m-1\}$, then*

$$h(t_{-1}) \geq (L + \frac{m-1}{Lc}) \cdot \alpha.$$

Proof. Since M is truthful and individually rational, inequality (2) applies, and

$$h(t_{-1}) \geq \int_0^\infty w(x, t_{-1}) \, dx$$
$$\geq \int_0^{\frac{\alpha}{Lc}} w(x, t_{-1}) \, dx + \int_{\frac{\alpha}{Lc}}^\alpha w(x, t_{-1}) \, dx + \int_\alpha^{m\alpha} w(x, t_{-1}) \, dx.$$

Apply M to vector (x, t_{-1}). By the local efficiency of w, job m should be assigned to the quickest machine. So for $x < \alpha$, $w(x, t_{-1}) \geq l_m$. When $x < \frac{\alpha}{Lc}$, all the jobs should be assigned to the machine with speed x for a makespan less than α/c. Otherwise the makespan is at least α, contradicting M is c-approximate. Since $w_i(m\alpha, t_{-i}) \geq 1$ for some $i \in \{1, \ldots, m-1\}$, monotonicity implies $w_i(x, t_{-1}) \geq 1$ for all $x \in (\alpha, m\alpha)$. Since $x \in (\alpha, m\alpha)$ is unique in vector (x, t_{-1}), we get $w(x, t_{-1}) \geq 1$ by anonymity. Thus

$$h(t_{-1}) \geq \int_0^{\frac{\alpha}{Lc}} L \, dx + \int_{\frac{\alpha}{Lc}}^\alpha m \, dx + \int_\alpha^{m\alpha} 1 \, dx \;=\; (L + \frac{m-1}{Lc})\alpha. \qquad \square$$

Let $t' = (1, m\alpha, \ldots, m\alpha, \alpha)$. Applying M to t', Lemma 4 implies

$$h(t'_{-1}) - h(t'_{-m}) \;\leq\; L + (\alpha - 1)w_1(t') \;\leq\; L \cdot \alpha.$$

Since $t'_{-1} = t_{-1}$, this implies

$$h(t_{-1}) \leq L \cdot \alpha + h(t'_{-m}). \tag{4}$$

Claim 6. *If M is c-approximate, then $h(t_{-1}) < L \cdot \alpha + f(m, c)$, where $f(m, c) = \gamma^{m-1}L + h(\gamma^{m-2}, \gamma^{m-1}, \ldots, \gamma, 1)$ and $\gamma = cL + \epsilon$ for some $0 < \epsilon < 1$.*

Proof. Define speed vector $t^{(i)} = (\gamma^{i-1}, \gamma^{i-2}, \ldots, \gamma, 1, m\alpha, \ldots, m\alpha)$ of length m for $i \geq 2$, where $\gamma = cL + \epsilon$ for some $0 < \epsilon < 1$.

Let us consider the allocation M makes to machine 1 for bid vector $t^{(i)}$. The speed of machine 1 is $\gamma^{i-1} \geq \gamma$ for $i \geq 2$. The speed of machine i is 1. The makespan of allocating all jobs to machine i is L while the makespan of allocating at least one job to machine 1 is at least $\gamma = cL + \epsilon$. Since M is c-approximate, this means $w_1(t^{(i)}) = 0$. Using Lemma 4, we have

$$h(t^{(i)}_{-1}) - h(t^{(i)}_{-m}) \;\leq\; t^{(i)}_1 L + (t^{(i)}_m - t^{(i)}_1)w_1(t^{(i)}) \;=\; \gamma^{i-1}L.$$

Since $t^{(i)}_{-m} = t^{(i+1)}_{-1}$, this implies $h(t^{(i)}_{-1}) - h(t^{(i+1)}_{-1}) \leq \gamma^{i-1}L$ for $i \in \{2, \ldots, m-1\}$. Summing up these inequalities on all i, we have

$$h(t^{(2)}_{-1}) - h(t^{(m)}_{-1}) \leq L \sum_{i=2}^{m-1} \gamma^{i-1} < \gamma^{m-1}L$$

Since $t'_{-m} = t^{(2)}_{-1}$, we get $h(t'_{-m}) < \gamma^{m-1}L + h(t^{(m)}_{-1}) = f(m, c)$. Plugging this into (4) yields

$$h(t_{-1}) < L \cdot \alpha + f(m, c). \tag{5}$$

$\square$

To complete the proof of Theorem 1, consider speed vector t with $\alpha = \frac{Lc}{m-1}f(m, c)$. If mechanism M does not allocate all jobs to machine m, then $w_i(t) \geq 1$ for some $i \in \{1, \ldots, m-1\}$. Then Claim 5 implies that $h(t_{-1}) \geq \alpha \cdot L + f(m, c)$, contradicting (5). So M must allocate all jobs to machine m in this case, yielding a makespan of $(2m - 1)\alpha$ while the makespan of the schedule that assigns job j to machine j for all j is $m\alpha$. Thus, M is c-approximate for some $c \geq 2 - 1/m$. $\square$

4   Characterizing Scalable Mechanisms on Two Machines

In this section, we will show that for two machines, there is just one deterministic scalable allocation that can be made truthful, envy-free, individually rational, and anonymous. This allocation turns out to be the same allocation as the VCG mechanism.

Lemma 7. *Let w be a deterministic and scalable allocation function for 2 machines. For some $k > 1$, if $w(x, a) > 0$ for all $a > 0$ and $x < ka$, then there is some $g(k) > 0$ such that*

$$\int_a^{ka} w(x,a) \; dx \geq \int_{\frac{a}{k}}^a w(a,x) \; dx + g(k) \cdot a.$$

Proof. For $a < x < ka$, let $x = \frac{a^2}{t}$.

$$\int_a^{ka} w(x,a) \; dx = \int_a^{\frac{a}{k}} w(\frac{a^2}{t}, a)(-\frac{a^2}{t^2}) \; dt \qquad \text{(integrate by substitution)}$$

$$= \int_{\frac{a}{k}}^a \frac{a^2}{t^2} w(a,t) \; dt \qquad\qquad \text{(w is scalable)}$$

$$= \int_{\frac{a}{k}}^{\frac{k+1}{2k}a} \frac{a^2}{t^2} w(a,t) \; dt + \int_{\frac{k+1}{2k}a}^a \frac{a^2}{t^2} w(a,t) \; dt$$

For $\frac{a}{k} < t < \frac{k+1}{2k}a$ and $k > 1$, we have $\frac{a^2}{t^2} \geq a^2/(\frac{k+1}{2k}a)^2 = \frac{4k^2}{(k+1)^2} > 1$. For $\frac{k+1}{2k}a < t < a$, we have $\frac{a^2}{t^2} \geq 1$. Therefore,

$$\int_a^{ka} w(x,a) \; dx \geq \frac{4k^2}{(k+1)^2} \int_{\frac{a}{k}}^{\frac{k+1}{2k}a} w(a,t) \; dt + \int_{\frac{k+1}{2k}a}^a w(a,t) \; dt$$

$$= \left(\frac{4k^2}{(k+1)^2} - 1 \right) \int_{\frac{a}{k}}^{\frac{k+1}{2k}a} w(a,t) \; dt + \int_{\frac{a}{k}}^a w(a,t) \; dt$$

We also have

$$\int_{\frac{a}{k}}^{\frac{k+1}{2k}a} w(a,t) \; dt \; = \; \int_{\frac{a}{k}}^{\frac{k+1}{2k}a} w(1, \frac{t}{a}) \; dt \; = \; a \int_{\frac{1}{k}}^{\frac{k+1}{2k}} w(1,y) \; dy.$$

The first equality follows the scalability of w and we get the second equality by substituting t with ay. Since $w(x, a) > 0$ for all $a > 0$ and $x < ka$, we have $w(1, y) > 0$ for $y > 1/k$. In sum, $\int_a^{ka} w(x,a) \; dx \geq \int_{\frac{a}{k}}^a w(a,x) \; dx + g(k) \cdot a$, where $g(k) = \left(\frac{4k^2}{(k+1)^2} - 1 \right) \int_{\frac{1}{k}}^{\frac{k+1}{2k}} w(1,y) \; dy > 0$. $\qquad \square$

Theorem 2. *Let $M = (w, p)$ be deterministic, truthful, envy-free, individually rational, and anonymous. If w is scalable, then for two machines of different speeds, w allocates all jobs to the quickest machine.*

Proof. Let L denote the total length of jobs. First, consider two machines of speed $t_1 = 1$ and $t_2 = a$ ($a > 1$). Since M is truthful, envy-free, and anonymous, by Lemma 4, we have

$$h(a) - h(1) \leq L + (a-1)L = L \cdot a \tag{6}$$

Since w is individually rational, $h(1) \geq \int_0^\infty w(x, 1) \geq 0$. We will show that $w(ka, a) = 0$ for any $k > 1$. For a contradiction, assume $w(ra, a) > 0$ for some $r > 1$. Let k be such that $w(x, a) > 0$ for $x < ka$ and $w(x, a) = 0$ for $x > ka$. By monotonicity, such a k exists. By the assumption that $w(ra, a) > 0$ for some $r > 1$, we know that $k > 1$. Since w is scalable, we have for any $x > ka$, $w(y, a) = w(a, x) = L$ if $y/a = a/x$, i.e. $y = a^2/x < a/k$. Therefore,

$$
\begin{aligned}
h(a) &\geq \int_0^\infty w(x, a) \, dx \\
&= \int_0^{\frac{a}{k}} L \, dx + \int_{\frac{a}{k}}^a w(x, a) \, dx + \int_a^{ka} w(x, a) \, dx \\
&\geq \frac{L}{k}a + \int_{\frac{a}{k}}^a w(x, a) \, dx + \int_{\frac{a}{k}}^a w(a, x) \, dx + g(k) \cdot a \quad \text{(Lemma 7)} \\
&\geq \frac{L}{k}a + \int_{\frac{a}{k}}^a (w(x, a) + w(a, x)) \, dx + g(k) \cdot a \\
&= \frac{L}{k}a + \int_{\frac{a}{k}}^a L \, dx + g(k) \cdot a \\
&= (L + g(k))a \tag{7}
\end{aligned}
$$

Take $a > h(1)/g(k)$. We have $h(a) > aL + h(1)$ from (7). This contradicts (6). ☐

5   Payments for Known Allocation Rules

Although the VCG mechanism is truthful, envy-free, individually rational, and anonymous, it does not have a good approximation guarantee for makespan. In this section, we show that the *LPT** algorithm that guarantees a 3-approximation of makespan [13] admits no truthful, envy-free, individually rational, and anonymous payments. We also prove a similar result for randomized mechanisms in the full paper: the randomized 2-approximation algorithm in [2,3], whose expected allocation is monotone decreasing and locally efficient, admits no payment function that make it simultaneously truthful-in-expectation, envy-free-in-expectation, individually rational, and anonymous. We can prove similar results with similar proofs for the allocations in [1,5,9].

Let w_i^j be the workload of machine i before job j is assigned. Assume the jobs are indexed so that $l_1 \geq l_2 \geq \ldots \geq l_m$. The *LPT** algorithm is presented as Algorithm 1. Note that this algorithm rounds the speeds and hence is not scalable.

Algorithm 1. *LPT** Algorithm

1: Define rounded speed of machine i to be $s_i := 2^{\lceil \log b_i \rceil}$.
2: **for** $j = 1$ to m **do**
3:    Assign job j to machine i that minimizes $(w_i^j + l_j) \cdot s_i$.
4: **end for**
5: Among machines of same rounded speed, reorder bundles on these machines so that a machine with smaller bid gets more jobs.

Theorem 3. *There is no payment function that will make LPT* simultaneously truthful, envy-free, individually rational, and anonymous.*

Proof. Let w denote the allocation of the *LPT** algorithm. For a contradiction, assume there exists a payment function p such that mechanism $M = (w, p)$ is truthful, envy-free, individually rational, and anonymous.

Apply M to the problem of two jobs with lengths $l_1 = 2$ and $l_2 = 1$, and two machines with speeds $t_1 = 1, t_2 = a$ where $a > 1$ and a is a power of 2. By Lemma 4, we have

$$h(a) - h(1) \leq 3 + (a-1) \cdot 3 = 3a. \tag{8}$$

Since M is individually rational, we also have

$$h(a) \geq \int_0^\infty w(x, a) \; dx$$

$$\geq \int_0^{\frac{a}{4}} w(x, a) \; dx + \int_{\frac{a}{4}}^a w(x, a) \; dx + \int_a^{2a} w(x, a) \; dx$$

$$\geq \frac{a}{4} w(\frac{a}{4}, a) + \int_{\frac{a}{4}}^a w(x, a) \; dx + a \cdot w(2a, a),$$

where the last inequality follows the monotonicity of w. Since a is a power of 2, the *LPT** algorithm ensures $w(\frac{a}{4}, a) = 3$ and $w(2a, a) = 1$. Since w is locally efficient, for any $\frac{a}{4} < x < a$, a machine with speed x gets at least job one, i.e., $w(x, a) \geq 2$. Therefore, $h(a) \geq \frac{a}{4} \cdot 3 + \frac{3a}{4} \cdot 2 + a \cdot 1 = \frac{13}{4}a$. Now take $a = 8h(1)$, we have $h(a) \geq \frac{13}{4}a = 26h(1)$ and $h(a) \leq h(1) + 3a = 25h(1)$ from (8), a contradiction.   □

References

1. Andelman, N., Azar, Y., Sorani, M.: Truthful approximation mechanisms for scheduling selfish related machines. In: 22nd Annual Symposium on Theoretical Aspects of Computer Science, pp. 69–82 (2005)
2. Archer, A.: Mechanisms for discrete optimization with rational agents. PhD thesis, Ithaca, NY, USA (2004)
3. Archer, A., Tardos, E.: Truthful mechanisms for one-parameter agents. In: Proceedings of the 42nd Annual Symposium on Foundations of Computer Science, pp. 482–491 (2001)
4. Ashlagi, I., Dobzinski, S., Lavi, R.: An optimal lower bound for anonymous scheduling mechanisms. In: Proceedings of the 10th ACM Conference on Electronic Commerce, pp. 169–176 (2009)
5. Christodoulou, G., Kovács, A.: A deterministic truthful PTAS for scheduling related machines. In: Proceedings of the Twenty-First Annual ACM-SIAM Symposium on Discrete Algorithms, pp. 1005–1016 (2010)
6. Cohen, E., Feldman, M., Fiat, A., Kaplan, H., Olonetsky, S.: Envy-free makespan approximation: extended abstract. In: Proceedings of the 11th ACM Conference on Electronic Commerce, pp. 159–166 (2010)
7. Cohen, E., Feldman, M., Fiat, A., Kaplan, H., Olonetsky, S.: On the interplay between incentive compatibility and envy freeness (2010),
http://arxiv.org/abs/1003.5328
8. Cohen, E., Feldman, M., Fiat, A., Kaplan, H., Olonetsky, S.: Truth and envy in capacitated allocation games (2010), http://arxiv.org/abs/1003.5326
9. Dhangwatnotai, P., Dobzinski, S., Dughmi, S., Roughgarden, T.: Truthful approximation schemes for single-parameter agents. In: Proceedings of 49th Annual Symposium on Foundations of Computer Science, pp. 15–24 (2008)
10. Dubins, L., Spanier, E.: How to cut a cake fairly. American Mathematical Monthly 68, 1–17 (1961)
11. Foley, D.: Resource allocation and the public sector. Yale Economic Essays 7, 45–98 (1967)
12. Hochbaum, D.S., Shmoys, D.B.: A polynomial approximation scheme for scheduling on uniform processors: Using the dual approximation approach. SIAM J. Comput. 17, 539–551 (1988)
13. Kovács, A.: Fast monotone 3-approximation algorithm for scheduling related machines. In: Brodal, G.S., Leonardi, S. (eds.) ESA 2005. LNCS, vol. 3669, pp. 616–627. Springer, Heidelberg (2005)
14. Kovács, A.: Tighter approximation bounds for LPT scheduling in two special cases. J. of Discrete Algorithms 7, 327–340 (2009)
15. Mu'Alem, A.: On multi-dimensional envy-free mechanisms. In: Proceedings of the 1st International Conference on Algorithmic Decision Theory, pp. 120–131 (2009)

A Truthful Mechanism for Value-Based Scheduling in Cloud Computing

Navendu Jain[1], Ishai Menache[1], Joseph (Seffi) Naor[2,*,**],
and Jonathan Yaniv[2,**]

[1] Extreme Computing Group, Microsoft Research, Redmond, WA
[2] Computer Science Department, Technion, Haifa, Israel

Abstract. We introduce a novel pricing and resource allocation approach for batch jobs on cloud systems. In our economic model, users submit jobs with a value function that specifies willingness to pay as a function of job due dates. The cloud provider in response allocates a subset of these jobs, taking into advantage the *flexibility* of allocating resources to jobs in the cloud environment. Focusing on social-welfare as the system objective (especially relevant for private or in-house clouds), we construct a resource allocation algorithm which provides a small approximation factor that approaches 2 as the number of servers increases. An appealing property of our scheme is that jobs are allocated non-preemptively, i.e., jobs run in one shot without interruption. This property has practical significance, as it avoids significant network and storage resources for checkpointing. Based on this algorithm, we then design an *efficient* truthful-in-expectation mechanism, which significantly improves the running complexity of black-box reduction mechanisms that can be applied to the problem, thereby facilitating its implementation in real systems.

1 Introduction

Cloud computing offers easily accessible computing resources of variable size and capabilities. This paradigm allows applications to rent computing resources and services on-demand, benefiting from dynamic allocation and the economy of scale of large data centers. Cloud computing providers, such as Microsoft, Amazon and Google, are offering cloud hosting of user applications under a utility pricing model. The most common purchasing options are pay-as-you-go (or *on-demand*) schemes, in which users pay per-unit resource (e.g., a virtual machine) per-unit time (e.g., per hour). The advantage of this pricing approach is in its simplicity, in the sense that users pay for the resources they get. However, such an approach suffers from two shortcomings. First, the user pays for computation as if it were a tangible commodity, rather than paying for desired performance. To exemplify this point, consider a finance firm which has to process the daily stock exchange

---

* Supported in part by the Google Inter-university center for Electronic Markets and Auctions, and by ISF grants 1366/07 and 954/11.
** Part of this work was done while visiting Microsoft Research.

G. Persiano (Ed.): SAGT 2011, LNCS 6982, pp. 178–189, 2011.

data with a deadline of an hour before the next trading day. Such a firm does not care about allocation of servers over time as long as the job is finished by its due date. At the same time, the cloud can deliver higher value to users by knowing user-centric valuation for the limited resources being contended for. This form of value-based scheduling, however, is not supported by pay-as-you-go pricing. Second, current pricing schemes lack a market feedback signal that prevents users from submitting unbounded amounts of work. Thus, users are not incentivized to respond to variation in resource demand and supply.

In this paper, we propose a novel pricing model for cloud environments, which focuses on *quality* rather than *quantity*. Specifically, we incorporate the significance of the completion time of a job, rather than the exact number of servers that the job gets at any given time. In our economic model, customers specify the overall amount of resources (server or virtual machine hours) which they require for their job, and how much they are willing to pay for these resources as a function of due date. For example, a particular customer may submit a job at 9am, specifying that she needs a total of 1000 server hours, and is willing to pay $100 if she gets them by 5pm and $200 if she gets them by 2pm. This framework is especially relevant for batch jobs (e.g., financial analytics, image processing, search index updates) that are carried out until completion. Under our scheme, the cloud determines the *scheduling* of resources according to the submitted jobs, the users' willingness to pay and its own capacity constraints. This entire approach raises fundamental issues in mechanism design, as users may try to game the system by reporting false values and potentially increasing their utility. Hence, any algorithmic solution should *incentivize* users to report their true values (or willingness to pay) for the different job due dates.

Pricing in shared computing systems such as cloud computing can have diverse objectives, such as maximizing profits or optimizing system-related metrics (e.g., delay or throughput). We focus in this work on maximizing the social welfare, i.e., the sum of users' values. This objective is especially relevant for private or in-house clouds, such as a government cloud, or enterprize computing clusters.

Our results. We design an efficient truthful-in-expectation mechanism for a new scheduling problem, called the Bounded Flexible Scheduling (BFS) problem, which is directly motivated by the cloud computing paradigm. A cloud containing C servers receives a set of job requests with heterogeneous demand and values per deadline (or due date), where the objective is to maximize the social welfare, i.e., the sum of the values of the scheduled jobs. The scheduling of a job is *flexible*, i.e., it can be allocated a different number of servers per time unit and in a possibly preemptive (non-contiguous) manner, under parallelism thresholds. The parallelism threshold represents the job's limitations on parallelized execution. For every job j, we denote by k_j the maximum number of servers that can be allocated to job j in any given time unit. The maximal parallelism thresholds across jobs, denoted by k, is assumed to be much smaller than the cloud capacity C, as typical in practical settings.

No approximation algorithm is known for the BFS problem. When relaxing the parallelism threshold constraint, our model coincides with the problem of maximizing the profit of preemptively scheduling jobs on a single server. Lawler [9] gives an optimal solution in pseudo-polynomial time via dynamic programming to this problem, implying also an FPTAS for it. However, his algorithm cannot be extended to the case where jobs have parallelization limits.

Our first result is an LP-based approximation algorithm for BFS that gives an approximation factor of $\alpha \triangleq \left(1 + \frac{C}{C-k}\right)(1 + \varepsilon)$ to the optimal social welfare for every $\varepsilon > 0$. With the gap between k and C being large, the approximation factor approaches 2. The running time of the algorithm, apart from solving the linear program, is polynomial in the number of jobs, the number of time slots and $\frac{1}{\varepsilon}$. The design of the algorithm proceeds through several steps. We first consider the natural LP formulation for the BFS problem. Since this LP has a very large integrality gap, we strengthen it by incorporating additional constraints that decrease the this gap. We proceed by defining a reallocation algorithm that converts any solution of the LP to a value-equivalent canonical form, in which the number of servers allocated per job does not decrease over the execution period of the job. Our approximation algorithm then decomposes the optimal solution in canonical form to a relatively small number of feasible BFS solutions, with their average social welfare being an α-approximation (thus, at least one of them is an α-approximation). An appealing property of our scheme is that jobs are allocated non-preemptively, i.e., jobs run in one shot without interruption. This property has practical significance, as it avoids significant network and storage resources for checkpointing the intermediate state of jobs that are distributed across multiple servers running in parallel.

The approximation algorithm that we develop is essential for constructing an efficient truthful-in-expectation mechanism that preserves the α-approximation. To obtain this result, we slightly modify the approximation algorithm to get an exact decomposition of an optimal fractional solution. This decomposition is then used to simulate (in expectation) a "fractional" VCG mechanism, which is truthful. The main advantage of our mechanism is that the allocation rule requires only a *single* execution of the approximation algorithm, whereas known black-box reductions that can be applied invoke the approximation algorithm many times, providing only a polynomial bound on the number of invocations. At the end of the paper, we discuss the process of computing the charged payments.

Related Work. We compare our results to known work in algorithmic mechanism design and scheduling. An extensive amount of work has been carried out in these fields, starting with the seminal paper of Nisan and Ronen [10] (see also [11] for a survey book). Of relevance to our work are papers which introduce *black-box* schemes of turning approximation algorithms to incentive compatible mechanisms, while maintaining the approximation ratio of the algorithm. Specifically, Lavi and Swamy [7] show how to construct a truthful-in-expectation mechanism for packing problems that are solved through LP-based approximation algorithms. Dughmi and Roughgarden [6] prove that packing problems that have an FPTAS solution can be turned into a truthful-in-expectation mechanism which is also an FPTAS. We note that there are several papers that combine scheduling and mechanism design (e.g., [8,1]), mostly focusing on makespan minimization.

Scheduling has been a perpetual field of research in operations research and computer science (see e.g., [5,3,4,12,9] and references therein). Of specific relevance to our work are [4,12], which consider variations of the interval-scheduling problem. These papers utilize a decomposition technique for their solutions, which we extend to a more complex model in which the amount of resources allocated to a job can change over time.

2   Definitions and Notation

In the Bounded Flexible Scheduling (**BFS**) problem, a cloud provider is in charge of a cloud containing a fixed number of C servers. The time axis is divided into T time slots $\mathcal{T} = \{1, 2, \ldots T\}$. The cloud provider receives requests from n clients, denoted by $\mathcal{J} = \{1, 2, \ldots n\}$, where each client has a job that needs to be executed. We will often refer to a client either as a player or by the job belonging to her. The cloud provider can choose to reject some of the job requests, for instance if allocating other jobs increases its profit. In this model, the cloud can gain profit only by fully completing a job.

Every job j is described by a tuple $\langle D_j, k_j, v_j \rangle$. The first parameter D_j, the *demand* of job j, is the total amount of demand units required to complete the job, where a demand unit corresponds to a single server being assigned to the job for a single time slot. Parallel execution of a job is allowed, that is, the job can be executed on several servers in parallel. In this model we assume that the additional overhead due to parallelism is negligible. However, parallel execution of a job is limited by a threshold k_j, which is the maximal number of servers that can be simultaneously assigned to job j in a single time slot. We assume that $k \triangleq \max_j \{k_j\}$ is substantially smaller than the total capacity C, i.e., $k \ll C$.

Let $v_j : \mathcal{T} \to \mathbb{R}^{+,0}$ be the *valuation function* of job j. That is, $v_j(t)$ is the value gained by the owner of job j if job j is completed at time t. The valuation function v_j is naturally assumed to be monotonically non-increasing in t. The goal is to maximize the sum of values of the jobs that are scheduled by the cloud. In this paper, two types of valuation functions will be of specific interest to us:

• **Deadline Valuation Functions:** Here, players have a deadline d_j which they need to meet. Formally, $v_j(t)$ is a step down function, which is equal to a constant scalar v_j until the deadline d_j and 0 afterwards.

• **General Valuation Functions:** The functions $v_j(t)$ are arbitrary monotonically non-increasing functions.

For simplicity of notation, when discussing the case of general valuation functions, we will set $d_j = T$ for every player. Define $\mathcal{T}_j = \{t \in \mathcal{T} : t \le d_j\}$ as the set of time slots in which job j can be executed and $\mathcal{J}_t = \{j \in \mathcal{J} : t \le d_j\}$ as the set of jobs that can be executed at time t.

A *mapping* $y_j : \mathcal{T}_j \to [0, k_j]$ is an assignment of servers to job j per time unit, which does not violate the parallelism threshold k_j[1]. A mapping which fully executes job j is called an *allocation*. Formally, an allocation $a_j : \mathcal{T}_j \to [0, k_j]$ is a mapping for job j with $\sum_t a_j(t) = D_j$. Denote by $\mathcal{A}_j$ the set of allocations a_j which fully execute job j and let $\mathcal{A} = \bigcup_{j=1}^n \mathcal{A}_j$. Let $s(y_j) = \min\{t : y_j(t) > 0\}$ and $e(y_j) = \max\{t : y_j(t) > 0\}$ denote the start and end times of a mapping y_j, respectively. Specifically, for an allocation a_j, $e(a_j)$ is the time in which job j is completed when the job is allocated according to a_j, and $v_j(e(a_j))$ is the value gained by the owner of job j. We will often use $v_j(a_j)$ instead of $v_j(e(a_j))$ to shorten notations.

3 Approximation Algorithm for BFS

In this section we present an algorithm for BFS that approximates the social welfare, i.e., the sum of values gained by the players. When discussing the approximation algorithm, we assume that players bid truthfully. In Section 4, we describe a payment scheme that gives players an incentive to bid truthfully. We begin this section by describing an LP relaxation for the case of deadline valuation functions and continue by presenting a canonical solution form in which all mappings are Monotonically Non Decreasing (MND) mappings, defined later. This result is then generalized for general valuation functions (Section 3.2). Finally, we give a decomposition algorithm (Section 3.3) which yields an α-approximation to the optimal social welfare of BFS.

3.1 LP Relaxation of BFS with Deadline Valuation Functions

Linear Relaxation. Consider the following relaxed linear program. Every variable $y_j(t)$ for $t \in \mathcal{T}_j$ in (LP-D) denotes the number of servers assigned to j at time t. We use y_j to denote the mapping induced by the variables $\{y_j(t)\}_{t \in \mathcal{T}_j}$ and x_j as the completed fraction of job j.

$$\text{(LP-D)}\quad \max \sum_{j=1}^n v_j x_j$$

$$\text{s.t.}\quad \sum_{t \in \mathcal{T}_j} y_j(t) = D_j \cdot x_j \qquad \forall j \in \mathcal{J} \tag{1}$$

$$\sum_{j \in \mathcal{J}_t} y_j(t) \leq C \qquad \forall t \in \mathcal{T} \tag{2}$$

$$0 \leq y_j(t) \leq k_j x_j \qquad \forall j \in \mathcal{J}, t \in \mathcal{T}_j \tag{3}$$

$$0 \leq x_j \leq 1 \qquad \forall j \in \mathcal{J} \tag{4}$$

---

[1] For tractability, we assume that the assignment y_j is a continuous decision variable. In practice, non-integer allocations will have to be translated to integer ones, for example by processor sharing within each time interval.

Reallocate(y)
1. While y contains non-MND mappings
    1.1. Let j be a job generating a maximal(a, b)-violation according to $\preceq$
    1.2. **ReallocationStep(y, j, a, b)**

ReallocationStep(y, j, a, b)
1. Let j' be a job such that $y_{j'}(a) < y_{j'}(b)$
2. $\mathcal{T}_{\max} = \{t \in [a, b] : y_{j'}(t) = y_{j'}(b)\}$
3. $\delta = \max\{y_{j'}(t) : t \in [a, b] \setminus \mathcal{T}_{\max}\}$
4. $\Delta = \min\left\{\frac{y_j(a) - y_j(b)}{1 + |\mathcal{T}_{\max}|}, \frac{y_{j'}(b) - y_{j'}(a)}{1 + |\mathcal{T}_{\max}|}, y_{j'}(b) - \delta\right\}$
5. Reallocate as follows:
    5.1. $y_{j'}(t) \leftarrow y_{j'}(t) - \Delta$ for every $t \in \mathcal{T}_{\max}$
    5.2. $y_{j'}(a) \leftarrow y_{j'}(a) + \Delta \cdot |\mathcal{T}_{\max}|$
    5.3. $y_j(a) \leftarrow y_j(a) - \Delta \cdot |\mathcal{T}_{\max}|$
    5.4. $y_j(t) \leftarrow y_j(t) + \Delta$ for every $t \in \mathcal{T}_{\max}$

Constraints (1) and (2) are job demand and capacity constraints. Typically, the parallelized execution constraints would take the form $0 \leq y_j(t) \leq k$. However, the integrality gap in this case can be as high as $\Omega(n)$. Intuitively, (3) "prevents" us from getting bad mappings which do not correspond to feasible allocations. That is, if we would have extended a mapping y_j (disregarding capacity constraints) by dividing every entry in y_j by x_j, we would have exceeded the parallelization threshold of job j. Before continuing, we mention that there is a strong connection between the choice of (3) and the *configuration LP* for the BFS problem. In fact, (3) can be viewed as an efficient way of implementing the configuration LP. We leave the details to the full version of this paper.

MND Mappings and the Reallocation Algorithm. We now present a canonical solution form of solutions for (LP-D), in which all mappings are monotonically non decreasing (defined next). This canonical form will allow us to construct an approximation algorithm for BFS with a good approximation factor.

Definition 1. *A **monotonically non-decreasing (MND)** mapping (allocation) $y_j : \mathcal{T}_j \to [0, k_j]$ is a mapping (allocation) which is monotonically non-decreasing in the interval $[s(y_j), e(y_j)]$.*

MND mappings propose implementation advantages, such as the allocation algorithm being non-preemptive, as well as theoretical advantages which will allow us to construct a good approximation algorithm for BFS. We first present the main result of this subsection:

Theorem 1. *There is a poly(n, T) time algorithm that transforms any feasible solution y of (LP-D) to an equivalent solution that obtains the same social welfare as y, in which all mappings are MND mappings.*

This theorem is a result of the following *reallocation algorithm*. Let y be a feasible solution to (LP-D). To simplify arguments, we add an additional "idle" job which is allocated whenever there are free servers. This allows us to assume without loss of generality that in every time slot, all C servers are in use. We present a reallocation algorithm that transforms the mappings in y to MND mappings. The reallocation algorithm will swap between assignments of jobs to servers, without changing the completed fraction of every job (x_j), such that no completion time of a job will be delayed. Since the valuation functions are deadline valuation functions, the social welfare of the resulting solution will be equal to the social welfare matching y. Specifically, an optimal solution to (LP-D) will remain optimal. We introduce some definitions and notations prior to the description of the reallocation algorithm.

Definition 2. *Job j generates an (a,b)-**violation**, $a < b$, if $y_j(a) > y_j(b) > 0$. Violations are weakly ordered according to a binary relation $\preceq$ over $\mathcal{T} \times \mathcal{T}$:*

$$(a,b) \preceq (a',b') \quad \Leftrightarrow \quad b < b' \text{ or } (b = b') \wedge (a \leq a') \tag{5}$$

Note that there can be several maximal pairs (a,b) according to $\preceq$.

Given a solution y to (LP-D), our goal is to eliminate all (a,b)-violations in y and consequently remain with only MND mappings, keeping y a feasible solution to (LP-D). The reallocation algorithm works as follows: In every step we try to eliminate one of the maximal (a,b)-violations, according to the order induced by $\preceq$. Let j be the job generating this maximal (a,b)-violation. The main observation is that there must be some job j' with $y_{j'}(a) < y_{j'}(b)$, since in every time slot all C servers are in use. We apply a *reallocation step*, which tries to eliminate this violation by shifting workload of job j from a to later time slots (b in particular), and by doing the opposite to j'. To be precise, we increase y_j in time slots in $\mathcal{T}_{\max}$ (line 2) by a value $\Delta > 0$ (line 4), and increase $y_{j'}(a)$ by the amount we decreased from other variables. We note that if we do not decrease $y_{j'}$ for all time slots in $\mathcal{T}_{\max}$, we will generate $(\tilde{a}, b)$-violations for $a < \tilde{a}$ and therefore the reallocation algorithm may not terminate.

We choose Δ such that after calling the reallocation step either: 1. $y_j(a) = y_j(b)$ 2. $y_{j'}(a) = y_{j'}(b)$ 3. The size of $\mathcal{T}_{\max}$ increases. In the second case, if the (a,b)-violation hasn't been resolved, there must be a different job j'' with $y_{j''}(a) < y_{j''}(b)$, and therefore we can call the reallocation step again. In the third case, we simply expand $\mathcal{T}_{\max}$ and recalculate Δ. The reallocation algorithm repeatedly applies the reallocation step, choosing the maximal (a,b)-violation under $\preceq$, until all mappings become MND mappings. The following lemma guarantees the correctness of the reallocation algorithm.

Lemma 1. *Let y be a feasible solution of (LP-D) and let j be a job generating a maximal (a,b)-violation over $\preceq$. Denote by $\tilde{y}$ the vector y after calling ReallocationStep(y, j, a, b) and let $\left(\tilde{a}, \tilde{b}\right)$ be the maximal violation in $\tilde{y}$. Then:*

1. *$\tilde{y}$ is a feasible solution of (LP-D).*
2. *$\left(\tilde{a}, \tilde{b}\right) \preceq (a, b)$*
3. *No new (a,b)-violations are added to $\tilde{y}$.*

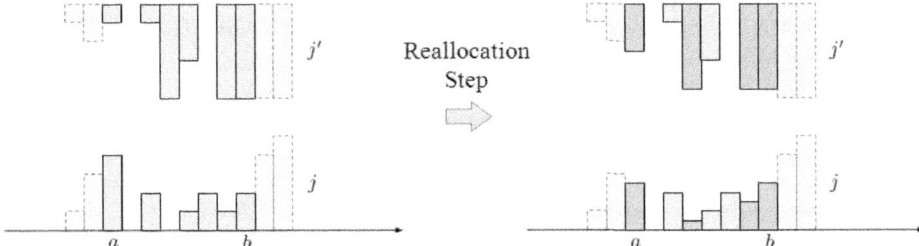

Fig. 1. Resolving an (a, b)-violation generated by j, with $\mathcal{T}_{\max} = \{b, b - 1, b - 4\}$. X-axis represents time

The reallocation algorithm runs in $poly\,(n, T)$ time. To show this, consider a potential function which is the total number of violations. The reallocation algorithm resolves at least one violation after at most nT calls to the reallocation step. The maximal initial number of such violations is bounded by $O\left(nT^2\right)$ and a reallocation step can be efficiently implemented, proving the statement.

3.2   Extension to General Valuation Functions

To extend the results presented so far to the case of general valuation functions, we expand (LP-D) by splitting every player into T subplayers, one for each end time, each associated with a deadline valuation function. Formally, every player j will be substituted by T subplayers $j^1, j^2, \ldots j^T$, all with the same demand and parallelization bound as j. For ease of notation, we denote by $y_j^e\,(t)$ the variables in the linear program matching subplayer j^e, and use similar superscript notations henceforth. For every subplayer j^e, we set $v_j^e = v_j\,(e)$ and $d_j^e = e$. Apart from demand and capacity constraints, we include constraint (3) for every subplayer j^e and add an additional set of constraints:

$$\sum_{e \in \mathcal{T}} x_j^e \leq 1 \qquad \forall j \in \mathcal{J} \qquad (6)$$

This is indeed a relaxation of BFS (we can map an allocation a_j in a BFS solution to the subplayer $j^{e(a_j)}$). The reallocation algorithm does not change the values x_j^e, thus it will not violate (6). We note that these results can be extended to the case where valuation functions are non-monotone. From this point on, we refer to (**LP**) as the relaxed linear program for general valuation functions, after adding (6). When applying results to deadline valuation functions settings, every player j will be viewed as a single subplayer j^{d_j} (making (6) redundant).

3.3   Decomposing an Optimal MND Fractional Solution

The approximation algorithm presented in this section constructs a set of feasible solutions to BFS from a fractional optimal solution to (LP) given in the canonical

---

Coloring Algorithm($\mathcal{S}$)
1. Sort the MND allocations $a \in \mathcal{S}$ according to $e(a)$ in descending order.
2. For every MND allocation a in this order
    2.1 Color a in some color c such that c remains a feasible integral solution.

MND form. The algorithm is similar to the coloring algorithm used in [4,12] for the weighted job interval scheduling problem. The first step of the algorithm constructs a multiset $\mathcal{S} \subset \bigcup_{j=1}^{n} \mathcal{A}_j$ of allocations out of an optimal solution of (LP) given in MND form and then divides the allocations in $\mathcal{S}$ into a set of feasible solutions to BFS.

Step I: Construction of $\mathcal{S}$. Let N be a large number to be determined later. Consider a job j which is substituted by a set of subplayers $j^1, j^2, \ldots, j^T$ (or a single subplayer j^{d_j} for the case of deadline valuation functions). Let y be an optimal solution of (LP) after applying the reallocation algorithm. For every subplayer j^e, let a_j^e be the allocation corresponding to y_j^e, defined: $a_j^e(t) = \frac{y_j^e(t)}{x_j^e}$ for every $t \in T_j$. Note that a_j^e is an allocation by the definition of x_j^e and by (3). We construct $\mathcal{S}$ as follows: Let $\bar{x}_j^e$ denote the value x_j^e rounded up to the nearest integer multiplication of $\frac{1}{N}$. For every subplayer j^e, add $N\bar{x}_j^e$ copies of a_j^e to $\mathcal{S}$.

Step II: Coloring Allocations. The coloring algorithm will color copies of MND allocations in $\mathcal{S}$ such that any set of allocations with identical color will induce a feasible integral solution to BFS. Let $1, 2, ..., COL$ denote the set of colors used by the coloring algorithm, described above. We use $a \in c$ to represent that an allocation a is colored in color c. Given a color c, let $c(t) = \sum_{a \in c} a(t)$ denote the total load of MND allocations colored in c at time t. The following two lemmas prove that the number of colors used is relatively small. This allows us to construct an α-approximation algorithm in Theorem 2.

Lemma 2. *Consider an iteration after some allocation $a \in \mathcal{S}$ is colored. Then, for every color c, $c(t)$ is monotonically non-decreasing in the range $[1, e(a)]$.*

Proof (Sketch). Since the sum of MND vectors is MND. Proof by induction. □

Lemma 3. *If $COL = N \cdot \left(1 + \frac{C}{C-k}\right)\left(1 + \frac{nT}{N}\right)$, then when the algorithm handles an allocation $a \in \mathcal{S}$ there is always a free color c in which a can be colored.*

Proof (Sketch). Consider the point when a copy of an allocation a_j is colored. Job j is associated with at most $N\left(1 + \frac{nT}{N}\right) - 1$ copies other than a_j (since we rounded up the values x_j^e. For any color c that cannot be used due to capacity constraints we must have: $c(e(a_j)) \geq C - k$. Thus, there is always a free color in which a_j can be colored. □

Theorem 2. *There is a poly $\left(n, T, \frac{1}{\varepsilon}\right)$ time approximation algorithm that given an optimal solution to (LP) returns an $\alpha \triangleq \left(1 + \frac{C}{C-k}\right)(1 + \varepsilon)$ approximation to BFS for every $\varepsilon > 0$.*

Proof (Sketch). Set $N = \frac{nT}{\varepsilon}$. The social welfare obtained by the best color out of the COL colors is at least: $\frac{N \cdot \sum_{je} v_j \bar{x}_j^e}{COL} \geq \frac{N \cdot OPT^*}{COL} = \frac{OPT^*}{\alpha}$ $\qquad\qquad$ □

4 Truthfulness-in-Expectation

Up until now we have assumed that players report their true valuation functions to the cloud provider and that prices are charged accordingly. However, in reality, players may choose to untruthfully report a valuation function b_j which differs from their true valuation function v_j if they may gain from it. In this section, we construct an efficient mechanism that charges costs from players such that reporting their valuation function untruthfully cannot benefit them. Unlike known black-box reductions for constructing such mechanisms, our construction calls the approximation algorithm only *once*, significantly improving the complexity of the mechanism.

We begin by introducing the common terminology used in mechanism design. Every participating player chooses a type out of a known type space. In our model, players choose a valuation function v_j out of the set of monotonically non-increasing valuation functions (or deadline valuation functions) to represent its true type. Denote by V_j the set of types from which player j can choose and let $V = V_1 \times \cdots \times V_n$. For a vector v, denote by v_{-j} the vector v restricted to entries of players other than j and denote V_{-j} accordingly. Let $\mathcal{O}$ denote the set of all possible outcomes of the mechanism and let $v_j(o)$ for $o \in O$ represents the value gained by player j under outcome o. A *mechanism* $\mathcal{M} = (f, p)$ consists of an allocation rule $f : V \rightarrow \mathcal{O}$ and a pricing rule $p_j : V \rightarrow \mathbb{R}$ for each player j. Players report a bid type $b_j \in V_j$ to the mechanism, which can be different from their true type v_j. The mechanism, given a reported type vector $b = (b_1, \ldots, b_n)$ computes an outcome $o = f(b)$ and charges $p_j(b)$ from each player. Each player strives to maximize its utility: $u_j(b) = v_j(o) - p_j(b)$, where o_j in our model is the allocation according to which job j is allocated, if at all. Mechanisms such as this, where the valuation function does not consist of a single scalar are called *multi-parameter* mechanisms. Our goal is to construct a multi-parameter mechanism where players benefit by declaring their true type. Another desired property is that players do not lose when truthfully reporting their values.

Definition 3. *A deterministic mechanism is **truthful** if for any player j, reporting its true type maximizes $u_j(b)$. That is, given any bid $b_j \in V_j$ and any $v_{-j} \in V_{-j}$, we have:*

$$u_j((v_j, v_{-j})) \geq u_j((b_j, v_{-j})) \tag{7}$$

*where $v_j \in V_j$ is the true type of player j. A randomized mechanism is **truthful-in-expectation** if for any player j, reporting its true type maximizes the expected value of $u_j(b)$. That is, (7) holds in expectation.*

Definition 4. *A mechanism is **individually rational (IR)** if $u_j(v)$ does not receive negative values when player j bids truthfully, for every j and $v_{-j} \in V_{-j}$.*

4.1   The Fractional VCG Mechanism

We start by giving a truthful, IR *fractional* mechanism that can return a fractional allocation, that is, allocate fractions of jobs according to (LP):

1. Given reported types $b_j : \mathcal{T} \to \mathbb{R}^{+,0}$, Solve (LP) and get an optimal solution y^*. Let $o \in \mathcal{O}$ be the outcome matching y^*.
2. Charge $p_j(b) = h_j(o_{-j}) - \sum_{i \neq j} b_i(o_i)$ from every player j, where h_j is any function independent of o_j.

This is the well known VCG mechanism. Recall that (LP) maximizes the social welfare, i.e., the sum of values gained by all players. Assuming all other players act truthful, player j gains $u_j(b) = OPT^* - h_j(o_{-j})$ by bidding truthfully and therefore the mechanism is optimal, since deviating can only decrease $\sum_i v_i(o)$. Note that by dividing both valuation functions and charged prices by some constant, the fractional VCG mechanism remains truthful. This will be useful later on. Individual rationality of the fractional VCG mechanism is obtained by setting the functions h_j according to Clarke's pivot rule [11].

4.2   A New Efficient Truthful-in-Expectation Mechanism

Lavi and Swamy [7] give a black-box reduction for combinatorial auction packing problems from constructing a truthful-in-expectation mechanism to finding an approximation algorithm that verifies an integrality gap of the "natural" LP for the problem. Their reduction finds an exact decomposition of the optimal fractional solution (scaled down by some constant β) into a distribution over feasible integer solutions. By sampling a solution out of this distribution and charging payments according to the fractional VCG mechanism (scaled down by β), they obtain truthfulness-in-expectation. The downside of the reduction given in [7] is that the approximation algorithm A is used as a separation oracle for an additional linear program used as part of the reduction, making their construction inefficient. We follow along the lines of [7] in order to construct a truthful-in-expectation mechanism for the BFS problem, and show how to achieve the same results as [7] by calling our approximation algorithm once.

Recall that the algorithm from Theorem 2 constructs a set of feasible solutions to BFS out of an optimal solution to LP. Ideally, we would have wanted to replace the exact decomposition found by [7] with the output of our decomposition algorithm (by drawing one of the colors uniformly). However, this does not work since our decomposition is not an exact one, because the values x_j^e have been rounded up to $\bar{x}_j^e$ prior to the construction of $\mathcal{S}$.

To overcome this issue, we use a simple alternative technique to round the entries in x to integer multiplications of $\frac{1}{N}$. We construct a vector $\tilde{x}$ such that $\mathbb{E}\left[\tilde{x}_j^e\right] = x_j^e$ for every subplayer j^e, as follows: Assume that $x_j^e = \frac{q}{N} + r$ for $q \in \mathbb{N}$ and $0 \leq r < \frac{1}{N}$. Then, set $\tilde{x}_j^e = \frac{q+1}{N}$ with probability $N \cdot r$ and $\tilde{x}_j^e = \frac{q}{N}$ otherwise. Note that $\mathbb{E}\left[\tilde{x}_j^e\right] = x_j^e$ as required. Now, we construct $\mathcal{S}$ out of $\tilde{x}$ and call the coloring algorithm. By uniformly drawing one of the colors c and scheduling jobs according to the allocations colored in c, we obtain an expected welfare of: $\mathbb{E}\left[\frac{N}{COL}\sum_{j^e}v_j\tilde{x}_j^e\right] = \frac{OPT^*}{\alpha}$. By charging fractional VCG prices, scaled down by α, we obtain truthfulness-in-expectation. Notice that this mechanism is not individually rational, since unallocated jobs may be charged. Lavi and Swamy [7] solve this problem by showing how to modify the pricing rule so that the mechanism will be individually rational. Notice that the number of colors used by the coloring algorithm must always be COL, even though it is an upper bound on the number of colors needed. Otherwise, players might benefit from reporting their valuation functions untruthfully by effecting the number of solutions.

Theorem 3. *There is a truthful-in-expectation, individually rational mechanism for BFS that provides an expected α-approximation of the optimal social welfare.*

Finally, we discuss the process of computing the payments p_j (b). Note that to directly calculate the payments charged by VCG, one must solve a linear program for every player j. [2] describes an implicit pricing scheme that requires only a single invocation of the approximation algorithm to construct both an allocation rule and pricing rules of a truthful-in-expectation mechanism. This result can be plugged into our mechanism, thus decreasing the number of calls to our approximation algorithm to one. However, their scheme induces a mechanism that is only individually rational in expectation (specifically, it may charge negative prices) and causes a multiplicative (constant) loss to social welfare.

References

1. Archer, A., Tardos, É.: Truthful mechanisms for one-parameter agents. In: FOCS, pp. 482–491 (2001)
2. Babaioff, M., Kleinberg, R., Slivkins, A.: Truthful mechanisms with implicit payment computation. In: EC, pp. 43–52 (2010)
3. Bar-Noy, A., Bar-Yehuda, R., Freund, A., Naor, J., Schieber, B.: A unified approach to approximating resource allocation and scheduling. JACM 48, 1069–1090 (2001)
4. Bar-Noy, A., Guha, S., Naor, J., Schieber, B.: Approximating the throughput of multiple machines in real-time scheduling. SIAM Journal of Computing 31(2), 331–352 (2001)
5. Brucker, P.: Scheduling Algorithms, 4th edn. Springer, Heidelberg (2004)
6. Dughmi, S., Roughgarden, T.: Black-box randomized reductions in algorithmic mechanism design. In: FOCS, pp. 775–784 (2010)
7. Lavi, R., Swamy, C.: Truthful and near-optimal mechanism design via linear programming. In: FOCS, pp. 595–604 (2005)
8. Lavi, R., Swamy, C.: Truthful mechanism design for multi-dimensional scheduling via cycle monotonicity. In: EC (2007)
9. Lawler, E.L.: A dynamic programming algorithm for preemptive scheduling of a single machine to minimize the number of late jobs. Annals of Oper. Research 26, 125–133 (1991)
10. Nisan, N., Ronen, A.: Algorithmic mechanism design. In: STOC (1999)
11. Nisan, N., Roughgarden, T., Tardos, É., Vazirani, V.V.: Algorithmic game theory. Cambridge University Press, Cambridge (2007)
12. Phillips, C.A., Uma, R.N., Wein, J.: Off-line admission control for general scheduling problems. In: SODA, pp. 879–888 (2000)

Random Bimatrix Games
Are Asymptotically Easy to Solve*
(A Simple Proof)

Panagiota N. Panagopoulou[2] and Paul G. Spirakis[1,2]

[1] Computer Engineering and Informatics Department, Patras University, Greece
[2] Research Academic Computer Technology Institute, Greece
{panagopp,spirakis}@cti.gr

Abstract. We focus on the problem of computing approximate Nash equilibria and well-supported approximate Nash equilibria in random bimatrix games, where each player's payoffs are bounded and independent random variables, not necessarily identically distributed, but with common expectations. We show that the completely mixed uniform strategy profile, i.e. the combination of mixed strategies (one per player) where each player plays with equal probability each one of her available pure strategies, is with high probability a $\sqrt{\frac{\ln n}{n}}$-Nash equilibrium and a $\sqrt{\frac{3\ln n}{n}}$-well supported Nash equilibrium, where n is the number of pure strategies available to each player. This asserts that the completely mixed, uniform strategy profile is an *almost Nash equilibrium* for random bimatrix games, since it is, with high probability, an ϵ-well-supported Nash equilibrium where ϵ tends to zero as n tends to infinity.

1 Introduction

Non-cooperative game theory has been extensively used in understanding the phenomena observed when decision makers interact. A non-cooperative game in strategic form consists of a set of players, and, for each player, a set of strategies available to her as well as a payoff function mapping each strategy profile (i.e. each combination of strategies, one for each player) to a real number that captures the preferences of the player over the possible outcomes of the game. The most important solution concept in non-cooperative game theory is the notion of *Nash equilibrium* [12]; this is a strategy profile from which no player would have an incentive to unilaterally deviate, i.e. no player could increase her payoff by choosing another strategy while the rest of the players persevered their strategies.

Despite the certain existence of such equilibria [12], the problem of finding any Nash equilibrium even for games involving only two players has been recently

---

* This work has been partially supported by the European Union under contract numbers ICT-2008-215270 (FRONTS), ICT-2009-5-257782 (ABC4Trust), and ERC/StG-210743 (RIMACO).

G. Persiano (Ed.): SAGT 2011, LNCS 6982, pp. 190–199, 2011.

proved to be complete in the PPAD (polynomial parity argument, directed version) class [3], introduced by [13]. Given this fact, the importance of the computation of *approximate* Nash equilibria, also referred to as ϵ-Nash equilibria, became clear. An ϵ-Nash equilibrium is a strategy profile such that no deviating player could achieve a payoff higher than the one that the specific profile gives her, plus ϵ.

A stronger notion of approximate Nash equilibria is the ϵ-*well-supported Nash equilibria*; these are strategy profiles such that each player plays with non-zero probability only strategies that are approximately best-responses, i.e., strategies giving the player a payoff that is no less than the one that the specific profile gives her, minus ϵ.

In this work, we focus on the problem of computing an ϵ-Nash equilibrium as well as an ϵ-well supported Nash equilibrium of a random $n \times n$ bimatrix game. In a random game, as considered in this work, the entries of each player's payoff matrix are random variables which are drawn independently from some probability distribution on the interval $[0, 1]$. We do not require that these random variables should be identically distributed, but we assume that, in each matrix, all entries have the same mean (μ_A for the first player and μ_B for the second player).

We show that the *completely mixed, uniform strategy profile* where each player plays with equal probability each of her n available pure strategies is, with high probability, a $\sqrt{\frac{\ln n}{n}}$-Nash equilibrium and a $\sqrt{\frac{3 \ln n}{n}}$-well supported Nash equilibrium. This implies that the simple uniform randomization over the set of pure strategies of each player yields an *almost Nash equilibrium* profile, in the sense that it is, with high probability, an ϵ-well-supported Nash equilibrium with ϵ tending to zero as the number of available strategies tends to infinity.

Related Work. [12] introduced the concept of Nash equilibria in non-cooperative games and proved that any game possesses at least one such equilibrium; however, the computational complexity of finding a Nash equilibrium used to be a wide open problem for several years. A well-known algorithm for computing a Nash equilibrium of a game with 2 players is the Lemke-Howson algorithm [10], however it has exponential worst-case running time in the number of available pure strategies [14].

Recently, [5] showed that the problem of computing a Nash equilibrium in a game with 4 or more players is PPAD-complete; this result was later extended to games with 3 players [6]. Eventually, [3] proved that the problem is PPAD-complete for bimatrix games in which each player has n available pure strategies.

In [11], following similar techniques as in [1], it was shown that, for any bimatrix game and for any *constant* $\epsilon > 0$, there exists an ϵ-Nash equilibrium with only logarithmic support (in the number n of available pure strategies). This result directly yields a quasi-polynomial ($n^{O(\ln n)}$) algorithm for computing such an approximate equilibrium. Moreover, as pointed out in [1], no algorithm that examines supports smaller than about $\ln n$ can achieve an approximation better than $\frac{1}{4}$.

In [4] it was shown that the problem of computing a $\frac{1}{n^{\Theta(1)}}$-Nash equilibrium is PPAD-complete, and that bimatrix games are unlikely to have a fully polynomial time approximation scheme (unless PPAD $\subseteq$ P). However, it was conjectured that it is unlikely that finding an ϵ-Nash equilibrium is PPAD-complete when ϵ is an absolute constant. Until know, the best polynomial-time algorithm for computing an ϵ-Nash equilibrium is due to [15] and achieves an approximation factor $\epsilon = 0.3393$.

[2] analyzed a Las Vegas algorithm for finding a Nash equilibrium in 2-player random games. In their model however the matrices entries were considered to be identically distributed (drawn either from the uniform distribution on some interval with zero mean, or from the standard Normal distribution $N(0, 1)$). The randomized algorithm proposed in [2] always finds an equilibrium, and an involved analysis of its time complexity shows that it runs in polynomial time (namely $O(n^3 \ln \ln n)$) with high probability.

Here we show that the simple strategy profile where each player plays with equal probability each of her available pure strategies is possibly an almost Nash equilibrium for the random games we consider. Our model of a random bimatrix game is less restrictive than the one considered in [2], since it allows for non-identically distributed payoffs. The results presented here give a clear and straightforward method for computing almost Nash equilibrium strategies in random games, without searching among possible supports or computing the probabilities that the pure strategies of each player should be played with.

2   Games and Nash Equilibria

2.1   Notation

For an integer n, let $[n] = \{1, 2, \ldots, n\}$. For a $n \times 1$ vector $\mathbf{x}$ we denote by $x_1, x_2, \ldots x_n$ the components of $\mathbf{x}$ and by $\mathbf{x}^T$ the transpose of $\mathbf{x}$. We denote by $\mathbf{e}_i$ the column vector with a 1 at the ith coordinate and 0 elsewhere; the size of $\mathbf{e}_i$ will be clear from the context. For an $n \times m$ matrix A, we denote a_{ij} the element in the i-th row and j-th column of A. Let $\mathbb{P}^n$ be the set of all probability vectors in n dimensions, i.e. $\mathbb{P}^n \equiv \{\mathbf{x} \in \mathbb{R}^n : \sum_{i=1}^n x_i = 1 \text{ and } x_i \geq 0 \text{ for all } i \in [n]\}$. For an event E in a sample space, denote $\Pr\{E\}$ the probability of event E occurring. For a random variable X that follows the probability distribution D, denote $\mathcal{E}[X]$ the *expectation* of X (according to the probability distribution D).

2.2   Bimatrix Games

A *noncooperative game* $\Gamma = \langle N, (S_i)_{i \in N}, (u_i)_{i \in N} \rangle$ consists of (i) a finite set of *players* N, (ii) a nonempty finite set of *pure strategies* S_i for each player $i \in N$ and (iii) a *payoff function* $u_i : \times_{i \in N} S_i \to \mathbb{R}$ for each player $i \in N$.

Bimatrix games [9, 10] are a special case of 2-player games (i.e. $|N| = 2$) such that the payoff functions can be described by two real $n \times m$ matrices A and B, where $n = |S_1|$ and $m = |S_2|$. More specifically, the n rows of matrices A and B

represent the n pure strategies of the first player (also called *the row player*), and the m columns represent the pure strategies of the second player (*the column player*). Then, when the row player chooses strategy i and the column player chooses strategy j, the former gets payoff a_{ij} while the latter gets payoff b_{ij}. Based on this, bimatrix games are denoted by $\Gamma = \langle A, B \rangle$.

A *mixed strategy* for player $i \in N$ is a probability distribution on the set of her pure strategies S_i. In a bimatrix game $\Gamma = \langle A, B \rangle$, a mixed strategy for the row player can be expressed as a probability vector $\mathbf{x} \in \mathbb{P}^n$ while a mixed strategy for the column player can be expressed as a probability vector $\mathbf{y} \in \mathbb{P}^m$. A *strategy profile* $(\mathbf{x}, \mathbf{y})$ is a combination of strategies, one for each player. In a given strategy profile $(\mathbf{x}, \mathbf{y})$ the players get *expected payoffs* $\mathbf{x}^T A \mathbf{y}$ (row player) and $\mathbf{x}^T B \mathbf{y}$ (column player). The *support* of a player in a strategy profile is the subset of her pure strategies that are assigned strictly positive probability.

We define the completely mixed, uniform strategy profile as the strategy profile where each player plays with equal probability each of her available pure strategies:

Definition 1 (Completely mixed, uniform strategy profile). *Consider a $n \times m$ bimatrix game $\Gamma = \langle A, B \rangle$. The completely mixed, uniform strategy profile is the strategy profile $(\mathbf{x}, \mathbf{y})$ such that*

$$x_i = \frac{1}{n} \quad \forall i \in [n] \quad and \quad y_j = \frac{1}{m} \quad \forall j \in [m] \ .$$

2.3 Nash Equilibria and Approximate Nash Equilibria

A Nash equilibrium [12] for a game Γ is a combination of (pure or mixed) strategies, one for each player, such that no player could increase her payoff by unilaterally changing her strategy.

Definition 2 (Nash equilibrium). *A strategy profile $(\mathbf{x}, \mathbf{y})$ is a Nash equilibrium for the $n \times m$ bimatrix game $\Gamma = \langle A, B \rangle$ if*

1. *For all pure strategies $i \in [n]$ of the row player, $\mathbf{e}_i^T A \mathbf{y} \leq \mathbf{x}^T A \mathbf{y}$ and*
2. *For all pure strategies $j \in [m]$ of the column player, $\mathbf{x}^T B \mathbf{e}_j \leq \mathbf{x}^T B \mathbf{y}$.*

Equivalently,

Definition 3 (Nash equilibrium). *A strategy profile $(\mathbf{x}, \mathbf{y})$ is a Nash equilibrium for the $n \times m$ bimatrix game $\Gamma = \langle A, B \rangle$ if*

1. *For all pure strategies $i \in [n]$ of the row player,*

$$x_i > 0 \Rightarrow \mathbf{e}_i^T A \mathbf{y} \geq \mathbf{e}_k^T A \mathbf{y} \quad \forall k \in [n]$$

2. *For all pure strategies $j \in [m]$ of the column player,*

$$y_j > 0 \Rightarrow \mathbf{x}^T B \mathbf{e}_j \geq \mathbf{x}^T B \mathbf{e}_k \quad \forall k \in [m] \ .$$

For $\epsilon > 0$, an ϵ-Nash equilibrium (or an ϵ-approximate Nash equilibrium) is a combination of (pure or mixed) strategies, one for each player, such that no player could increase her payoff more than ϵ by unilaterally changing her strategy:

Definition 4 (ϵ-Nash equilibrium). *For any $\epsilon > 0$ a strategy profile $(\mathbf{x}, \mathbf{y})$ is an ϵ-Nash equilibrium for the $n \times m$ bimatrix game $\Gamma = \langle A, B \rangle$ if*

1. *For all pure strategies $i \in [n]$ of the row player, $\mathbf{e}_i^T A \mathbf{y} \leq \mathbf{x}^T A \mathbf{y} + \epsilon$ and*
2. *For all pure strategies $j \in [m]$ of the column player, $\mathbf{x}^T B \mathbf{e}_j \leq \mathbf{x}^T B \mathbf{y} + \epsilon$.*

A stronger notion of approximate Nash equilibria was introduced in [7, 5]: For $\epsilon > 0$, an ϵ-well supported Nash equilibrium (or a well-supported ϵ-approximate Nash equilibrium) is a combination of (pure or mixed) strategies, one for each player, such that each player plays only approximately best-response pure strategies with non-zero probability:

Definition 5 (ϵ-well-supported Nash equilibrium). *For any $\epsilon > 0$ a strategy profile $(\mathbf{x}, \mathbf{y})$ is an ϵ-well-supported Nash equilibrium for the $n \times m$ bimatrix game $\Gamma = \langle A, B \rangle$ if*

1. *For all pure strategies $i \in [n]$ of the row player,*

$$x_i > 0 \Rightarrow \mathbf{e}_i^T A \mathbf{y} \geq \mathbf{e}_k^T A \mathbf{y} - \epsilon \quad \forall k \in [n]$$

2. *For all pure strategies $j \in [m]$ of the column player,*

$$y_j > 0 \Rightarrow \mathbf{x}^T B \mathbf{e}_j \geq \mathbf{x}^T B \mathbf{e}_k - \epsilon \quad \forall k \in [m] \ .$$

Observe that every ϵ-well-supported Nash equilibrium is also an ϵ-Nash equilibrium, but the converse need not be true. However, as pointed out in [4], for every $\epsilon > 0$, given an $\frac{\epsilon}{8n}$-Nash equilibrium we can compute in polynomial time an ϵ-well-supported Nash equilibrium.

We define an *almost Nash equilibrium* as an ϵ-well-supported Nash equilibrium such that ϵ tends to zero as the number of the available pure strategies of the players tends to infinity:

Definition 6 (Almost Nash equilibrium). *A strategy profile $(\mathbf{x}, \mathbf{y})$ is an almost Nash equilibrium for the $n \times n$ bimatrix game $\Gamma = \langle A, B \rangle$ if*

1. *For all pure strategies $i \in [n]$ of the row player,*

$$x_i > 0 \Rightarrow \mathbf{e}_i^T A \mathbf{y} \geq \mathbf{e}_k^T A \mathbf{y} - \varepsilon(n) \quad \forall k \in [n]$$

2. *For all pure strategies $j \in [n]$ of the column player,*

$$y_j > 0 \Rightarrow \mathbf{x}^T B \mathbf{e}_j \geq \mathbf{x}^T B \mathbf{e}_k - \varepsilon(n) \quad \forall k \in [n] \ .$$

3. $\lim_{n \to \infty} \varepsilon(n) = 0$.

2.4   Random Bimatrix Games

As pointed out in [4], the set of Nash equilibria of a bimatrix game remains precisely the same if we multiply all entries of a matrix by a positive constant or if we add the same constant to each entry. Therefore, it suffices to consider bimatrix games with normalized matrices so as to study their complexity. We adopt the normalization used in [11]: we assume that the value of each entry in the matrices is lower bounded by 0 and upper bounded by 1. Such games are referred to as *positively normalized* [4]. Furthermore, in this work we focus on *random* bimatrix games:

Definition 7 (Random bimatrix game). *A $n \times m$ random bimatrix game $\Gamma = \langle A, B \rangle$ is $n \times m$ bimatrix game such that*

1. *all elements of matrix A are independent random variables, each taking a value in the interval $[0, 1]$ and with expectation $\mu_A \in [0, 1]$, and*
2. *all elements of matrix B are independent random variables, each taking a value in the interval $[0, 1]$ and with expectation $\mu_B \in [0, 1]$.*

Note that, according to the above definition, the entries of each matrix of a bimatrix game need not be identically distributed; it suffices that they all have the same expectation.

3   Approximate Nash Equilibria in Random Games

In the following, we deal with the problem of computing an ϵ-Nash equilibrium of a random $n \times n$ bimatrix game $\Gamma = \langle A, B \rangle$. We show that the completely mixed, uniform strategy profile is, with high probability, a $\sqrt{\frac{\ln n}{n}}$-Nash equilibrium. For the proof we will use the following lemma:

Lemma 1 ([8]). *If $Y_1, \ldots Y_m$, $Z_1, \ldots, Z_n$ are independent random variables with values in the interval $[a, b]$, and if $\bar{Y} = (Y_1 + \cdots + Y_m)/m$ and $\bar{Z} = (Z_1 + \cdots + Z_n)/n$, then for $t > 0$*

$$\Pr\left\{\bar{Y} - \bar{Z} - \left(\mathcal{E}[\bar{Y}] - \mathcal{E}[\bar{Z}]\right) \geq t\right\} \leq \exp\left(-\frac{2t^2}{\left(\frac{1}{m} + \frac{1}{n}\right)(b - a)^2}\right).$$

Theorem 1. *Let $\Gamma = \langle A, B \rangle$ be a $n \times m$ random bimatrix game and let $(\mathbf{x}, \mathbf{y})$ be the completely mixed, uniform strategy profile for Γ. Then, for any $\epsilon > 0$,*

$$\Pr\left\{(\mathbf{x}, \mathbf{y}) \text{ is an } \epsilon\text{-Nash equilibrium}\right\} \geq 1 - \left(n \exp(-2m\epsilon^2) + m \exp(-2n\epsilon^2)\right).$$

Proof. Fix $r \in [n]$. Then

$$\mathbf{e}_r^T A \mathbf{y} = \sum_{j \in [m]} a_{rj} y_j = \frac{1}{m} \sum_{j \in [m]} a_{rj},$$

$$\mathbf{x}^T A \mathbf{y} = \sum_{i \in [n]} \sum_{j \in [m]} a_{ij} x_i y_j = \frac{1}{nm} \sum_{i \in [n]} \sum_{j \in [m]} a_{ij}.$$

Now,

$$
\begin{aligned}
\mathbf{e}_r^T A \mathbf{y} - \mathbf{x}^T A \mathbf{y} &= \frac{1}{m} \sum_{j \in [m]} a_{rj} - \frac{1}{nm} \sum_{i \in [n]} \sum_{j \in [m]} a_{ij} \\
&= \frac{1}{m} \sum_{j \in [m]} a_{rj} - \frac{1}{nm} \sum_{j \in [m]} a_{rj} - \frac{1}{nm} \sum_{i \in [n], i \neq r} \sum_{j \in [m]} a_{ij} \\
&= \frac{n-1}{nm} \sum_{j \in [m]} a_{rj} - \frac{1}{nm} \sum_{i \in [n], i \neq r} \sum_{j \in [m]} a_{ij} \ ,
\end{aligned}
$$

and therefore, for any $\epsilon > 0$,

$$
\Pr\left\{ \mathbf{e}_r^T A \mathbf{y} - \mathbf{x}^T A \mathbf{y} \geq \epsilon \right\} =
$$
$$
\Pr\left\{ \frac{\sum_{j \in [m]} a_{rj}}{m} - \frac{\sum_{i \in [n], i \neq r} \sum_{j \in [m]} a_{ij}}{(n-1)m} \geq \frac{n}{n-1}\epsilon \right\} .
$$

Now we can apply Lemma 1: all random variables are mutually independent, $a = 0$ and $b = 1$, $t = \frac{n}{n-1}\epsilon$. Moreover,

$$
\mathcal{E}\left[\frac{1}{m} \sum_{j \in [m]} a_{rj} \right] = \frac{1}{m} \cdot m \cdot \mu_A = \mu_A
$$
$$
\mathcal{E}\left[\frac{1}{(n-1)m} \sum_{i \in [n], i \neq r} \sum_{j \in [m]} a_{ij} \right] = \frac{1}{(n-1)m} \cdot (n-1)m \cdot \mu_A = \mu_A \ .
$$

So

$$
\begin{aligned}
\Pr\left\{ \mathbf{e}_r^T A \mathbf{y} - \mathbf{x}^T A \mathbf{y} \geq \epsilon \right\} &\leq \exp\left(-\frac{2\frac{n^2}{(n-1)^2}\epsilon^2}{\left(\frac{1}{m} + \frac{1}{(n-1)m}\right)(1-0)^2} \right) \\
&= \exp\left(-2\frac{nm}{n-1}\epsilon^2 \right) \\
&\leq \exp(-2m\epsilon^2) \ .
\end{aligned}
$$

Now,

$$
\begin{aligned}
\Pr\left\{ \exists r \in [n] : \mathbf{e}_r^T A \mathbf{y} - \mathbf{x}^T A \mathbf{y} \geq \epsilon \right\} &\leq \sum_{r \in [n]} \Pr\left\{ \mathbf{e}_r^T A \mathbf{y} - \mathbf{x}^T A \mathbf{y} \geq \epsilon \right\} \\
&\leq n\exp(-2m\epsilon^2) \ .
\end{aligned}
$$

Similarly, we can show that for any $c \in [m]$,

$$
\Pr\left\{ \mathbf{x}^T B \mathbf{e}_c - \mathbf{x}^T B \mathbf{y} \geq \epsilon \right\} \leq \exp(-2n\epsilon^2)
$$

and hence

$$\Pr\left\{\exists c \in [m] : \mathbf{x}^T B \mathbf{e}_c - \mathbf{x}^T B \mathbf{y} \geq \epsilon\right\} \leq \sum_{c \in [m]} \Pr\left\{\mathbf{x}^T B \mathbf{e}_c - \mathbf{x}^T B \mathbf{y} \geq \epsilon\right\}$$
$$\leq m \exp(-2n\epsilon^2) \ .$$

Therefore, for any $\epsilon > 0$,

$$\Pr\left\{(\mathbf{x}, \mathbf{y}) \text{ is an } \epsilon\text{-Nash equilibrium}\right\} \geq 1 - \left(n \exp(-2m\epsilon^2) + m \exp(-2n\epsilon^2)\right),$$

as needed.   □

Corollary 1. *Consider a $n \times n$ random bimatrix game Γ. Then the completely mixed, uniform strategy profile is, with high probability, a $\sqrt{\frac{\ln n}{n}}$-Nash equilibrium for Γ.*

Proof. Let $(\mathbf{x}, \mathbf{y})$ denote the uniform strategy profile for Γ. By Theorem 1,

$$\Pr\left\{(\mathbf{x}, \mathbf{y}) \text{ is an } \epsilon\text{-Nash equilibrium}\right\} \geq 1 - 2\left(n \exp(-2n\epsilon^2)\right) \ .$$

Setting $\epsilon = \sqrt{\frac{\ln n}{n}}$, it follows that the uniform strategy profile is a $\sqrt{\frac{\ln n}{n}}$-Nash equilibrium for $\langle A, B \rangle$ with probability

$$\Pr\left\{(\mathbf{x}, \mathbf{y}) \text{ is a } \sqrt{\frac{\ln n}{n}}\text{-Nash equilibrium}\right\} \geq 1 - 2n \exp\left(-2n\frac{\ln n}{n}\right) = 1 - \frac{2}{n} \ .\square$$

4 Well-Supported Nash Equilibria in Random Games

In this section we show that the completely mixed, uniform strategy profile is, with high probability, a $\sqrt{\frac{3 \ln n}{n}}$-well supported Nash equilibrium for a $n \times n$ random bimatrix game.

Theorem 2. *Let $\Gamma = \langle A, B \rangle$ be a $n \times n$ random bimatrix game. Then the completely mixed, uniform strategy profile is a $\sqrt{\frac{3 \ln n}{n}}$-well supported Nash equilibrium for Γ, with high probability.*

Proof. By definition, the completely mixed, uniform strategy profile $(\mathbf{x}, \mathbf{y})$ is an ϵ-well supported Nash equilibrium if and only if, for all $i, j \in [n]$,

$$\mathbf{e}_i^T A \mathbf{y} \geq \mathbf{e}_j^T A \mathbf{y} - \epsilon$$
$$\mathbf{e}_j^T A \mathbf{y} - \mathbf{e}_i^T A \mathbf{y} \leq \epsilon$$
$$\frac{1}{n} \sum_{k=1}^{n} a_{jk} - \frac{1}{n} \sum_{k=1}^{n} a_{ik} \leq \epsilon$$

and

$$\mathbf{x}^T B \mathbf{e}_i \geq \mathbf{x}^T B \mathbf{e}_j - \epsilon$$
$$\mathbf{x}^T B \mathbf{e}_j - \mathbf{x}^T B \mathbf{e}_i \leq \epsilon$$
$$\frac{1}{n} \sum_{k=1}^{n} b_{kj} - \frac{1}{n} \sum_{k=1}^{n} b_{ki} \leq \epsilon .$$

Now fix $i, j \in [n]$. Then

$$\mathcal{E}\left[\frac{1}{n} \sum_{k=1}^{n} a_{jk}\right] = \mathcal{E}\left[\frac{1}{n} \sum_{k=1}^{n} a_{ik}\right] = \frac{1}{n} \cdot n \cdot \mu_A = \mu_A$$

$$\mathcal{E}\left[\frac{1}{n} \sum_{k=1}^{n} b_{kj}\right] = \mathcal{E}\left[\frac{1}{n} \sum_{k=1}^{n} b_{ki}\right] = \frac{1}{n} \cdot n \cdot \mu_B = \mu_B ,$$

therefore, by Lemma 1,

$$\Pr\left\{\frac{1}{n} \sum_{k=1}^{n} a_{jk} - \frac{1}{n} \sum_{k=1}^{n} a_{ik} \geq \epsilon\right\} \leq \exp\left(-\frac{2\epsilon^2}{\left(\frac{1}{n} + \frac{1}{n}\right)(1-0)^2}\right) = \exp\left(-n\epsilon^2\right) ,$$

and similarly

$$\Pr\left\{\frac{1}{n} \sum_{k=1}^{n} b_{kj} - \frac{1}{n} \sum_{k=1}^{n} b_{ki} \geq \epsilon\right\} \leq \exp\left(-n\epsilon^2\right) .$$

Thus

$$\Pr\{(\mathbf{x}, \mathbf{y}) \quad \text{is an } \epsilon\text{-well-supported Nash equilibrium}\}$$
$$= 1 - \Pr\left\{\exists i, j : \mathbf{e}_i^T A \mathbf{y} \geq \mathbf{e}_j^T A \mathbf{y} - \epsilon \quad \text{or} \quad \mathbf{x}^T B \mathbf{e}_i \geq \mathbf{x}^T B \mathbf{e}_j - \epsilon\right\}$$
$$\geq 1 - \binom{n}{2} \cdot 2 \cdot \exp\left(-n\epsilon^2\right)$$
$$= 1 - n(n-1)\exp\left(-n\epsilon^2\right) .$$

Setting $\epsilon = \sqrt{\frac{3 \ln n}{n}}$ we get

$$\Pr\{(\mathbf{x}, \mathbf{y}) \quad \text{is an } \epsilon\text{-well-supported Nash equilibrium}\}$$
$$\geq 1 - n(n-1)\exp\left(-n\frac{3 \ln n}{n}\right)$$
$$\geq 1 - n^2 \cdot \frac{1}{n^3}$$
$$= 1 - \frac{1}{n} .$$

Therefore the completely mixed, uniform strategy profile $(\mathbf{x}, \mathbf{y})$ is a $\sqrt{\frac{3 \ln n}{n}}$-well supported Nash equilibrium for Γ with high probability. $\qquad\square$

Corollary 2. *The completely mixed, uniform strategy profile is an almost Nash equilibrium for random bimatrix games, with high probability.*

References

[1] Althöfer, I.: On sparse approximations to randomized strategies and convex combinations. Linear Algebra and Applications 199, 339–355 (1994)
[2] Bárány, I., Vempala, S., Vetta, A.: Nash equilibria in random games. In: Proceedings of the 46th Annual IEEE Symposium on Foundations of Computer Science (FOCS 2005), pp. 123–131 (2005)
[3] Chen, X., Deng, X.: Settling the complexity of 2-player Nash-equilibrium. In: Proceedings of the 47th Annual IEEE Symposium on Foundations of Computer Science, FOCS 2006 (2005)
[4] Chen, X., Deng, X., Teng, S.-H.: Computing Nash equilibria: Approximation and smoothed complexity. In: Electronic Colloquium on Computational Complexity, ECCC (2006)
[5] Daskalakis, C., Goldberg, P., Papadimitriou, C.: The complexity of computing a Nash equilibrium. In: Proceedings of the 38th Annual ACM Symposium on Theory of Computing (STOC 2006), pp. 71–78 (2006)
[6] Daskalakis, C., Papadimitriou, C.: Three-player games are hard. In: Electronic Colloquium on Computational Complexity, ECCC (2005)
[7] Goldberg, P., Papadimitriou, C.: Reducibility among equilibrium problems. In: Proceedings of the 38th Annual ACM Symposium on Theory of Computing (STOC 2006), pp. 61–70 (2006)
[8] Hoeffding, W.: Probability inequalities for sums of bounded random variables. Journal of the American Statistical Association 58, 13–30 (1963)
[9] Lemke, C.E.: Bimatrix equilibrium points and mathematical programming. Management Science 11, 681–689 (1965)
[10] Lemke, C.E., Howson, J.T.: Equilibrium points of bimatrix games. J. Soc. Indust. Appl. Math. 12, 413–423 (1964)
[11] Lipton, R.J., Markakis, E., Mehta, A.: Playing large games using simple startegies. In: Proceedings of the 4th ACM Conference on Electronic Commerce (EC 2003), pp. 36–41 (2003)
[12] Nash, J.: Noncooperative games. Annals of Mathematics 54, 289–295 (1951)
[13] Papadimitriou, C.H.: On inefficient proofs of existence and complexity classes. In: Proceedings of the 4th Czechoslovakian Symposium on Combinatorics (1991)
[14] Savani, R., von Stengel, B.: Exponentially many steps for finding a nash equilibrium in a bimatrix game. In: Proceedings of the 45th Annual IEEE Symposium on Foundations of Computer Science (FOCS 2004), pp. 258–267 (2004)
[15] Tsaknakis, H., Spirakis, P.G.: An optimization approach for approximate Nash equilibria. In: Deng, X., Graham, F.C. (eds.) WINE 2007. LNCS, vol. 4858, pp. 42–56. Springer, Heidelberg (2007)

Complexity of Rational and Irrational Nash Equilibria

Vittorio Bilò[1,*] and Marios Mavronicolas[2,**]

[1] Dipartimento di Matematica "Ennio De Giorgi", Università del Salento,
Provinciale Lecce-Arnesano, P.O. Box 193, 73100 Lecce, Italy
vittorio.bilo@unisalento.it
[2] Department of Computer Science, University of Cyprus, Nicosia CY-1678, Cyprus
mavronic@cs.ucy.ac.cy

Abstract. We introduce two new decision problems, denoted as ∃ RA-
TIONAL NASH and ∃ IRRATIONAL NASH, pertinent to the rational-
ity and irrationality, respectively, of *Nash equilibria* for (finite) strategic
games. These problems ask, given a strategic game, whether or not it
admits *(i)* a *rational* Nash equilibrium where all probabilities are rational numbers, and *(ii)* an *irrational* Nash equilibrium where at least
one probability is irrational, respectively. We are interested here in the
complexities of ∃ RATIONAL NASH and ∃ IRRATIONAL NASH.

Towards this end, we study two other decision problems, denoted as
NASH-EQUIVALENCE and NASH-REDUCTION, pertinent to some mutual properties of the sets of Nash equilibria of two given strategic games
with the same number of players. NASH-EQUIVALENCE asks whether
the two sets of Nash equilibria coincide; we identify a restriction of
its complementary problem that witnesses ∃ RATIONAL NASH. NASH-
REDUCTION asks whether or not there is a so called *Nash reduction* (a
suitable map between corresponding strategy sets of players) that yields
a Nash equilibrium of the former game from a Nash equilibrium of the
latter game; we identify a restriction of it that witnesses ∃ IRRATIONAL
NASH.

As our main result, we provide two distinct reductions to simultaneously show that *(i)* NASH-EQUIVALENCE is co-$\mathcal{NP}$-hard and ∃ RATIONAL
NASH is $\mathcal{NP}$-hard, and *(ii)* NASH-REDUCTION and ∃ IRRATIONAL NASH
are $\mathcal{NP}$-hard, respectively. The reductions significantly extend techniques
previously employed by Conitzer and Sandholm [6, 7].

1 Introduction

Motivation, Framework and Techniques. Understanding the complexity of
algorithmic problems pertinent to equilibria in (finite) strategic games is one

---

* Partially supported by the PRIN 2008 research project COGENT "Computational
and game-theoretic aspects of uncoordinated networks" funded by the Italian Ministry of University and Research. Part of the work of this author was done while
visiting Università di L'Aquila, Italy and University of Cyprus, Cyprus.
** Partially supported by research funds at the University of Cyprus. Part of the work
of this author was done while visiting Università di L'Aquila, Italy.

G. Persiano (Ed.): SAGT 2011, LNCS 6982, pp. 200–211, 2011.
© Springer-Verlag Berlin Heidelberg 2011

of the most intensively studied topics in *Algorithmic Game Theory* today (see, for example, [3–9, 15, 16, 19] and references therein). Much of this research has focused on *Nash equilibria* [17, 18], perhaps the most influential equilibrium concept in all of *Game Theory* ever. In the wake of the complexity results for *search problems* about Nash equilibria, a series of breakthrough results [4, 8] shows that, even for *two-player* games, computing an (exact) Nash equilibrium is complete for $\mathcal{PPAD}$ [19], a complexity class to capture the computation of discrete fixed points; so also is the problem of computing an *approximate* Nash equilibrium for games with any number of players.

In this work, we study the complexity of *decision problems* related to Nash equilibria. The celebrated result of John Nash [17, 18] shows that every (finite) game admits a *mixed* Nash equilibrium; so, it trivializes the decision problem for (the existence of) mixed Nash equilibria, while it simultaneously opens up a wide avenue for studying the complexity of decision problems for Nash equilibria with certain properties (e.g., *pure*). To the best of our knowledge, Gilboa and Zemel [13] were the *first* to present complexity results (more specifically, $\mathcal{NP}$-hardness results) about mixed Nash equilibria (and *correlated equilibria*) for games represented in explicit form; they identified some $\mathcal{NP}$-hard decision problems about the existence of (mixed) Nash equilibria with certain properties for *two-player* strategic games.

Much later, Conitzer and Sandholm [6, 7] provided a very notable *unifying reduction*, henceforth abbreviated as CS-reduction, to show that all decision problems from [13] and many more are $\mathcal{NP}$-hard. The CS-reduction [6, 7] yields a two-player game out of a CNF formula ϕ; it is then shown that the game has a Nash equilibrium with certain properties (in addition to some fixed *pure* Nash equilibrium) if and only if ϕ is satisfiable. Hence, deciding the properties is $\mathcal{NP}$-hard. The CS-reduction uses literals, variables and clauses from the formula ϕ together with a special strategy f as the strategies of each player. The essence is that *(i)* both players choosing f results in a pure Nash equilibrium, and *(ii)* a player could otherwise improve (by switching to f) unless both players only randomize over literals. More important, a Nash equilibrium where both players only randomize over literals is possible (and has certain properties) if and only if ϕ is satisfiable.

In this paper, we extend the work from [6, 7, 13]. We study *for the first time* the complexity of *deciding* rationality and irrationality properties of mixed Nash equilibria. Recall that a Nash equilibrium is *rational* if all involved probabilities are rational and otherwise it is *irrational*; all *two-player* games have only rational Nash equilibria, while there are known three-player games with no rational Nash equilibrium (cf. [18]). The corresponding decision problems, denoted as ∃ RATIONAL NASH and ∃ IRRATIONAL NASH, respectively, ask if there is a rational (resp., irrational) Nash equilibrium[1]. So, both problems trivialize when restricted to two-player games but become non-trivial for games with at least three players. Since the CS-reduction [6, 7] applies to two-player games, it will

---

[1] We were inspired to study these problems by a corresponding question posed by E. Koutsoupias [14] to M. Yannakakis during his Invited Talk at *SAGT 2009*.

not be directly applicable to settling the complexity of ∃ RATIONAL NASH and
∃ IRRATIONAL NASH.

Contribution and Significance. Gross plan: Establishing the $\mathcal{NP}$-hardness of
the problems ∃ RATIONAL NASH and ∃ IRRATIONAL NASH is achieved via the
following plan:

1. Identify suitable decision problems witnessing ∃ RATIONAL NASH and ∃
   IRRATIONAL NASH that make no reference to rationality or irrationality.
   These will be NASH-EQUIVALENCE and NASH-REDUCTION, respectively;
   on input a pair of strategic games SG and $\widehat{SG}$ with the same number of players
   $r \geq 2$, they inquire about some mutual properties of their Nash equilibria.
2. Use CS-like reductions to simultaneously show that both problems NASH-
   EQUIVALENCE and ∃ RATIONAL NASH (resp., NASH-REDUCTION and ∃
   IRRATIONAL NASH) are $\mathcal{NP}$-hard.

The problems NASH-EQUIVALENCE and ∃ RATIONAL NASH: NASH-
EQUIVALENCE asks whether the sets of Nash equilibria of SG and $\widehat{SG}$ coincide. Fixing $\widehat{SG}$ to some *gadget* game yields the restricted problem NASH-
EQUIVALENCE($\widehat{SG}$); so, NASH-EQUIVALENCE($\widehat{SG}$) takes the single input SG.
Note that *(i)* if the set of Nash equilibria for $\widehat{SG}$ is a subset of those for SG *and*
(ii) $\widehat{SG}$ has no rational Nash equilibrium, then SG has a rational Nash equilibrium if and only if the set of Nash equilibria for SG and $\widehat{SG}$ do *not* coincide. So,
the existence of a rational Nash equilibrium is a witness to the *non-equivalence*
of SG and $\widehat{SG}$. So, if NASH-EQUIVALENCE($\widehat{SG}$) is $\mathcal{NP}$-hard, then so is ∃ RA-
TIONAL NASH. We show that NASH-EQUIVALENCE($\widehat{SG}$) is $\mathcal{NP}$-hard for an
arbitrary but *fixed* strategic game $\widehat{SG}$ (Proposition 2 and Theorem 1). Fixing
$\widehat{SG}$ to admit no rational Nash equilibrium yields that ∃ RATIONAL NASH is
$\mathcal{NP}$-hard (Proposition 4 and Theorem 2).

The problems NASH-REDUCTION and ∃ IRRATIONAL NASH: NASH-
REDUCTION asks whether there is a Nash reduction from SG to $\widehat{SG}$. Roughly
speaking, a *Nash reduction* consists of a family of surjective functions, one per
player, mapping the strategy set of each player in SG to the strategy set of the
same player in $\widehat{SG}$. Note that any family of surjective functions induces a map
from mixed profiles for SG to mixed profiles for $\widehat{SG}$ in the natural way: probabilities to different strategies of a player in SG that map to the same strategy (of
the player) in $\widehat{SG}$ are added up. However, a Nash reduction must, in addition,
preserve at least one Nash equilibrium: there must be a Nash equilibrium σ
for SG that maps to a Nash equilibrium $\widehat{\sigma}$ for $\widehat{SG}$. Note that *(i)* if there is a
Nash reduction from SG to $\widehat{SG}$, *and (ii)* SG has only rational Nash equilibria,
then $\widehat{SG}$ has at least one rational Nash equilibrium. So, if $\widehat{SG}$ is chosen to have
no rational Nash equilibrium, then *either* there is no Nash reduction from SG
to $\widehat{SG}$ *or* SG has an irrational Nash equilibrium. Hence, the inexistence of an
irrational Nash equilibrium is a witness to the inexistence of a Nash reduction

from SG to $\widehat{\text{SG}}$. So, if NASH-REDUCTION($\widehat{\text{SG}}$) is $\mathcal{NP}$-hard, then so is $\exists$ IRRA-TIONAL NASH. We show that NASH-REDUCTION($\widehat{\text{SG}}$) is $\mathcal{NP}$-hard for a *fixed* strategic game $\widehat{\text{SG}}$ which *(a)* is constant-sum with sum $r \cdot u$, *(b)* has a unique Nash equilibrium which is *(b/i)* fully mixed and in which *(b/ii)* the utility of each player is u (Proposition 6 and Theorem 3). Fixing the gadget $\widehat{\text{SG}}$ so that, in addition, it admits no rational Nash equilibrium yields that $\exists$ IRRATIONAL NASH is $\mathcal{NP}$-hard (Proposition 8 and Theorem 4).

To the best of our knowledge, our complexity results for $\exists$ RATIONAL NASH and $\exists$ IRRATIONAL NASH are the *first $\mathcal{NP}$-hardness* results for a decision problem inquiring the existence of a combinatorial object involving rational (resp., irrational) numbers; no such $\mathcal{NP}$-hard problems are listed in [12].

Other Related Work. Etessami and Yannakakis [9] study the related search problem of *computing* an approximation to a Nash equilibrium (for games with at least three players) within a specified precision: a rational point that differs from an (irrational) Nash equilibrium by at most ϵ in every coordinate (which is different than computing an ϵ-approximate Nash equilibrium, which may be very far from an actual Nash equilibrium). It is shown [9, Theorem 4] that placing this problem in $\mathcal{NP}$ would imply the breakthrough result that the SQUARE-ROOT-SUM [11] problem in also in $\mathcal{NP}$, which is a long-standing open problem in Complexity Theory. It is also shown [9, Theorem 18] that the same problem is complete for the complexity class $\mathcal{FIXP}$ introduced there.

Fiat and Papadimitriou [10] use a new gadget based on a generalized *rock-paper-scissors game* in a reduction arguably simpler than the CS-reduction [6, 7]; thereby, they prove [10, Theorem 5] that it is $\mathcal{NP}$-hard to decide if a two-player game where players are not *expectation-maximizers* has a Nash equilibrium. (The existence result of Nash [17, 18] does not apply to such games.) Austrin et al [2] prove that approximate variants of several $\mathcal{NP}$-hard problems about Nash equilibria from [6, 7] are as hard as finding a hidden clique of size $O(\log n)$ in the random graph $G(n, 1/2)$.

2   Definitions and Preliminaries

Background from Game Theory. A *game* is a triple $\text{SG} = \langle [r], \{\Sigma_i\}_{i \in [r]}, \{U_i\}_{i \in [r]} \rangle$, where: *(i)* $[r] = \{1, \ldots, r\}$ is a finite set of *players* with $r \geq 2$, and *(ii)* for each player $i \in [r]$, Σ_i is the set of *strategies* for player i, and U_i is the *utility function* $U_i : \times_{k \in [r]} \Sigma_k \to \mathbb{R}$ for player i. For any integer $r \geq 2$, denote as r-$\mathcal{SG}$ the set of r-*player games*; so, $\mathcal{SG} = \bigcup_{r \geq 2} r$-$\mathcal{SG}$ is the set of all games.

Denote $\Sigma = \times_{k \in [r]} \Sigma_k$. A *profile* is a tuple $\mathbf{s}$ of r strategies, one for each player. For a profile $\mathbf{s}$, the vector $U(\mathbf{s}) = \langle U_1(\mathbf{s}), \ldots, U_r(\mathbf{s}) \rangle$ is called the *utility vector*. The game SG is *constant-sum* if there is a constant c such that for any profile $\mathbf{s}$, $\sum_{i \in [r]} U_i(\mathbf{s}) = c$. A *partial profile* $\mathbf{s}_{-i}$ is a tuple of $r - 1$ strategies, one for each player other than i; it results by eliminating the strategy s_i of player $i \in [r]$ from the profile $\mathbf{s}$. For a profile $\mathbf{s}$ and a strategy $t_i \in \Sigma_i$ of player i, denoted as $\mathbf{s}_{-i} \diamond t_i$ the profile obtained by substituting strategy t_i for strategy s_i in the profile $\mathbf{s}$. Denote $\Sigma_{-i} = \times_{k \in [r] | k \neq i} \Sigma_k$.

A **mixed strategy** for player $i \in [r]$ is a probability distribution σ_i on her strategy set Σ_i; so, a mixed strategy for player i is a probability distribution on her strategies: a function $\sigma_i : \Sigma_i \to [0,1]$ such that $\sum_{s \in \Sigma_i} \sigma_i(s) = 1$. Denote as $\mathsf{Support}(\sigma_i)$ the set of strategies $s \in \Sigma_i$ such that $\sigma_i(s) > 0$. The mixed strategy $\sigma_i : \Sigma_i \to [0,1]$ is **rational** if all values of σ_i are rational numbers; else, it is **irrational**.

A **mixed profile** $\boldsymbol{\sigma} = (\sigma_i)_{i \in [r]}$ is a tuple of mixed strategies, one for each player. A mixed profile $\boldsymbol{\sigma}$ induces a probability measure $\mathbb{P}_{\boldsymbol{\sigma}}$ on the set of profiles in the natural way. Say that the profile $\mathbf{s}$ is **enabled** in the mixed profile $\boldsymbol{\sigma}$, and write $\mathbf{s} \sim \boldsymbol{\sigma}$, if $\mathbb{P}_{\boldsymbol{\sigma}}(\mathbf{s}) > 0$; note that for a profile $\mathbf{s}$, $\mathbb{P}_{\boldsymbol{\sigma}}(\mathbf{s}) = \prod_{k \in [r]} \sigma_k(s_k)$. Under the mixed profile $\boldsymbol{\sigma}$, the utility of each player becomes a random variable. So, associated with the mixed profile $\boldsymbol{\sigma}$ is the **expected utility** for each player $i \in [r]$, denoted as $\mathsf{U}_i(\boldsymbol{\sigma})$ and defined as the expectation according to $\mathbb{P}_{\boldsymbol{\sigma}}$ of her utility for a profile $\mathbf{s}$ enabled in the mixed profile $\boldsymbol{\sigma}$; so, $\mathsf{U}_i(\boldsymbol{\sigma}) = \mathbb{E}_{\mathbf{s} \sim \boldsymbol{\sigma}}(\mathsf{U}_i(\mathbf{s})) = \sum_{\mathbf{s} \in \Sigma(\mathsf{SG})} \mathbb{P}_{\boldsymbol{\sigma}}(\mathbf{s}) \cdot \mathsf{U}_i(\mathbf{s}) = \sum_{\mathbf{s} \in \Sigma(\mathsf{SG})} \left(\prod_{k \in [r]} \sigma_k(s_k) \right) \cdot \mathsf{U}_i(\mathbf{s})$. A **partial mixed profile** $\boldsymbol{\sigma}_{-i}$ is a tuple of $r-1$ mixed strategies, one for each player other than i. For a mixed profile $\boldsymbol{\sigma}$ and a mixed strategy τ_i of player $i \in [r]$, denote as $\boldsymbol{\sigma}_{-i} \diamond \tau_i$ the mixed profile obtained by substituting the mixed strategy τ_i for the mixed strategy σ_i in the mixed profile $\boldsymbol{\sigma}$. A mixed profile is **rational** if all of its mixed strategies are rational; else, it is **irrational**. So, a profile is the degenerate case of a rational mixed profile where all rational probabilities are either 0 or 1.

A **pure Nash equilibrium**, or **Nash equilibrium** for short, is a profile $\mathbf{s} \in \Sigma$ such that for each player $i \in [r]$, for each strategy $t_i \in \Sigma_i$, $\mathsf{U}_i(\mathbf{s}) \geq \mathsf{U}_i(\mathbf{s}_{-i} \diamond t_i)$. A **mixed Nash equilibrium** is a mixed profile $\boldsymbol{\sigma}$ such that for each player $i \in [r]$, for each mixed strategy τ_i, $\mathsf{U}_i(\boldsymbol{\sigma}) \geq \mathsf{U}_i(\boldsymbol{\sigma}_{-i} \diamond \tau_i)$. For the mixed profile $\boldsymbol{\sigma}$ to be a mixed Nash equilibrium, it is (necessary and) sufficient that for each player $i \in [r]$, for each strategy $t_i \in \Sigma_i$, $\mathsf{U}_i(\boldsymbol{\sigma}) \geq \mathsf{U}_i(\boldsymbol{\sigma}_{-i} \diamond t_i)$. For a strategic game SG, denote as $\mathcal{NE}(\mathsf{SG})$ the set of Nash equilibria for SG. Nash equilibria will classified as **pure**, **mixed rational** (or **rational** for short) and **mixed irrational** (or **irrational** for short) in the natural way.

Background from Complexity Theory. A *decision problem* Π is identified with a set of *positive instances* encoded over $\{0,1\}$; so, $\Pi \subseteq \{0,1\}^*$. The class $\mathcal{NP}$ is the set of all decision problems for which there is a deterministic, polynomial-time algorithm V such that for every instance w, w is a positive instance if and only if there is a *certificate* $\mathbf{c} \in \{0,1\}^{\mathsf{p}(|w|)}$ (for some polynomial $\mathsf{p} : \mathbb{N} \to \mathbb{N}$) such that V *accepts* the input $\langle w, \mathbf{c} \rangle$. The class co-$\mathcal{NP}$ is the set of all complements of decision problems in $\mathcal{NP}$. Say that the decision problem Π' *polynomially reduces to* the decision problem Π, denoted as $\Pi' \leq_{\mathsf{P}} \Pi$, if there is a polynomial-time function $\mathsf{f} : \{0,1\} \to \{0,1\}^*$ such that w is a positive instance for Π_1 if and only if w is a positive instance for Π_2. The decision problem Π is $\mathcal{NP}$-*hard* if for every decision problem $\Pi' \in \mathcal{NP}$, $\Pi' \leq_{\mathsf{P}} \Pi$.

A boolean formula ϕ is in *Conjunctive Normal Form*, abbreviated as CNF, if $\phi = \bigwedge_{1 \leq i \leq k} \bigvee_{1 \leq j \leq l} \ell_{ij}$, where ℓ_{ij} is a *literal*: a boolean variable or its negation; so, ϕ is a *conjunction* of *clauses*. Denote $\mathcal{C}(\phi) = \{ \bigvee_{j \in [l]} \ell_{ij} \mid i \in [k] \}$, the set of clauses in ϕ. Denote as $\mathsf{V}(\phi)$ and $\mathsf{L}(\phi)$ the sets of variables and literals, respectively, in the formula ϕ, with $|V(\phi)| = n$ and $|L(\phi)| = 2n$. The function $v : L(\phi) \rightarrow V(\phi)$ gives the variable corresponding to a literal; so, for a variable $v \in V(\phi)$, $v(\ell) = v(\overline{\ell}) = v$. An *assignment* is a function from $V(\phi)$ to $\{0, 1\}$; so, an assignment is represented by a tuple of literals $\langle \ell_1, \ldots, \ell_n \rangle$ where for each index $j \in [n]$, $v(\ell_j) = v_j$ and $\ell_j = 1$ (under the assignment). The formula ϕ is *satisfiable* if there is a *satisfying* assignment: one that makes ϕ equal to 1. The decision problem CNF SAT is identified with the set of satisfiable boolean formulas in CNF. We recall that CNF SAT is $\mathcal{NP}$-hard.

Some Decision Problems about Nash Equilibria.
Decision problems about Nash equivalence. A pair of games $\widehat{SG}$ and SG are ***Nash-equivalent*** if $\mathcal{NE}(\widehat{SG}) = \mathcal{NE}(SG)$. So, this leads to the following decision problem:

NASH-EQUIVALENCE
> INSTANCE: Two games $\widehat{SG}$ and SG from $r\text{-}\mathcal{SG}$, for some integer $r \geq 2$.
> QUESTION. Are $\widehat{SG}$ and SG Nash-equivalent?

The next decision problem is a restriction of NASH-EQUIVALENCE; it is parameterized by some *fixed* game $\widehat{SG}$ from $r\text{-}\mathcal{SG}$, for some integer $r \geq 2$. The game $\widehat{SG}$ will be called a ***gadget game***.

NASH-EQUIVALENCE $(\widehat{SG})$
> INSTANCE: A game SG from $r\text{-}\mathcal{SG}$.
> QUESTION: Are $\widehat{SG}$ and SG Nash-equivalent?

So, NASH-EQUIVALENCE $(\widehat{SG}) \leq_P$ NASH-EQUIVALENCE.
Decision problems about Nash reductions. Consider now a pair of games $\widehat{SG}, SG \in r\text{-}\mathcal{SG}$. Assume that for each player $i \in [r]$, $|\Sigma_i| \geq |\widehat{\Sigma}_i|$. A ***surjective mapping*** from SG to $\widehat{SG}$ is a family of *surjective functions* $\mathcal{H} = \{h_i\}_{i \in [r]}$ where for each player $i \in [r]$, $h_i : \Sigma_i \rightarrow \widehat{\Sigma}_i$; so, a surjective mapping maps strategies of each player $i \in [r]$ in the game SG to strategies of the same player in the game $\widehat{SG}$ in a surjective way. Note that a surjective mapping $\mathcal{H}$ from SG to $\widehat{SG}$ induces a corresponding surjective mapping from the set of profiles Σ for the game SG to the set of profiles $\widehat{\Sigma}$ for the game $\widehat{SG}$. In turn, it induces a mapping $\mathcal{H}$ from the set of mixed profiles for SG to the set of mixed profiles for $\widehat{SG}$ as follows. Consider a mixed profile $\boldsymbol{\sigma}$ for SG. Then, $\boldsymbol{\sigma}$ maps to the mixed profile $\widehat{\boldsymbol{\sigma}} = \mathcal{H}(\boldsymbol{\sigma})$, where for each player $i \in [r]$, for each strategy $\widehat{s} \in \widehat{\Sigma}_i$, $\widehat{\sigma}_i(\widehat{s}) = \sum_{s \in \Sigma_i | h_i(s) = \widehat{s}} \sigma_i(s)$.

A ***Nash reduction*** from the game SG to the game $\widehat{SG}$ is a surjective mapping from SG to $\widehat{SG}$ such that there is a Nash equilibrium $\boldsymbol{\sigma}$ for the game SG for which the mixed profile $\mathcal{H}(\boldsymbol{\sigma})$ is a Nash equilibrium for the game $\widehat{SG}$. Say that the game SG ***Nash-reduces to*** the game $\widehat{SG}$ if there is a Nash reduction from SG to $\widehat{SG}$. So, this leads to the following decision problem:

NASH-REDUCTION
>   INSTANCE: Two games $\widehat{SG}$ and SG from r-$\mathcal{SG}$, for some integer $r \geq 2$.
>   QUESTION: Does SG Nash-reduce to $\widehat{SG}$?

The next decision problem is a restriction of NASH-REDUCTION; it is parameterized by some *fixed* game $\widehat{SG}$ from r-$\mathcal{SG}$, for some integer $r \geq 2$. The game $\widehat{SG}$ will be called a ***gadget game***.

NASH-REDUCTION $(\widehat{SG})$
>   INSTANCE: A game $\widehat{SG}$ from r-$\mathcal{SG}$.
>   QUESTION: Does SG Nash-reduce to $\widehat{SG}$?

So, NASH-REDUCTION $(\widehat{SG}) \leq_P$ NASH-REDUCTION.
Rationality and Irrationality Problems: To the best of our knowledge, the following problems are new.

∃ RATIONAL NASH
>   INSTANCE: A game SG.
>   QUESTION: Does SG have a rational Nash equilibrium?

∃ IRRATIONAL NASH
>   INSTANCE: A game SG.
>   QUESTION: Does SG have an irrational Nash equilibrium?

3   Complexity of NASH-EQUIVALENCE and ∃ RATIONAL NASH

The Reduction. Given a CNF formula ϕ, construct the game $SG(\phi) = \left\langle [r], \{\Sigma_i\}_{i \in [r]}, \{U_i\}_{i \in [r]} \right\rangle$:

- For each player $i \in [2]$, $\Sigma_i := \widehat{\Sigma}_i \cup L(\phi) \cup V(\phi) \cup C(\phi)$; for each player $i \in [r] \setminus [2]$, $\Sigma_i := \widehat{\Sigma}_i \cup \{\delta\}$. Roughly speaking, each strategy in $\widehat{\Sigma}_i$ with $i \in [r]$ is inherited from the gadget game $\widehat{SG}$, while δ is some new strategy; the remaining strategies come from the formula ϕ. Players 1 and 2 are *special*; they are the only players whose sets of strategies are influenced by the formula ϕ.
- To specify the utility functions, we need some notation. First, π denotes a permutation on $[r]$. Second, denote $\overline{u} = \max \left\{ \widehat{U}_i(s) \right\}_{i \in [r], s \in \widehat{\Sigma}}$ and $\underline{u} = \min \left\{ \widehat{U}_i(s) \right\}_{i \in [r], s \in \widehat{\Sigma}}$; so, $\overline{u}$ and $\underline{u}$ are the maximum and minimum utilities, respectively, of a player in the strategic game $\widehat{SG}$.

  Fix now a profile $\mathbf{s} = \langle s_1, \ldots, s_r \rangle$ from $\Sigma = \Sigma_1 \times \ldots \times \Sigma_r$. Use $\mathbf{s}$ to partition $[r]$ into $\widehat{\mathcal{P}}(\mathbf{s}) = \{i \in [r] \mid s_i \in \widehat{\Sigma}_i\}$ and $\mathcal{P}(\mathbf{s}) = \{i \in [r] \mid s_i \notin \widehat{\Sigma}_i\}$; loosely speaking, $\widehat{\mathcal{P}}(\mathbf{s})$ and $\mathcal{P}(\mathbf{s})$ are the sets of players choosing and not choosing strategies inherited from $\widehat{SG}$, respectively. The utility vector $U(\mathbf{s})$ is depicted in the following table, where $v \in V(\phi)$, $\ell_1, \ell_2, \ell \in L(\phi)$ and $c \in C(\phi)$.

Case	Condition on the profile $\mathbf{s} = \langle s_1, \ldots, s_r \rangle$	Utility vector $\mathsf{U}(\mathbf{s})$
(1)	$\mathbf{s} = \langle \ell_1, \ell_2, \delta, \ldots, \delta \rangle$ with $\ell_1 \neq \overline{\ell}_2$	$\langle \overline{u} + 1, \ldots, \overline{u} + 1 \rangle$
(2)	$\mathbf{s} = \langle \ell_1, \ell_2, \delta, \ldots, \delta \rangle$ with $\ell_1 = \overline{\ell}_2$	$\langle \underline{u} - 1, \ldots, \underline{u} - 1 \rangle$
(3)	$\mathbf{s} = \langle v, \ell, \delta, \ldots, \delta \rangle$ with $\mathsf{v}(\ell) \neq v$	$\langle \overline{u} + 2, \underline{u} - 1, \ldots, \underline{u} - 1 \rangle$
(4)	$\mathbf{s} = \langle v, \ell, \delta, \ldots, \delta \rangle$ with $\mathsf{v}(\ell) \neq v$	$\langle \overline{u} + 2 - n, \underline{u} - 1, \ldots, \underline{u} - 1 \rangle$
(5)	$\mathbf{s} = \langle \mathsf{c}, \ell, \delta, \ldots, \delta \rangle$ with $\ell \notin \mathsf{c}$	$\langle \overline{u} + 2, \underline{u} - 1, \ldots, \underline{u} - 1 \rangle$
(6)	$\mathbf{s} = \langle \mathsf{c}, \ell, \delta, \ldots, \delta \rangle$ with $\ell \in \mathsf{c}$	$\langle \overline{u} + 2 - n, \underline{u} - 1, \ldots, \underline{u} - 1 \rangle$
(7)	For each $i \in [r]$, $s_i \in \widehat{\Sigma}_i$	$\widehat{\mathsf{U}}(\langle s_1, \ldots, s_r \rangle)$
(8)	$\widehat{\mathcal{P}}(\mathbf{s}) \neq \emptyset$ and $\mathcal{P}(\mathbf{s}) \neq \emptyset$	$\mathsf{U}_i(\mathbf{s}) = \overline{u} + 1$ if $i \in \widehat{\mathcal{P}}(\mathbf{s})$ or $\underline{u} - 1$ if $i \in \mathcal{P}(\mathbf{s})$
(9)	$\mathbf{s} = \pi(\mathbf{t})$, where $\mathbf{t}$ falls in one of the Cases (1) through (8)	$\pi\left(\widehat{\mathsf{U}}(\mathbf{t})\right)$
(10)	None of the above	$\mathsf{U}_i(\mathbf{s}) = \underline{u} - 1$ for each $i \in [r]$

Clearly, the construction of $\mathsf{SG}(\phi)$ from ϕ is carried out in polynomial time. For brevity, we shall also write SG, L, V and $\mathcal{C}$ for $\mathsf{SG}(\phi)$, $\mathsf{L}(\phi)$, $\mathsf{V}(\phi)$ and $\mathcal{C}(\phi)$.

From the construction of the utility functions, we observe that for each player $i \in [r] \setminus [2]$, $\mathsf{U}_i(\mathbf{s}) \leq \overline{u} + 1$. We also observe that $\sum_{i \in [2]} \mathsf{U}_i(\mathbf{s}) \leq 2(\overline{u} + 1)$ with $\sum_{i \in [2]} \mathsf{U}_i(\mathbf{s}) = 2(\overline{u} + 1)$ if and only if $\mathbf{s} = \langle \ell_1, \ell_2, \delta, \ldots, \delta \rangle$ with $\ell_1 \neq \overline{\ell}_2$. This implies that for any property P satisfied by at least one pure profile enabled in a mixed profile $\boldsymbol{\sigma}$, $\mathbb{E}_{\mathbf{s} \sim \boldsymbol{\sigma}}\left(\sum_{i \in [2]} \mathsf{U}_i(\mathbf{s}) \mid \mathbf{s} \text{ satisfies } \mathsf{P} \right) \leq 2(\overline{u} + 1)$.

For a fixed player $i \in [r]$, denote $\sigma_i(\widehat{\Sigma}_i) = \sum_{s \in \widehat{\Sigma}_i} \sigma_i(\sigma)$ and $\sigma_i(\mathsf{L}) = \sum_{\ell \in \mathsf{L}} \sigma_i(\ell)$; so, $\sigma_i(\widehat{\Sigma}_i)$ and $\sigma_i(\mathsf{L})$ are the probability masses put on strategies from $\widehat{\Sigma}_i$ and on literals from L, respectively. We prove:

Proposition 1. *The following conditions are equivalent:* (1) *ϕ is satisfiable,* (2) *$\mathcal{NE}(\mathsf{SG}) \neq \mathcal{NE}(\widehat{\mathsf{SG}})$, and* (3) *$\mathsf{SG}$ admits a rational Nash equilibrium in $\mathcal{NE}(\mathsf{SG}) \setminus \mathcal{NE}(\widehat{\mathsf{SG}})$.*

The equivalence of Conditions (1) and (2) in Proposition 1 yields:

Proposition 2. *Fix an arbitrary game $\widehat{\mathsf{SG}}$. Then, $\overline{\mathsf{CNF\ SAT}} \leq_{\mathrm{P}}$* NASH-EQUIVALENCE $(\widehat{\mathsf{SG}})$.

There are already known three-players games with no rational Nash equilibrium (cf. [18]). As an additional example, consider the gadget game $\widehat{\mathsf{SG}}_1 = \left\langle [3], \{\widehat{\Sigma}_i\}_{i \in [3]}, \{\widehat{\mathsf{U}}_i\}_{i \in [3]} \right\rangle$, where for each player $i \in [3]$, $\widehat{\Sigma}_i = \{0, 1\}$. The utilities, given in the style $\mathbf{s} = \langle s_1, s_2, s_3 \rangle \rightarrow \langle \mathsf{U}_1(\mathbf{s}), \mathsf{U}_2(\mathbf{s}), \mathsf{U}_3(\mathbf{s}) \rangle$, are defined as follows: $\langle 0, 0, 0 \rangle \rightarrow \langle 1, 3, 3 \rangle$, $\langle 0, 0, 1 \rangle \rightarrow \langle 0, 1, 2 \rangle$, $\langle 0, 1, 0 \rangle \rightarrow \langle 3, 2, 2 \rangle$, $\langle 0, 1, 1 \rangle \rightarrow \langle 2, 3, 1 \rangle$, $\langle 1, 0, 0 \rangle \rightarrow \langle 2, 3, 1 \rangle$, $\langle 1, 0, 1 \rangle \rightarrow \langle 2, 0, 3 \rangle$, $\langle 1, 1, 0 \rangle \rightarrow \langle 1, 0, 3 \rangle$, $\langle 1, 1, 1 \rangle \rightarrow \langle 0, 1, 2 \rangle$.

Proposition 3. *The gadget game $\widehat{\mathsf{SG}}_1$ has a unique Nash equilibrium $\boldsymbol{\sigma}$, and $\boldsymbol{\sigma}$ is irrational.*

Using $\widehat{SG}_1$ as the gadget game in the reduction, Condition (3) of Proposition 1 becomes: (3') SG admits a rational Nash equilibrium. So, the equivalence of Conditions (1) and (3) implies:

Proposition 4. CNF SAT $\leq_P$ ∃ RATIONAL NASH.

Since CNF SAT is $\mathcal{NP}$-hard, Propositions 2 and 4 immediately imply:

Theorem 1. NASH-EQUIVALENCE($\widehat{SG}$) *is co-$\mathcal{NP}$-hard.*

Theorem 2. ∃ RATIONAL NASH *is $\mathcal{NP}$-hard.*

4   Complexity of NASH-REDUCTION and ∃ IRRATIONAL NASH

The Reduction. We shall consider a game $\widehat{SG} = \left\langle [r], \left\{\widehat{\Sigma}_i\right\}_{i\in[r]}, \left\{\widehat{U}_i\right\}_{i\in[r]} \right\rangle$ with the following properties: (P1) $\widehat{SG}$ is constant-sum with constant $c = ur$, and (P2) $\widehat{SG}$ has a unique Nash equilibrium $\widehat{\boldsymbol{\sigma}}$ which is fully mixed and such that $\widehat{U}_i(\widehat{\boldsymbol{\sigma}}) = u$ for each player $i \in [r]$.

For each player $i \in [r]$ in the game $\widehat{SG}$, set $\widehat{\Sigma}_i := \{1, \ldots, m_i\}$; so, for each player $i \in [r]$, for each strategy $j \in [m_i]$, $\widehat{\sigma}_i(j)$ is the probability that player i chooses strategy j in a mixed profile $\widehat{\boldsymbol{\sigma}}$.

Consider now a boolean formula ϕ in CNF with $|V(\phi)| = n \geq m := \max_{i\in[r]}\{m_i\}$ variables. Set $V(\phi) := \{1, 2, \ldots, n\}$; so, the boolean variables in ϕ are numbered $1, 2, \ldots, n$. For each player $i \in [r]$ and for each variable $k \in [n]$, define

$$g_i(\ell_k) = g_i(\bar{\ell}_k) = \begin{cases} k & \text{if } k \in [m_i - 1] \\ m_i & \text{if } k \geq m_i \end{cases}.$$

Moreover, for each player $i \in [r]$ and for each $j \in [m_i]$ define

$$n_i(j) = \begin{cases} 1 & \text{if } j \in [m_i - 1] \\ n - m_i + 1 & \text{if } j = m_i \end{cases}.$$

For a player $i \in [r]$, denote $s(i) = (i + 1) \bmod r$; so, $s(i)$ is the *successor modulo* r of player $i \in [r]$.

Construct the strategic game $SG(\phi) = \left\langle [r], \{\Sigma_i\}_{i\in[r]}, \{U_i\}_{i\in[r]} \right\rangle$ as follows.

- For each player $i \in [r]$, $\Sigma_i := V(\phi) \cup L(\phi) \cup C(\phi) \cup \{f\}$.
- Fix a profile $\mathbf{s}$ from $\Sigma = \Sigma_1 \times \ldots \times \Sigma_r$. Denote $\mathbf{f} = \langle f, \ldots, f \rangle$. Denote $P_f(\mathbf{s}) = \{i \in [r] \mid s_i = f\}$; so, $|P_f(\mathbf{s})| = r$ if and only if $\mathbf{s} = \mathbf{f}$. The utility vector $U(\mathbf{s})$ is depicted in the following table.

Case	Condition on the profile $s = \langle s_1, \ldots, s_r \rangle$	Utility vector $U(s)$
(1)	$s = \langle \ell_1, \ldots, \ell_r \rangle$ with (i) $\ell_i \in L(\phi)$ for each $i \in [r]$ & (ii) $\ell_i \neq \overline{\ell_j}$ for each distinct pair $i, j \in [r]$	$\widehat{U}\left(\langle s_{g(\ell_1)}, \ldots, s_{g(\ell_r)} \rangle\right)$
(2)	(i) $s_i = v \in V(\phi)$ for some $i \in [r]$, (ii) $s_j \in L(\phi)$ for each $j \in [r] \setminus \{i\}$ & (iii) $v(s_{s(i)}) \neq v$	$U_i(s) = u + 1$ & $U_j(s) = u - 2$ for each $j \in [r] \setminus \{i\}$
(3)	(i) $s_i = v \in V(\phi)$ for some $i \in [r]$, (ii) $s_j \in L(\phi)$ for each $j \in [r] \setminus \{i\}$ & (iii) $v(s_{s(i)}) = v$	$U_i(s) = u + 1 - \dfrac{n_i(g_i(s_{s(i)}))}{\widehat{\sigma}_{s(i)}(g_i(s_{s(i)}))}$ & $U_j(s) = u - 2$ for each $j \in [r] \setminus \{i\}$
(4)	(i) $s_i = c \in C(\phi)$ for some $i \in [r]$, (ii) $s_j \in L(\phi)$ for each $j \in [r] \setminus \{i\}$ & (iii) $s_{s(i)} \notin c$	$U_i(s) = u + 1$ & $U_j(s) = u - 2$ for each $j \in [r] \setminus \{i\}$
(5)	(i) $s_i = c \in C(\phi)$ for some $i \in [r]$, (ii) $s_j \in L(\phi)$ for each $j \in [r] \setminus \{i\}$ & (iii) $s_{s(i)} \in c$	$U_i(s) = u + 1 - \dfrac{n_i(g_i(s_{s(i)}))}{\widehat{\sigma}_{s(i)}(g_i(s_{s(i)}))}$ & $U_j(s) = u - 2$ for each $j \in [r] \setminus \{i\}$
(6)	$|P_f(s)| = r$	$U_i(s) = u - 1$ for each $i \in r$
(7)	$0 < |P_f(s)| < r$	$U_i(s) = u$ if $s_i = f$ $U_j(s) = u - 2$ if $s_i \neq f$
(8)	None of the above	$U_i(s) = u - 1$ for each $i \in [r]$

Clearly, the construction of $SG(\phi)$ from ϕ is carried out in polynomial time. For brevity, we shall also write SG, L, V and C for $SG(\phi)$, $L(\phi)$, $V(\phi)$ and $C(\phi)$. Here is an observation to use later:

For each profile s, $\sum_{i \in [r]} U_i(s) \leq r \cdot u$ with $\sum_{i \in [r]} U_i(s) = r \cdot u$ if and only if $s = \langle \ell_1, \ldots, \ell_r \rangle$ with (i) $\ell_i \in L$ for each $i \in [r]$, and (ii) $\ell_i \neq \overline{\ell_j}$ for each distinct pair $i, j \in [r]$. (To see this, note that by Case (1) in the definition of the utility vectors $\sum_{i \in [r]} U_i(\langle \ell_1, \ldots, \ell_r \rangle) = \sum_{i \in [r]} \widehat{U}_i(\langle s_{g(\ell_1)}, \ldots, s_{g(\ell_r)} \rangle) = r \cdot u$, since $\widehat{SG}$ is constant-sum with $c = r \cdot u$.) This implies that for any property P satisfied by at least one pure profile enabled in a mixed profile σ, it holds that $\mathbb{E}_{s \sim \sigma}\left(\sum_{i \in [r]} U_i(s) \mid s \text{ satisfies } P\right) \leq r \cdot u$.

Let $\mathcal{H}$ be the surjective mapping from SG to $\widehat{SG}$ defined as follows: for each player $i \in [r]$,

$$h_i(s) = \begin{cases} g_i(\ell) \text{ if } s = \ell, \\ * \quad \text{otherwise,} \end{cases}$$

where $*$ means a *don't care* situation. We prove:

Proposition 5. *Consider a gadget game $\widehat{SG}$ satisfying Properties (P1) and (P2). Then, the following conditions are equivalent: (1) ϕ is satisfiable, (2) SG admits a Nash equilibrium $\sigma \neq f$, and (3) SG Nash-reduces to $\widehat{SG}$.*

The equivalence of Conditions (1) and (2) in Proposition 5 yields:

Proposition 6. *Consider a gadget game* $\widehat{\mathsf{SG}}$ *satisfying Properties* (P1) *and* (P2). *Then,* $\mathsf{CNF\ SAT} \leq_{\mathrm{P}} \mathsf{NASH\text{-}REDUCTION}(\widehat{\mathsf{SG}})$.

We shall now exhibit a gadget game satisfying Properties (P1) and (P2). Consider the gadget game $\widehat{\mathsf{SG}}_2 = \left\langle [3], \left\{ \widehat{\Sigma}_i \right\}_{i \in [3]}, \left\{ \widehat{\mathsf{U}}_i \right\}_{i \in [3]} \right\rangle$, where for each player $i \in [3]$, $\widehat{\Sigma}_i = \{0, 1\}$. The utilities, given in the style $\mathbf{s} = \langle s_1, s_2, s_3 \rangle \rightarrow \langle \mathsf{U}_1(\mathbf{s}), \mathsf{U}_2(\mathbf{s}), \mathsf{U}_3(\mathbf{s}) \rangle$, are defined as follows: $\langle 0, 0, 0 \rangle \rightarrow \langle 2, 4, -3 \rangle$, $\langle 0, 0, 1 \rangle \rightarrow \langle -1, 5, -1 \rangle$, $\langle 0, 1, 0 \rangle \rightarrow \langle 3, -4, 4 \rangle$, $\langle 0, 1, 1 \rangle \rightarrow \langle 2, 0, 1 \rangle$, $\langle 1, 0, 0 \rangle \rightarrow \langle -3, 6, 0 \rangle$, $\langle 1, 0, 1 \rangle \rightarrow \langle 4, -6, 5 \rangle$, $\langle 1, 1, 0 \rangle \rightarrow \langle -1, -1, 5 \rangle$, $\langle 1, 1, 1 \rangle \rightarrow \left\langle \frac{3863 - 173\sqrt{471}}{74}, \frac{6877 + 107\sqrt{471}}{1850}, \frac{-1323 + 57\sqrt{471}}{25} \right\rangle$.

Note that $\widehat{\mathsf{SG}}_2$ is constant-sum, with constant equal to r. We prove:

Proposition 7. *The gadget game* $\widehat{\mathsf{SG}}_2$ *has a single Nash equilibrium* $\widehat{\boldsymbol{\sigma}}$, *which is irrational, fully mixed and in which for each player* $i \in [3]$, $\widehat{\mathsf{U}}_i(\widehat{\boldsymbol{\sigma}}) = 1$.

Using $\widehat{\mathsf{SG}}_2$ as the gadget game in the reduction, Condition (3) in Proposition 5 implies that SG has an irrational Nash equilibrium. Hence, Condition (2) is equivalent to the Condition (2') SG admits an irrational Nash equilibrium. So, the equivalence of Conditions (1) and (2) implies:

Proposition 8. $\mathsf{CNF\ SAT} \leq_{\mathrm{P}} \exists\ \mathsf{IRRATIONAL\ NASH}$.

Since $\mathsf{CNF\ SAT}$ is $\mathcal{NP}$-hard, Propositions 6 and 8 immediately imply:

Theorem 3. $\mathsf{NASH\text{-}REDUCTION}(\widehat{\mathsf{SG}})$ *is* $\mathcal{NP}$-*hard.*

Theorem 4. $\exists\ \mathsf{IRRATIONAL\ NASH}$ *is* $\mathcal{NP}$-*hard.*

5 Epilogue

Our work initiates the study of the complexity of decision problems about the rationality and irrationality of Nash equilibria. It remains open to determine the extent of strategic games for which $\mathcal{NP}$-hardness holds for $\exists$ RATIONAL NASH and $\exists$ IRRATIONAL NASH. For example, what happens if we restrict to *win-lose games* [1]: games where all utilities are 0 or 1? What is the smallest value for the utilities that suffices for $\mathcal{NP}$-hardness? A wide avenue for further research concerns the extent of the decision problems and the games for which $\mathcal{NP}$-hardness holds.

What is the relation between $\exists$ IRRATIONAL NASH and the SQUARE-ROOT-SUM problem [11]? Is $\exists$ IRRATIONAL NASH SQUARE-ROOT-SUM-hard (cf. [9])?

Irrational Nash equilibria are a manifestation of a familiar phenomenon in science and engineering, where the quantities of interest are solutions to nonlinear equations, so that they can be irrational. Are there other manifestations where deciding irrationality is $\mathcal{NP}$-hard?

References

1. Abbott, T., Kane, D., Valiant, P.: On the Complexity of Two-Player Win-Lose Games. In: Proceedings of the 46th Annual IEEE Symposium on Foundations of Computer Sciences, pp. 113–122 (October 2005)
2. Austrin, P., Braverman, M., Chlamtáč, E.: Inapproximability of NP-complete Variants of Nash Equilibrium, arXiv:1104.3760v1, April 19 (2001)
3. Borgs, C., Chayes, J., Immorlica, N., Tauman Kalai, A., Mirrokni, V., Papadimitriou, C.H.: The Myth of the Folk Theorem. Games and Economic Behavior 70(1), 34–43 (2010)
4. Chen, X., Deng, X., Teng, S.H.: Settling the Complexity of Computing Two-Player Nash Equilibria. Journal of the ACM 56(3) (2009)
5. Codenotti, B., Stefanovic, D.: On the Computational Complexity of Nash Equilibria for $(0, 1)$ Bimatrix Games. Information Processing Letters 94(3), 145–150 (2005)
6. Conitzer, V., Sandholm, T.: Complexity Results about Nash Equilibria. In: Proceedings of the 18th Joint Conference on Artificial Intelligence, pp. 765–771 (August 2003)
7. Conitzer, V., Sandholm, T.: New Complexity Results about Nash Equilibria. Games and Economic Behavior 63(2), 621–641 (2008)
8. Daskalakis, C., Goldberg, P.W., Papadimitriou, C.H.: The Complexity of Computing a Nash Equilibrium. SIAM Journal on Computing 39(1), 195–259 (2009)
9. Etessami, K., Yannakakis, M.: On the Complexity of Nash Equilibria and Other Fixed Points. SIAM Journal on Computing 39(6), 2531–2597 (2010)
10. Fiat, A., Papadimitriou, C.H.: When Players are not Expectation Maximizers. In: Kontogiannis, S., Koutsoupias, E., Spirakis, P.G. (eds.) Algorithmic Game Theory. LNCS, vol. 6386, pp. 1–14. Springer, Heidelberg (2010)
11. Garey, M.R., Graham, R.L., Johnson, D.S.: Some NP-complete geometric problems. In: Proceedings of the 8th Annual ACM Symposium on Theory of Computing, pp. 10–22 (1976)
12. Garey, M.R., Johnson, D.S.: Computers and Intractability — A Guide to the Theory of $\mathcal{NP}$-Completeness. W.H. Freeman, New York (1979)
13. Gilboa, I., Zemel, E.: Nash and Correlated Equilibria: Some Complexity Considerations. Games and Economic Behavior 1(1), 80–93 (1989)
14. Koutsoupias, E.: Personal communication during. In: 2nd International Symposium on Algorithmic Game Theory, Paphos, Cyprus (October 2009)
15. Mavronicolas, M., Monien, B., Wagner, K.K.: Weighted Boolean Formula Games. In: Deng, X., Graham, F.C. (eds.) WINE 2007. LNCS, vol. 4858, pp. 469–481. Springer, Heidelberg (2007)
16. McLennan, A., Tourky, R.: Simple Complexity from Imitation Games. Games abd Economic Behavior 68(2), 683–688 (2010)
17. Nash, J.F.: Equilibrium Points in n-Person Games. Proceedings of the National Academy of Sciences of the United States of America 36, 48–49 (1950)
18. Nash, J.F.: Non-Cooperative Games. Annals of Mathematics 54(2), 286–295 (1951)
19. Papadimitriou, C.H.: On the Complexity of the Parity Argument and Other Inefficient Proofs of Existence. Journal of Computer and System Sciences 48(3), 498–532 (1994)

Diffusion in Social Networks with Competing Products[*]

Krzysztof R. Apt[1,2] and Evangelos Markakis[3]

[1] CWI, Amsterdam, The Netherlands
[2] University of Amsterdam
apt@cwi.nl
[3] Athens University of Economics and Business,
Dept. of Informatics, Athens, Greece
markakis@gmail.com

Abstract. We introduce a new threshold model of social networks, in which the nodes influenced by their neighbours can adopt one out of several alternatives. We characterize the graphs for which adoption of a product by the whole network is possible (respectively necessary) and the ones for which a unique outcome is guaranteed. These characterizations directly yield polynomial time algorithms that allow us to determine whether a given social network satisfies one of the above properties.

We also study algorithmic questions for networks without unique outcomes. We show that the problem of computing the minimum possible spread of a product is NP-hard to approximate with an approximation ratio better than $\Omega(n)$, in contrast to the maximum spread, which is efficiently computable. We then move on to questions regarding the behavior of a node with respect to adopting some (resp. a given) product. We show that the problem of determining whether a given node has to adopt some (resp. a given) product in all final networks is co-NP-complete.

1 Introduction

1.1 Background

Social networks have become a huge interdisciplinary research area with important links to sociology, economics, epidemiology, computer science, and mathematics. A flurry of numerous articles and recent books [10,6] shows the growing relevance of this field as it deals with such diverse topics as epidemics, spread of certain patterns of social behaviour, effects of advertising, and emergence of 'bubbles' in financial markets.

A large part of research on social networks focusses on the problem of *diffusion*, that is the spread of a certain event or information over the network, e.g., becoming infected or adopting a given product. In the remainder of the paper, we will use as a running example the adoption of a product, which is being marketed over a social network.

Two prevalent models have been considered for capturing diffusion: the *threshold models* introduced in [8] and [15] and the *independent cascade models* studied in [7]. In threshold models, which is the focus of our work, each node i has a threshold $\theta(i) \in (0, 1]$ and it decides to adopt a product when the total weight of incoming edges from

---

[*] A full version with all missing proofs is available at the authors' homepages.

G. Persiano (Ed.): SAGT 2011, LNCS 6982, pp. 212–223, 2011.

nodes that have already adopted a product reaches or exceeds $\theta(i)$. In a special case a node decides to adopt a product if at least the fraction $\theta(i)$ of its neighbours has done so. In cascade models, each node that adopts a product can activate each of his neighbours with a certain probability and each node has only one chance of activating a neighbour.

Most of research has focussed on the situation in which the players face the choice of adopting a specific product or not. In this setting, the algorithmic problem of choosing an initial set of nodes so as to maximize the adoption of a given product and certain variants of this were studied initially in [11] and in several publications that followed, e.g., [5,14].

When studying social networks from the point of view of adopting new products that come to the market, it is natural to lift the restriction of a single product. One natural example is when users choose among competing programs from providers of mobile telephones. Then, because of lower subscription costs, each owner of a mobile telephone naturally prefers to choose the same provider that his friends choose. In such situations, the outcome of the adoption process does not need to be unique. Indeed, individuals with a low 'threshold' can adopt any product a small group of their friends adopts. As a result this leads to different considerations than before.

In the presence of multiple products, diffusion has been investigated recently for cascade models in [2,4,12], where new approximation algorithms and hardness results have been proposed. For threshold models, an extension to two products has been recently proposed in [3], where the authors examine whether the algorithmic approach of [11] can be extended. Algorithms and hardness of approximation results are provided for certain variants of the diffusion process.

Game theoretic aspects have also been considered in the case of two products. In particular, the behavior of best response dynamics in infinite graphs is studied in [13], when each node has to choose between two different products. An extension of this model is studied in [9] with a focus on notions of compatibility and bilinguality, i.e., having the option to adopt both products at an extra cost so as to be compatible with all your neighbours.

1.2   Contributions

We study a new model of a social network in which nodes (agents) can choose out of *several* alternatives *and* in which various outcomes of the adoption process are possible. Our model combines a number of features present in various models of networks.

It is a threshold model and we assume that the threshold of a node is a fixed number as in [5] (and unlike [11,3], where they are random variables). This is in contrast to Hebb's model of learning in networks of neurons, the focus of which is on learning, leading to strengthening of the connections (here thresholds). In our context threshold should be viewed as a fixed 'resistance level' of a node to adopt a product. In contrast to the SIR model, see, e.g., [10], in which a node can be in only two states, in our model each node can choose out of several states (products). We also allow that not all nodes have exactly the same set of products to choose from, e.g. due to geographic or income restrictions some products may be available only to a subset of the nodes. If a node changes its state from the initial one, the new state (that corresponds to the adopted product) is final, as is the case with most of the related literature.

Our work consists of two parts. In the first part (Sections 3, 4, 5) we study three basic problems concerning this model. In particular, we find necessary and sufficient conditions for determining whether

- a specific product will possibly be adopted by all nodes.
- a specific product will necessarily be adopted by all nodes.
- the adoption process of the products will yield a unique outcome.

For each of these questions, we obtain a characterization with respect to the structure of the underlying graph.

In the second part (Section 6), we focus on networks that do not possess a unique outcome and investigate the complexity of various problems concerning the adoption process. We start with estimating the minimum and maximum number of nodes that may adopt a given product. Then we move on to questions regarding the behavior of a given node in terms of adopting a given product or some product from its list. We resolve the complexity of all these problems. As we show, some of these problems are efficiently solvable, whereas the remaining ones are either co-NP-complete or have strong inapproximability properties.

2 Preliminaries

Assume a fixed weighted directed graph $G = (V, E)$ (with no parallel edges and no self-loops), with $n = |V|$ and $w_{ij} \in [0, 1]$ being the weight of edge (i, j). Given a node i of G we denote by $N(i)$ the set of nodes from which there is an incoming edge to i. We call each $j \in N(i)$ a ***neighbour*** of i in G. We assume that for each node i such that $N(i) \neq \emptyset$, $\sum_{j \in N(i)} w_{ji} \leq 1$. Further, we have a ***threshold function*** θ that assigns to each node $i \in V$ a fixed value $\theta(i) \in (0, 1]$. Finally, we fix a finite set P of alternatives to which we shall refer as ***products***.

By a ***social network*** we mean a tuple (G, P, p, θ), where p is a function that assigns to each node of G a non-empty subset of P. The idea is that each node i is offered a non-empty set $p(i) \subseteq P$ of products from which it can make its choice. If $p(i)$ is a singleton, say $p(i) = \{t\}$, the node adopted the product t. Otherwise it can adopt a product if the total weight of incoming edges from neighbours that have already adopted it is at least equal to the threshold $\theta(i)$. To formalize the questions we want to address, we need to introduce a number of notions. Since G, P and θ are fixed, we often identify each social network with the function p.

Consider a binary relation $\rightarrow$ on social networks. Denote by $\rightarrow^*$ the reflexive, transitive closure of $\rightarrow$. We call a reduction sequence $p \rightarrow^* p'$ ***maximal*** if for no p'' we have $p' \rightarrow p''$. In that case we will say that p' is a ***final*** network, given the initial network p.

Definition 1. *Assume an initial social network p and a network p'. We say that*

- *p' is **reachable** (from p) if $p \rightarrow^* p'$,*
- *p' is **unavoidable** (from p) if for all maximal sequences of reductions $p \rightarrow^* p''$ we have $p' = p''$,*
- *p has a **unique outcome** if some social network is unavoidable from p.*

From now on we specialize the relation $\rightarrow$. Given a social network p, and a product $t \in p(i)$ for some node i with $N(i) \neq \emptyset$, we use the abbreviation $A(t,i)$ (for 'adoption condition of product t by node i') for

$$\sum_{j \in N(i) | p(j) = \{t\}} w_{ji} \geq \theta(i)$$

When $N(i) = \emptyset$, we stipulate that $A(t,i)$ holds for every $t \in p(i)$.

Definition 2.
- *We write $p_1 \rightarrow p_2$ if $p_2 \neq p_1$ and for all nodes i, if $p_2(i) \neq p_1(i)$, then $|p_1(i)| \geq 2$ and for some $t \in p_1(i)$*

$$p_2(i) = \{t\} \text{ and } A(t,i) \text{ holds in } p_1.$$

- *We say that node i in a social network p*
  - **adopted product** t *if $p(i) = \{t\}$,*
  - **can adopt product** t *if $t \in p(i)$, $|p(i)| \geq 2$, and $A(t,i)$ holds in p.*

In particular, a node i with no neighbours and more than one product in $p(i)$ can adopt any product that is a possible choice for it. Note that each modification of the function p results in assigning to a node i a singleton set. Thus, if $p_1 \rightarrow^* p_2$, then for all nodes i either $p_2(i) = p_1(i)$ or $p_2(i)$ is a singleton set.

One of the questions we are interested is whether a product t can spread to the whole network. We will denote this final network by $[t]$, where $[t]$ denotes the constant function p such that $p(i) = \{t\}$ for all nodes i. Furthermore, given a social network (G, P, p, θ) and a product $t \in P$ we denote by $G_{p,t}$ the weighted directed graph obtained from G by removing from it all edges to nodes i with $p(i) = \{t\}$. That is, in $G_{p,t}$ for all such nodes i we have $N(i) = \emptyset$ and for all other nodes the set of neighbours in $G_{p,t}$ and G is the same.

If each weight $w_{j,i}$ in the considered graph equals $\frac{1}{|N(i)|}$, then we call the corresponding social network ***equitable***. Hence in equitable social networks the adoption condition, $A(t,i)$, holds if at least a fraction $\theta(i)$ of the neighbours of i adopted in p product t.

Example 1. As an example for illustrating the definitions, consider the equitable social networks in Figure 1, where $P = \{t_1, t_2\}$ and where we mention next to each node the set of products available to it.

In the first social network, if $\theta(a) \leq \frac{1}{3}$, then the network in which node a adopts product t_1 is reachable, and so is the case for product t_2. If $\frac{1}{3} < \theta(a) \leq \frac{2}{3}$, then only the network in which node a adopts product t_1 is reachable. Further, if $\theta(a) > \frac{2}{3}$, then none of the above two networks is reachable. Finally, the initial network has a unique outcome iff $\frac{1}{3} < \theta(a)$.

For the second social network the following more elaborate case distinction lists the possible values of p in the final reachable networks.

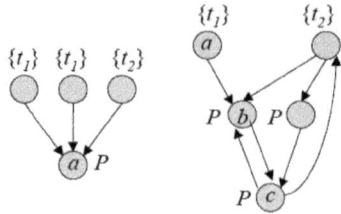

Fig. 1. Two examples of social networks

$$\theta(b) \le \tfrac{1}{3} \wedge \theta(c) \le \tfrac{1}{2} \qquad : (p(b) = \{t_1\} \vee p(b) = \{t_2\}) \wedge (p(c) = \{t_1\} \vee p(c) = \{t_2\})$$
$$\theta(b) \le \tfrac{1}{3} \wedge \theta(c) > \tfrac{1}{2} \qquad : (p(b) = \{t_1\} \wedge p(c) = P) \vee (p(b) = p(c) = \{t_2\})$$
$$\tfrac{1}{3} < \theta(b) \le \tfrac{2}{3} \wedge \theta(c) \le \tfrac{1}{2} : p(b) = p(c) = \{t_2\}$$
$$\tfrac{1}{3} < \theta(b) \wedge \theta(c) > \tfrac{1}{2} \qquad : p(b) = p(c) = P$$
$$\tfrac{2}{3} < \theta(b) \wedge \theta(c) \le \tfrac{1}{2} \qquad : p(b) = P \wedge p(c) = \{t_2\}$$

In particular, when $\tfrac{1}{3} < \theta(b) \le \tfrac{2}{3}$ and $\theta(c) \le \tfrac{1}{2}$, node b adopts product t_2 only *after* node c adopts it.

3   Reachable Outcomes

We start with providing necessary and sufficient conditions for a product to be reachable by all nodes. This is achieved by a structural characterization of graphs that allow products to spread to the whole graph, given the threshold function θ. In particular, we shall need the following notion.

Definition 3. *Given a threshold function θ we call a weighted directed graph θ-**well-structured** if for some function $level$ that maps nodes to natural numbers, we have that for all nodes i such that $N(i) \ne \emptyset$*

$$\sum_{j \in N(i) \mid level(j) < level(i)} w_{ji} \ge \theta(i). \tag{1}$$

In other words, a weighted directed graph is θ-well-structured if levels can be assigned to its nodes in such a way that for each node i such that $N(i) \ne \emptyset$, the sum of the weights of the incoming edges from lower levels is at least $\theta(i)$. We will often refer to the function $level$ as a **certificate** for the graph being θ-well-structured. Note that there can be many certificates for a given graph. Note also that θ-well structured graphs can have cycles. For instance, it is easy to check that the second social network in Figure 1 is θ-well structured when $\theta(i) \le \tfrac{1}{3}$ for every node i.

We have the following characterization.

Theorem 1. *Assume a social network (G, P, p, θ) and a product $top \in P$. A social network $(G, P, [top], \theta)$ is reachable from (G, P, p, θ) iff*

- *for all i, $top \in p(i)$,*
- *$G_{p,top}$ is θ-well-structured.*

Proof. ($\Rightarrow$) If for some node i we have $top \notin p(i)$, then i cannot adopt product top and $[top]$ is not reachable.

To establish the second condition consider a reduction sequence

$$p_1 \to p_2 \to \ldots \to p_m$$

starting in p and such that $p_m = [top]$.

Assign now to each node i the minimal k such that $p_{k+1}(i) = \{top\}$. We claim that this definition of the *level* function shows that $G_{p,top}$ is θ-well-structured. Consider a node i.

Case 1. $level(i) = 0$.

Then $p(i) = \{top\}$, so by the definition of $G_{p,top}$ we have $N(i) = \emptyset$ in $G_{p,top}$. Hence we do not need to argue about these nodes since we only need to ensure condition (1) for nodes with $N(i) \neq \emptyset$.

Case 2. $level(i) > 0$.

Suppose that $N(i) \neq \emptyset$ and that $level(i) = k$. By the definition of the reduction $\to$ the adoption condition $A(top, i)$ holds in p_k, i.e.,

$$\sum_{j \in N(i) | p_k(j) = \{top\}} w_{ji} \geq \theta(i).$$

But for each $j \in N(i)$ such that $p_k(j) = \{top\}$ we have by definition $level(j) < level(i)$. So (1) holds.

($\Leftarrow$) Consider a certificate function *level* showing that $G_{p,top}$ is θ-well-structured. Without loss of generality we can assume that the nodes in $G_{p,top}$ such that $N(i) = \emptyset$ are exactly the nodes of level 0. We construct by induction on the level m a reduction sequence $p \to^* p''$, such that for all nodes i we have $top \in p''(i)$ and for all nodes i of level $\leq m$ we have $p''(i) = \{top\}$.

Consider level 0. By definition of $G_{p,top}$, a node i is of level 0 iff it has no neighbours in G or $p(i) = \{top\}$. In the former case, by the first condition, $top \in p(i)$. So $p \to^* p''$, where the function p'' is defined by

$$p''(i) := \begin{cases} \{top\} & \text{if } level(i) = 0 \\ p(i) & \text{otherwise} \end{cases}$$

This establishes the induction basis.

Suppose the claim holds for some level m. So we have $p \to^* p'$, where for all nodes i we have $top \in p'(i)$ and for all nodes i of level $\leq m$ we have $p'(i) = \{top\}$.

Consider the nodes of level $m + 1$. For each such node i we have $top \in p'(i)$, $N(i) \neq \emptyset$ and

$$\sum_{j \in N(i) | level(j) < level(i)} w_{ji} \geq \theta(i).$$

By the definition of $G_{p,top}$ the sets of neighbours of i in G and $G_{p,top}$ are the same. By the induction hypothesis for all nodes j such that $level(j) < level(i)$ we have $p'(j) = \{top\}$.

So either node i adopted product top in p' or can adopt product top in p'. Hence $p' \to^* p''$, where the function p'' is defined by

$$p''(i) := \begin{cases} \{top\} & \text{if } level(i) = m+1 \\ p'(i) & \text{otherwise} \end{cases}$$

Consequently $p \to^* p''$, which establishes the induction step. We conclude $p \to^* [top]$.

Next we show that testing if a graph is θ-well-structured can be efficiently solved.

Theorem 2. *Given a weighted directed graph G and a threshold function θ, we can decide whether G is θ-well-structured in time $O(n^2)$.*

Proof. (Sketch) We claim that the following simple algorithm achieves this:

- Given a weighted directed graph G, first assign level 0 to all nodes with $N(i) = \emptyset$. If no such node exists, output that the graph is not θ-well-structured.
- Inductively, at step i, assign level i to each node for which condition (1) from Definition 3 is satisfied when considering only its neighbours that have been assigned levels $0, \ldots, i-1$.
- If by iterating this all nodes are assigned a level, then output that the graph is θ-well-structured. Otherwise, output that G is not θ-well-structured.

The above algorithm can be implemented in time $O(n^2 + |E|) = O(n^2)$, by using the adjacency list representation. To prove correctness, note that if the input graph is not θ-well-structured, then the algorithm will output No, as otherwise, at termination it would have constructed a level function for a non-θ-well-structured graph. For the reverse, suppose a graph G is θ-well-structured. The idea of the proof is to use a certificate function, in which all nodes are assigned the minimum possible level. We then prove by induction that this is precisely the level assignment produced by the algorithm and hence it outputs Yes. Due to lack of space, we omit the proof.

Finally, we end this section by observing that determining whether a network $[top]$ is reachable can also be solved efficiently.

Theorem 3. *Assume a social network (G, P, p, θ) and a product $top \in P$. There is an algorithm running in time $O(n^2)$ that determines whether the social network $(G, P, [top], \theta)$ is reachable.*

4   Unavoidable Outcomes

Next, we focus on the notion of unavoidable outcomes. We establish the following characterization.

Theorem 4. *Assume a social network (G, P, p, θ) and a product $top \in P$. A social network $(G, P, [top], \theta)$ is unavoidable iff*

- *for all i, if $N(i) = \emptyset$, then $p(i) = \{top\}$,*
- *for all i, $top \in p(i)$,*
- *$G_{p,top}$ is θ-well-structured.*

To prove this, we need first a few lemmas, the proofs of which we omit from this version.

Lemma 1. *Suppose that $p \rightarrow^* p'$ and for some node i we have $p'(i) = \{t\}$. Then for some node j such that $N(j) = \emptyset$ or $p(j)$ is a singleton, we have $t \in p(j)$.*

Intuitively, this means that each product eventually adopted can also be initially adopted (by a possibly different node).

Lemma 2. *Assume a social network (G, P, p, θ) and a product $top \in P$. Suppose that*

- *for all i, if $N(i) = \emptyset$ or $p(i)$ is a singleton, then $p(i) = \{top\}$.*

Then a unique outcome of (G, P, p, θ) exists.

Intuitively, this means that if initially only one product can be adopted, then a unique outcome of the social network exists.

Proof of Theorem 4: (Sketch) By Theorem 1 and Lemma 2.

In analogy to Theorem 3, we also have the following simple fact.

Theorem 5. *Assume a social network (G, P, p, θ) and a product $top \in P$. There is an algorithm, running in time $O(n^2)$, that determines whether the social network $(G, P, [top], \theta)$ is unavoidable.*

5   Unique Outcomes

We now consider the question of when does a network admit a unique outcome. To answer this, we introduce the following definitions.

Definition 4. *Given social networks p, p' based on the same graph we say that*

- *node i **can switch in** p' **given** p if i adopted in p' a product t and for some $t' \neq t$*

$$t' \in p(i) \wedge A(t', i) \text{ holds in } p',$$

- *p' is **ambivalent given** p if it contains a node that either can adopt more than one product or can switch in p' given p,*
- *the reduction $p \rightarrow p'$ is **fast** if for each node i, if i can adopt a product in p then i adopted a product in p'. Intuitively, $p \rightarrow p'$ is then a 'maximal' one-step reduction of p.*

Definition 5. *By the **contraction sequence** of a social network we mean the unique reduction sequence $p \rightarrow^* p'$ such that*

- *each of its reduction steps is fast,*
- *either $p \rightarrow^* p'$ is maximal or p' is the first network in the sequence $p \rightarrow^* p'$ that is ambivalent given p.*

We now formulate a characterization of social networks that admit a unique outcome. We omit the proof.

Theorem 6. *A social network admits a unique outcome iff its contraction sequence ends in a non-ambivalent social network.*

Corollary 1. *Assume a social network (G, P, p, θ) such that*

- *for all nodes i we have $\theta(i) > \frac{1}{2}$,*
- *for all i, if $N(i) = \emptyset$, then $p(i)$ is a singleton.*

Then (G, P, p, θ) admits a unique outcome.

The above corollary can be strengthened by assuming that the network is such that if $\theta(i) \leq \frac{1}{2}$ then $|N(i)| < 2$ or $|p(i)| = 1$. The reason is that the nodes for which $|N(i)| < 2$ or $|p(i)| = 1$ cannot introduce an ambivalence.

When for some node i, $\theta(i) \leq \frac{1}{2}$ holds and neither $|N(i)| < 2$ nor $|p(i)| = 1$, the equitable social network still may admit a unique outcome but it does not have to. For instance the second social network in Figure 1 admits a unique outcome for the last three alternatives (explained in Example 1), while for the first two is does not.

Theorem 6 also yields an algorithm to test if a network has a unique outcome. The algorithm simply has to simulate the contraction sequence of a network and determine whether it ends in a non-ambivalent network. The statement of the algorithm and its analysis are omitted.

Theorem 7. *There exists a polynomial time algorithm, running in time $O(n^2 + n|P|)$, that determines whether a social network admits a unique outcome. Furthermore, if for all nodes i we have $\theta(i) > \frac{1}{2}$, there is a $O(n^2)$ algorithm.*

For all practical purposes we have $|P| << n$, so even for the general case the running time would typically be $O(n^2)$.

6   Product Adoption in Networks without Unique Outcomes

The results of the previous section reveal that many social networks will not admit a unique outcome. In this section, we consider some natural questions regarding product adoption that are of interest for such networks. We start with two optimization problems.

Suppose that a product *top* is neither unavoidable by all nodes nor reachable. We would like then to estimate the worst and best-case scenario for the spread of this product. That is, starting from a given initial network p, what is the minimum (resp. maximum) number of nodes that will adopt this product in a final network (recall that a final network is one that has been obtained from some initial network by a maximal sequence of reductions). Hence, the following two problems are of interest.

MIN-ADOPTION: Given a social network (G, P, p, θ) and a product top, find the minimum number of nodes that adopted top in a final network, starting from (G, P, p, θ).

MAX-ADOPTION: Given a social network (G, P, p, θ) and a product top, find the maximum number of nodes that adopted top in a final network, starting from (G, P, p, θ).

We show that these two problems are substantially different, the first being essentially inapproximable while the second efficiently solvable.

Theorem 8. *If n is the number of nodes of a network, then*

(i) It is NP-hard to approximate MIN-ADOPTION with an approximation ratio better than $\Omega(n)$.

(ii) The MAX-ADOPTION problem can be solved in $O(n^2)$ time.

Proof. (i) We give a reduction from the PARTITION problem, which is: given n positive rational numbers $(a_1, \ldots, a_n)$, is there a set S such that $\sum_{i \in S} a_i = \sum_{i \notin S} a_i$? Consider an instance I of PARTITION. WLOG, suppose we have normalized the numbers so that $\sum_{i=1}^{n} a_i = 1$. Hence the question is to decide whether there is a set S such that $\sum_{i \in S} a_i = \sum_{i \notin S} a_i = \frac{1}{2}$.

We build an instance of our problem with 3 products, namely $P = \{top, t, t'\}$, and with the graph shown in Figure 2. The number of nodes in the line that starts to the right of node e is $M = n^{O(1)}$, hence the reduction is of polynomial time. The weights in those edges is 1. The thresholds of the nodes are $\theta(a) = \theta(b) = \theta(c) = \theta(d) = \frac{1}{2}$, $\theta(e) = 1/2 + \epsilon$, for some $\epsilon > 0$ and for the nodes to the right of e we can set the thresholds to an arbitrary positive number in $(0, 1]$. Finally, for each node $i \in \{1, \ldots, n\}$, we set $w_{i,a} = w_{i,b} = a_i$. The weights of the other edges can be seen in the figure.

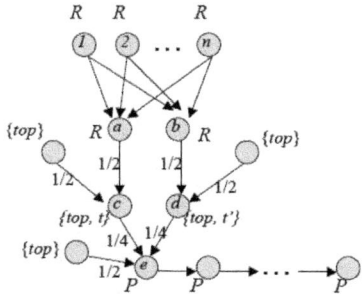

Fig. 2. The graph of the reduction with $P = \{top, t, t'\}$ and $R = \{t, t'\}$

We claim that if there exists a solution to I, then a final network exists where the number of nodes that adopted top equals 3, otherwise in all final networks the number of nodes that adopted top equals $M + 5$. This directly yields the desired result.

Suppose there is a solution S to I. Then we can have the nodes corresponding to the set S adopt t and the remaining nodes from $\{1, \ldots, n\}$ adopt t'. This implies that node a can adopt t and node b can adopt t'. Subsequently, node c can adopt t and node d can

222 K.R. Apt and E. Markakis

adopt t', which implies that node e cannot adopt any product. Hence a final network exists in which only 3 nodes adopted top.

For the reverse direction, suppose there is no solution to the PARTITION problem. Then, no matter how we partition the nodes $\{1, \ldots, n\}$, into 2 sets S, S', it will always be that for one of them, say S, we have $\sum_{i \in S} a_i > \frac{1}{2}$, whereas for the other we have $\sum_{i \in S'} a_i < \frac{1}{2}$. Thus in each final network, no matter which nodes from $\{1, \ldots, n\}$ adopted t or t', the nodes a and b adopted the same product. Suppose that nodes a and b both adopted t (the same applies if they both adopt t'). This in turn implies that node c adopted t and node d did not adopt t'. Thus, the node d could only adopt top. But then the only choice for node e was to adopt top and this propagates along the whole line to the right of e. This completes the proof of (i).

(ii) The algorithm for MAX-ADOPTION resembles the one used in the proof of Theorem 7. Given the product top, it suffices to start with the nodes that have already adopted the product and perform fast reductions but only with respect to top until no further adoption of top is possible.

We now move on to some decision problems that concern the behavior of a specific node in a given social network. We consider the following natural questions.

ADOPTION 1: (unavoidable adoption of some product)
Determine whether a given node has to adopt some product in all final networks.
ADOPTION 2: (unavoidable adoption of a given product)
Determine whether a given node has to adopt a given product in all final networks.
ADOPTION 3: (possible adoption of some product)
Determine whether a given node can adopt some product in some final network.
ADOPTION 4: (possible adoption of a given product)
Determine whether a given node can adopt a given product in some final network.

Theorem 9. *The complexity of the above problems is as follows:*

  (i) ADOPTION 1 is co-NP-complete.
 (ii) ADOPTION 2 is co-NP-complete.
(iii) ADOPTION 3 can be solved in $O(n^2 |P|)$ time.
(iv) ADOPTION 4 can be solved in $O(n^2)$ time.

The proofs of (i) and (ii) use the reduction given in the proof of Theorem 8. We omit the proof due to lack of space.

7 Conclusions and Future Work

We have introduced a diffusion model in the presence of multiple competing products and studied some basic questions. We have provided characterizations of the underlying graph structure for determining whether a product can spread or will necessarily spread to the whole graph, and of the networks that admit a unique outcome. We also studied the complexity of various problems that are of interest for networks that do not admit a unique outcome, such as the problems of computing the minimum or maximum number of nodes that will adopt a given product, or determining whether a given node has to adopt some (resp. a given) product in all final networks.

In the proposed model, one could also incorporate game theoretic aspects by considering a strategic game either between the nodes who decide which product to choose, or between the producers who decide to offer their products for free to some selected nodes. In the former case, a game theoretic analysis for players choosing between two products has been presented in [13]. An extension with the additional option of adopting both products has been considered in [9]. The latter case, with the producers being the players, has been recently studied in [1] in a different model than the threshold ones. We are particularly interested in analyzing the set of Nash equilibria in the presence of multiple products, as well as in introducing threshold behavior in the model of [1].

Acknowledgement. We would like to thank Berthold Vöcking for suggesting to us the first two problems in Section 6.

References

1. Alon, N., Feldman, M., Procaccia, A.D., Tennenholtz, M.: A note on competitive diffusion through social networks. Inf. Process. Lett. 110(6), 221–225 (2010)
2. Bharathi, S., Kempe, D., Salek, M.: Competitive influence maximization in social networks. In: Deng, X., Graham, F.C. (eds.) WINE 2007. LNCS, vol. 4858, pp. 306–311. Springer, Heidelberg (2007)
3. Borodin, A., Filmus, Y., Oren, J.: Threshold models for competitive influence in social networks. In: Saberi, A. (ed.) WINE 2010. LNCS, vol. 6484, pp. 539–550. Springer, Heidelberg (2010)
4. Carnes, T., Nagarajan, C., Wild, S.M., van Zuylen, A.: Maximizing influence in a competitive social network: A follower's perspective. In: Proc. 9th International Conference on Electronic Commerce (ICEC), pp. 351–360 (2007)
5. Chen, N.: On the approximability of influence in social networks. SIAM J. Discrete Math. 23(3), 1400–1415 (2009)
6. Easley, D., Kleinberg, J.: Networks, Crowds, and Markets. Cambridge University Press, Cambridge (2010)
7. Goldenberg, J., Libai, B., Muller, E.: Talk of the network: A complex systems look at the underlying process of word-of-mouth. Marketing Letters 12(3), 211–223 (2001)
8. Granovetter, M.: Threshold models of collective behavior. American Journal of Sociology 83(6), 1420–1443 (1978)
9. Immorlica, N., Kleinberg, J.M., Mahdian, M., Wexler, T.: The role of compatibility in the diffusion of technologies through social networks. In: ACM Conference on Electronic Commerce, pp. 75–83 (2007)
10. Jackson, M.: Social and Economic Networks. Princeton University Press, Princeton (2008)
11. Kempe, D., Kleinberg, J.M., Tardos, É.: Maximizing the spread of influence through a social network. In: Getoor, L., Senator, T.E., Domingos, P., Faloutsos, C. (eds.) KDD, pp. 137–146. ACM, New York (2003)
12. Kostka, J., Oswald, Y.A., Wattenhofer, R.: Word of mouth: Rumor dissemination in social networks. In: Shvartsman, A.A., Felber, P. (eds.) SIROCCO 2008. LNCS, vol. 5058, pp. 185–196. Springer, Heidelberg (2008)
13. Morris, S.: Contagion. The Review of Economic Studies 67(1), 57–78 (2000)
14. Mossel, E., Roch, S.: On the submodularity of influence in social networks. In: STOC, pp. 128–134 (2007)
15. Schelling, T.: Micromotives and Macrobehavior. Norton, New York (1978)

A Clustering Coefficient
Network Formation Game

Michael Brautbar and Michael Kearns

Computer and Information Science,
University of Pennsylvania,
3330 Walnut Street,
Philadelphia, PA 19104
{brautbar,mkearns}@cis.upenn.edu

Abstract. Social and other networks have been shown empirically to exhibit high edge clustering — that is, the density of local neighborhoods, as measured by the clustering coefficient, is often much larger than the overall edge density of the network. In social networks, a desire for tight-knit circles of friendships — the colloquial "social clique" — is often cited as the primary driver of such structure.

We introduce and analyze a new network formation game in which rational players must balance edge purchases with a desire to maximize their own clustering coefficient. Our results include the following:

- Construction of a number of specific families of equilibrium networks, including ones showing that equilibria can have rather general binary tree-like structure, including highly asymmetric binary trees. This is in contrast to other network formation games that yield only symmetric equilibrium networks. Our equilibria also include ones with large or small diameter, and ones with wide variance of degrees.
- A general characterization of (non-degenerate) equilibrium networks, showing that such networks are always sparse and paid for by low-degree vertices, whereas high-degree "free riders" always have low utility.
- A proof that for edge cost $\alpha \geq 1/2$ the Price of Anarchy grows linearly with the population size n while for edge cost α less than $1/2$, the Price of Anarchy of the formation game is bounded by a constant depending only on α, and independent of n. Moreover, an explicit upper bound is constructed when the edge cost is a "simple" rational (small numerator) less than $1/2$.
- A proof that for edge cost α less than $1/2$ the average vertex clustering coefficient grows at least as fast as a function depending only on α, while the overall edge density goes to zero at a rate inversely proportional to the number of vertices in the network.
- Results establishing the intractability of even weakly approximating best response computations.

Several of our results hold even for weaker notions of equilibrium, such as those based on link stability.

G. Persiano (Ed.): SAGT 2011, LNCS 6982, pp. 224–235, 2011.
© Springer-Verlag Berlin Heidelberg 2011

1 Introduction

The proliferation of large-scale social and technological networks over the last decade has given rise to an emerging science. One of the primary aims of the empirical branch of this new science is to quantify and examine the striking apparent structural commonalities that many of these large networks share, despite their differing origins, populations, and function. For example, one empirical narrative in this vein that is still unfolding is the claim that large-scale networks from social, economic, technological and other origins often share the properties of small diameter, heavy-tailed degree distributions, and high edge clustering.

Because of this, one of the primary goals of the theoretical branch of this new science is the formulation of simple models of network formation that can explain such apparent structural universalities. Interestingly, to date such efforts have mainly fallen into two categories. In the stochastic network formation literature, probabilistic models for network growth are proposed that exhibit one or more of the structural universals of interest in expectation or with high probability. In contrast, in the game-theoretic network formation links do not form randomly, but for a "reason" (rationality), and the interest is in the structural and other properties that can arise at population equilibrium. The game-theoretic models to date have primarily technological, rather than sociological, motivations, such as efficient routing concerns in communication networks (see [18,13] for good overviews of both approaches, as well as Related Work below).

In this paper we introduce and study a new network formation game explicitly motivated by an empirical phenomenon often cited in large social networks: the tendency for friendship to be transitive, or for friends of friends to be friends themselves [13,8]. In sociology and other fields, this notion is quantified by the *clustering coefficient* of a network, and a long series of studies has documented the fact that social networks routinely exhibit much larger clustering coefficients than would be expected from their overall edge density alone [19,13]. In social networks, homophily (the tendency for like to associate with like), the tendency for introductions to be made through mutual acquaintances, and a human desire for tight-knit cohorts are all cited as possible forces towards high clustering coefficients [12,8]. Given the frequent observation of clustering in social networks, and the long history of sociological and psychological theories regarding its origins in individuals, it is of interest to examine the consequences when clustering is considered the primary source of utility in a network formation game. In the same way that previous papers have taken abstract human or organizational desires, such as those of being well-connected or centrally placed in a network, and studied them as network formation games [14,3,11,10], here we do so for the notion of clustering.

We thus introduce and analyze a network formation game in which rational players must balance edge purchases, each of fixed cost, with a desire to maximize their own clustering coefficients. Like most of the prior work in formation games, we consider a unilateral, rather than bilateral, edge purchase model (Twitter rather than Facebook); such a model is appropriate for many, though obviously not all, social networks. Our results include the following:

- Construction of a number of specific families of equilibrium networks, including ones showing that equilibria can have rather general binary tree-like structure, including highly asymmetric binary trees. This is in contrast to other network formation games that yield only symmetric equilibrium networks. Our equilibria also include ones with large or small diameter, and ones with wide variance of degrees.
- A general characterization of (non-degenerate) equilibrium networks, showing that such networks are always sparse and paid for by low-degree vertices, whereas high-degree "free riders" always have low utility.
- A proof that for edge cost $\alpha \geq 1/2$ the Price of Anarchy grows linearly with the population size n while for edge cost α less than $1/2$, the Price of Anarchy of the formation game is bounded by a constant depending only on α, and independent of n. Moreover, an explicit upper bound is constructed when the edge cost is a "simple" rational (small numerator) less than $1/2$.
- A proof that for edge cost α less than $1/2$ the average vertex clustering coefficient grows at least as fast as a function depending only on α, while the overall edge density goes to zero at a rate inversely proportional to the number of vertices in the network.
- Results establishing the intractability of even weakly approximating best response computations.

Several of our results hold even for weaker notions of equilibrium, such as those based on link stability.

In the extended version of the paper we also consider other variants of the game, including a non-normalized version of clustering coefficient and bilateral edge purchases one [7].

2   Related Work

Models of social and technological networks can be roughly divided into two categories — stochastic generative models and game-theoretic models.

A stochastic generative model captures the dynamics of a specific stochastic process and characterizes the networks created in the limit of that process. Perhaps the most notable stochastic generative models are the preferential attachment model [4] and the small-world model [20]. In the preferential attachment model nodes arrive one at a time and each new node stochastically connects to a fixed number of previous nodes, where the probability of connecting to a specific node is proportional to that node's current degree in the network. Networks created by the model are known to have a limiting power-law degree distribution [5], a prominent property of various social networks. In contrast to the preferential attachment model, the small-world generative model assumes that all nodes are given in advance. In that model one starts with a ring lattice on the n nodes and rewires each edge independently with some fixed probability. Networks created this way are known to have low diameter and a large average clustering coefficient, for a large range of the rewiring probability [20]. While the preferential attachment model and the small-world model are able to only generate networks with some properties of real social networks, a recent model following similar lines as that of preferential attachment was shown to being able to generate networks with several more properties of real social networks [17].

A second approach to modeling social and technological networks is based on game theory. A node is equipped with a utility function that for each outcome of the game quantifies how good the outcome is for that node. The utility of a node is a function that depends on the structure of the outcome network and the cost of the edges the node purchased. Game theoretic formation models roughly divide into unilateral and bilateral games. In unilateral games a node can purchase an edge to another node without asking for that node's consent. In a bilateral game each edge is a result of mutual consent between the edge's endpoints. In both variants once the edge is constructed both parties can use it[1]. The seminal work of Fabrikant et al. [11] present an Internet routing latency game where the utility of a node is the sum of its shortest path distances to all other nodes plus the cost of the edges the node purchased. The game is perceived as a minimum latency game where a node's goal is to route packets quickly to their destination. The authors showed that regular trees are Nash Equilibrium (NE) networks of the game and raised the question whether the game has non-tree NE. Albers et al. [1] provide the first construction of a cyclic NE for the game using methods from finite affine spaces. Alon et al. [2] provide a combinatorial construction of a link stable network with diameter three for the routing game.

Bala and Goyal analyzed a general formation game where the utility of a node is a two-parameter function where the first parameter is the number of nodes a node is connected to in the outcome graph, and the second parameter is the number of edges the node bought [3]. Under a mild monotonicity condition on this utility function the authors showed that the Nash Equilibrium networks of the game are trees and the strict Nash Equilibrium networks are star-like (plus the empty network for some edge costs).

Borgs et al. [6] have recently introduced a unilateral network formation game motivated from affiliation networks. In their model a player can unilaterally initiate social events with a cost proportional to the number of invitees. Any two players that meet regularly at events will then form an (undirected) edge. The utility of a player is its degree in the network minus the cost of events he initiated. The authors show that the class of NE of the game contains sparse networks as well as power-law networks and that the average clustering coefficient of each NE network is bigger than the inverse of the average degree in that network.

Jackson and Wolinsky [14] were the first to introduce a general bilateral game, called the "connections model". The utility of a node in this game is a sum of discounted shortest path distances to all other nodes plus the cost of the edges adjacent to the node. The authors presented the notion of link-stability where no two nodes want to purchase a missing edge between them, and no node wants to unilaterally remove an adjacent edge. The authors presented a partial

---

[1] A third type of formation games, where edges are purchased unilaterally and can only be used by the purchasing party, is rarely considered in the literature.

characterization of all link stable networks of the game. A specific version of the game, where the edge cost is not uniform but depends on a metric on the network nodes, was further analyzed in [15]. The authors showed that for a specifically chosen discount factor for the utility of path lengths the link-stable networks of the metric version include regular networks, complete networks, chain, and star networks. However, the analysis is limited to specific values of the discount factor and no general characterization of equilibria networks is given.

Evan-Dar et al. [9] analyze a formation game for bipartite exchange economies. The network is bipartite containing buyers on one side and sellers on the other and edge purchases represent trading opportunities between its endpoint parties. The authors where able to provide a complete characterization of all NE of the network formation game which is rather exceptional in the literature.

The Price of Anarchy measure was introduced by [16] to quantify the inefficiency of NE networks with respect to a central designed solution. It is defined as the ratio between the best welfare (sum of node's utilities) of a network to the worst welfare of a NE network. The routing game presented by Fabrikant et al. was shown to have a low Price of Anarchy [11,1].

3   Preliminaries

The game we shall study, which we will refer to as the *CC game*, is a one-shot, full information game on n players that shall form the vertices of an undirected graph or network. The pure strategies of the game are the possible sets of undirected edges a player may purchase to the other $n - 1$ players. The price of all edges is the same and known in advance to all players. The edge price is denoted by α.

As in a number of previously studied network formation games, we consider edge purchases to be *unilateral* — a player may purchase an edge to any other party without consent from that party — but all players may potentially benefit from the edge purchases of others. In this sense edges are undirected, but we also need to keep track of who purchased each edge. Given the edge purchases of all players the outcome of the game yields a directed network on n nodes, denoted as G, where an edge from node u to node v is present if and only if u purchased an edge to v. Throughout we shall analyze both the properties of the directed graph G, as well as the undirected graph it induces.

We denote by I_v the set of nodes that purchased edges to v and by O_v the set of nodes v purchased an edge to. We denote the in-degree of v in G as $in\text{-}deg(v)$ and its out-degree as $out\text{-}deg(v)$. The total degree of v is defined as $deg(v) = in\text{-}deg(v) + out\text{-}deg(v)$.

We denote the number of triangles that v is part of in G by $\Delta(v)$. The number of triangles containing v in which the two other nodes both belong to I_v is denoted as $\Delta_I(v)$. Similarly, the number of triangles containing v where the two other nodes belong to O_v is denoted as $\Delta_O(v)$. The number of triangles containing v where one of the other nodes belongs to I_v and the remaining one belongs to O_v is denoted as $\Delta_{I,O}(v)$. These sets are all disjoint by definition and we have $\Delta(v) = \Delta_I(v) + \Delta_O(v) + \Delta_{I,O}(v)$.

The *clustering coefficient* of a node v in G is defined as the probability that two randomly selected neighbors of v are directly connected to each other: $CC(v) = \frac{\Delta(v)}{\binom{deg(v)}{2}}$ if $deg(v) \geq 2$, and 0 otherwise. In the CC game, players must balance their desire for high clustering coefficient against their edge expenditures. The utility of v in the game is defined to be $utility(v) = CC(v) - \alpha \cdot out\text{-}deg(v)$. When the edge cost $\alpha \geq 1$ all strategies for a node v are dominated by the strategy to purchase no edges at all, so we will assume from now on that $0 < \alpha < 1$. Most of our results will consider the natural case in which α is a constant not depending on the population size n — forming edges has a fixed cost — though we will occasionally discuss cases where α diminishes with increasing n. Some of our results will also depend on α being a rational number.

As in much of the related literature, our main interest in this paper is to study the properties of the *pure* Nash equilibrium (NE) networks of the CC game. For some of our results we shall slightly refine this notion to exclude some degenerate cases and thus focus on the interesting ones. Note that the empty network (no edge purchases) is a trivial NE with zero social welfare (total utility) that we will omit from consideration. We also ask that players who purchase edges have non-zero utility. Note that (at least) zero utility can always be obtained by purchasing *no edges*. This condition demands that the action taken by only a subset of the players (those buying edges) be better than only one of their many alternatives (buying no edges), and even then only in the case that the latter gives zero utility. It is thus a considerable weakening of the standard notion of a *strict* Nash equilibrium. We next codify these restrictions:

Definition 1. *A non-degenerate NE is a non-empty, pure Nash Equilibrium of the CC game in which out-deg(v) ≥ 1 implies utility(v) > 0 for all players v.*

The *social welfare* of a given network is defined as the sum of all players' utilities. The (non-degenerate) Price of Anarchy (PoA) is defined as the ratio of the highest social welfare of any directed network with n nodes to the worst social welfare of any non-degenerate NE.

4   A (Partial) Catalog of CC Game Nash Equilibria

We begin by constructing a number of families of non-degenerate NE of the CC game, focusing primarily on the network topologies that can arise at equilibrium. Each of these families is defined for arbitrarily large population size n, and has social welfare scaling linearly with n. We do not propose this catalog to be exhaustive; indeed it is interesting to see the diversity of structures that can arise at equilibrium, and we suspect there are others. Subsequent sections are devoted to the study of general properties of non-degenerate NE.

The first three constructions below are sufficiently simple that their equilibrium proofs can be established by straightforward calculations that we omit. We do provide the equilibrium proof for our last, and richest, construction.

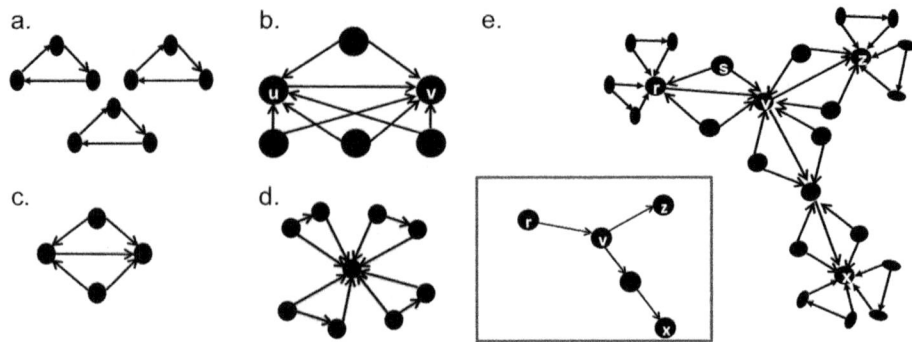

Fig. 1. A variety of Nash Equilibrium networks of the CC-Game: Disjoint Triangles NE (a), Popular Victims NE (b and c), Triangular Hub and Spokes NE (d), Binary Tree-Like NE (e)

Disjoint Triangles NE. Perhaps the simplest non-degenerate NE consists of $n/3$ disjoint triangles. The nodes in each group form a triangle by purchasing one edge each (Figure 1a). Clearly this structure is a non-degenerate NE for any $0 < \alpha < 1$; for n divisible by 3 it also maximizes the social welfare, a fact we shall use throughout.

Popular Victims NE. This non-degenerate NE shows a case where the most "popular" (highest degree) nodes suffer the lowest utility. Let $n \geq 4$. The construction is as follows: a player u connects to a player v, and each other node connects directly to both u and v by purchasing two edges (Figure 1b). When the edge cost is inversely proportional to n, $\alpha = \frac{2}{n-1} - \epsilon$, for any $\epsilon > 0$, this network is a non-degenerate equilibrium. To see this, notice that all players other than u and v are playing their best responses and get a positive utility provided $\alpha < \frac{1}{2}$. Node v cannot improve its utility since all nodes are connected to it. Last, if $\alpha < \frac{2}{n-1}$, u wouldn't want to remove the edge it purchased to v and therefore is playing its best response. Furthermore, u is getting a positive utility.

Note that this network is "paid for" by low-degree vertices, all of whom enjoy high utility, while the high-degree victims u and v suffer low utility. We shall show later that in fact this is a property of all non-degenerate NE.

Triangular Hub and Spokes NE. Consider the network shown in Figure 1d; it is easily verified that for edge cost $\alpha = \frac{1}{2} - \epsilon$, for any $\epsilon > 0$ this is a non-degenerate NE. Furthermore, this construction can be scaled up to make the "hub" node have arbitrarily high degree at the same (constant) edge cost, and disjoint copies of this construction of different size can be combined to form new non-degenerate NE. In this fashion we may create non-degenerate NE whose (total) degree distributions are effectively unconstrained.

Binary Tree-Like NE. We next construct a large family of non-degenerate NE obtained by the following construction. We take any rooted, directed binary tree T (with edges always oriented towards the leaves), where the root has out-degree of one, and replace each directed edge in T with a local gadget of the type given in Figure 1c. As an example of the construction, consider starting with the rooted, directed tree T on five vertices shown in Figure 1e (inset). The resulting network $G(T)$ is given in Figure 1e.

It is worth emphasizing that this construction yields a rather rich family of non-degenerate NE with a variety of asymmetries possible, which is somewhat unusual in the network formation game literature. At one extreme it contains connected, small diameter networks (constructed from balanced binary trees) and on the other extreme it contains connected, large diameter networks (constructed from path-like graphs). Since the argument that the construction does yield NE is considerably more involved than for our previous examples, a formal theorem is given. The proof is omitted due to lack of space and is given in the extended version of the paper [7].

Theorem 1. *For any rooted, directed binary tree T where the root has out-degree of one, let $G(T)$ be the directed network obtained by the construction described above. Then for any edge cost that is smaller than some constant independent of network size, $G(T)$ is a non-degenerate pure NE of the CC game.*

5 General Properties of CC Game Nash Equilibria

Given the apparent diversity and potential asymmetry of the NE of the CC game, what general statements might we hope to make about their topological and utility properties? Certain very basic and crude characterizations are easily obtained — for instance, the fact that any NE has at most $\frac{n}{\alpha}$ edges follows from the fact that each node can purchase at most $\frac{1}{\alpha}$ edges at equilibrium since all utilities are non-negative. Notice that this observation does not imply a non-trivial restriction on the total degree or utility of any individual node.

In this section, we prove a considerably stronger characterization motivated by the commonalities in the NE described in the last section. Namely, we prove that any (non-degenerate) NE is paid for by nodes of low total degree and high utility, while high-degree vertices are always victims of low utility. This characterization will then be applied in the following section to obtain non-trivial bounds on the Price of Anarchy for the CC game.

Theorem 2. *Let $0 < \alpha < \frac{1}{2}$. Then in any non-degenerate NE of the CC game:*

- *For any node v, if out-deg$(v) \geq 1$, then $deg(v) < \frac{3}{\alpha}$, and utility$(v) = c(\alpha) > 0$, where the strictly positive constant $c(\alpha)$ depends only on α, and not the population size n. Moreover, when $\frac{1}{\alpha}$ is integral, $c(\alpha) > \frac{\alpha^3}{9}$. [2] Thus, vertices purchasing an edge have low total degree and a positive, constant utility.*
- *For any node v with $deg(v) \geq \frac{3}{\alpha}$, utility$(v) < \frac{3}{\alpha(deg(v)-1)}$. Thus, high-degree vertices have low utility.*

---

[2] A similar bound holds for "simple" rational α; see the proof.

Proof. We start by proving the first part of the theorem. Let v be any node in a non-degenerate NE network that purchased an edge and has an in-degree of at least two (the claim is trivially true when in-degree of v is at most one). The upper bound on v's total degree is derived from the fact the v's utility is higher than what it could have gotten by purchasing no edges at all:

$$\frac{\Delta(v)}{\binom{deg(v)}{2}} - \alpha \cdot out\text{-}deg(v) \geq \frac{\Delta_I(v)}{\binom{in\text{-}deg(v)}{2}}.$$

Simplifying, we get

$$\Delta_I(v) \left(\frac{2}{deg(v)(deg(v) - 1)} - \frac{2}{in\text{-}deg(v)(in\text{-}deg(v) - 1)} \right) +$$

$$\frac{\Delta_{I,O}(v) + \Delta_O(v)}{\frac{deg(v)(deg(v)-1)}{2}} \geq \alpha \cdot out\text{-}deg(v).$$

Since $\frac{2}{deg(v)(deg(v)-1)} - \frac{2}{in\text{-}deg(v)(in\text{-}deg(v)-1)} < 0$, we get

$$\frac{\Delta_{I,O}(v) + \Delta_O(v)}{out\text{-}deg(v)\,deg(v)(deg(v) - 1)} > \frac{\alpha}{2}.$$

By using $\Delta_O(v) \leq \binom{out\text{-}deg(v)}{2}$ and $\Delta_{I,O}(v) \leq in\text{-}deg(v) \cdot out\text{-}deg(v)$, we get

$$\frac{in\text{-}deg(v)\,out\text{-}deg(v) + \frac{out\text{-}deg(v)(out\text{-}deg(v)-1)}{2}}{out\text{-}deg(v)\,deg(v)(deg(v) - 1)} > \frac{\alpha}{2},$$

so $\frac{1}{deg(v)} + \frac{1}{2deg(v)} > \frac{\alpha}{2}$, or alternatively, $deg(v) < \frac{3}{\alpha}$.

Next, we prove a lower bound on v's utility that follows from it being strictly positive (non-degeneracy). Recall that $utility(v) = \frac{\Delta(v)}{\binom{deg(v)}{2}} - \alpha\,out\text{-}deg(v) > 0$. Since $deg(v) < \frac{3}{\alpha}$, the RHS of the utility expression can only equal one out of a finite number possible of possibilities that depend only on α and not on n. In particular, for each α we can choose the worst possible value that still renders $utility(v)$ strictly positive. We denote that value by $c(\alpha)$.

Furthermore if $\frac{1}{\alpha}$ is integral, by taking a common denominator the left hand-side can be written as a strictly positive numerator divided by a denominator of $\frac{1}{\frac{1}{\alpha}\binom{deg(v)}{2}}$. Using $deg(v) < \frac{3}{\alpha}$, we get that v's utility is bigger than $\frac{\alpha^3}{9}$. (More generally note that if $\alpha = p/q$ for integers $p < q$ and thus rational, a similar argument yields a lower bound of $\frac{\alpha^3}{9p}$ on the utility, and thus "simple" α give constructive lower bounds.)

We next prove the second part of the theorem. Consider a node v with a total degree of at least $\frac{3}{\alpha}$. We saw earlier that a node that purchased edges has a degree of less than $\frac{3}{\alpha}$ so v could not have purchased edges at all. Moreover, a node u that purchased an edge to v has degree less than $\frac{3}{\alpha}$ and so v is part of less than $\frac{3}{\alpha}$ joint triangles with u. Therefore the total triangle count of node v is less than $\frac{1}{2}d \cdot \frac{3}{\alpha}$. Thus, v's utility is less than $\frac{\frac{1}{2}d\frac{3}{\alpha}}{\binom{d}{2}} = \frac{3}{\alpha(d-1)}$.

6 The Price of Anarchy

As has been mentioned, a disjoint union of triangles is a maximum social welfare NE, whereas all the specific families of NE given in Section 4 have a social welfare growing linearly with the population size n. In this section we prove that the non-degenerate Price of Anarchy is upper bounded by a function depending only on α, and not on n, for all $\alpha < 1/2$, and give an explicit expression for the upper bound when α is a "simple" rational (small numerator). This turns out to be a fairly straightforward consequence of the characterization given in Theorem 2. The proof can be found in the extended version of the paper [7].

Theorem 3. *For edge cost $\alpha \geq \frac{1}{2}$ the non-degenerate Price of Anarchy for the CC game is lower bounded by $\Omega(n(1-\alpha))$, and for edge cost $\alpha < \frac{1}{2}$ it is upper bounded by an expression that depends only on α. Moreover, when $\frac{1}{\alpha}$ is integral the Price of Anarchy is upper bounded by $\frac{36(1-\alpha)}{\alpha^4}$* [3]

While Theorem 3 upper bounds the non-degenerate Price of Anarchy independent of the population size n for $\alpha < 1/2$, it leaves open the question of the exact dependence on α and whether it is even real or not. Indeed, all specific constructions in Section 4 have a constant Price of Anarchy independent of α, even when α is a small numerator rational. We leave the resolution of this dependence as an open problem.

It is natural to ask how robust the results we have described so far are with respect to modifications of the equilibrium notion — especially in light of the results in the following section, where we will prove that even approximate best-response computations for the CC game are intractable. Indeed, it is for similar reasons that in other network formation games, researchers often consider weaker notions of equilibrium, such as *link stability* (which asks only that players cannot improve their utilities by adding or dropping a single edge purchase).

Notice that an equilibrium concept resilient only to the addition or removal of a single one edge already has a Price of Anarchy of $\Omega(n(1-\alpha))$ for any edge cost, since a network with one triangle and many isolated nodes is then in equilibrium no matter how small α is (a single edge purchase can never help). However, define *k-stability* to be the equilibrium concept in which players cannot benefit by switching from their assigned edge purchase set S to any other edge purchase set S' for which the symmetric set difference $|S - S'| \leq k$. (Thus standard link stability corresponds to 1-stability.) For any fixed value of k, computing best responses under k-stability becomes a computationally tractable problem, and for $k \geq 2$, all of our results can be shown to hold under this notion as well:

Theorem 4. *For all $k \geq 2$, Theorems 2 and 3 remain true when we replace NE by k-stability.*

The proof is omitted, but mainly involves technical modifications of the proof of the first part of Theorem 2 to consider the utility effects of dropping only the most beneficial edge purchases, rather than all edge purchases.

---

[3] A similar bound holds for "simple" (small numerator) rational α.

We end by noting that a low PoA implies that the average vertex clustering coefficient is high.

Corollary 1. *For edge cost $\alpha < \frac{1}{2}$ the average vertex clustering coefficient grows at least as some function $g(\alpha)$ independent of the network size, while the network's overall edge density goes to zero at a rate smaller or equal to $\frac{2\alpha}{n-1}$.*

7   Intractability of Best Responses

A natural question that arises in many complex network formation games is how difficult it can be to compute best responses, which would seem a prerequisite to reaching NE dynamically; for instance, best-response computation was shown to be NP-hard to compute for a routing formation game [11]. Here we show that best responses in the CC game are intractable even to approximate, thus motivating the weaker notion of k-stability in the last section.

Theorem 5. *Given a directed graph G and a node v in G (where G represents the edge purchases of the other nodes), the edge cost α (encoded as a rational number), and an integer $f \geq 1$, computing a strategy (set of edge purchases) for v with CC game utility at least $\frac{1}{f}$ of the best-response utility is not polynomial time computable, unless $P = NP$.*

The proof is given in the extended version of the paper [7].

One way to deal with the inapproximability of best response is to focus on computing best responses under k-stability, $k \geq 1$. Although the problem of computing best response under k stability for each node becomes tractable for fixed values of k, the corresponding dynamics doesn't always converge to a k-stable network, as shown in Figure 2. Therefore there is no simple solution to the inapproximability of best-responses.

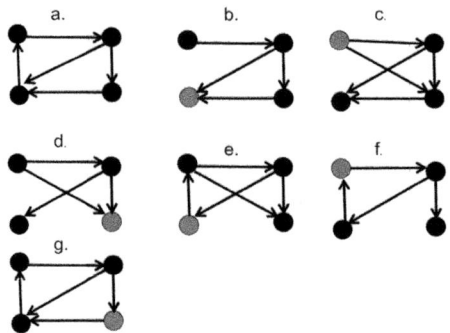

Fig. 2. Here we consider the weakest notion of best response, where a node can either add or remove at most one edge to improve its utility. Edge cost can be taken to be any $0 < \alpha < 1$. Figure 2a shows the initial network, and in each consecutive round the network is drawn after the node colored gray played its best-response under k-stability, $k \geq 1$. The dynamics returns to its initial configuration after six rounds as shown in Figure 2g, so it never converges to a k-stable network.

References

1. Albers, S., Eilts, S., Even-Dar, E., Mansour, Y., Roditty, L.: On nash equilibria for a network creation game. In: SODA, pp. 89–98 (2006)
2. Alon, N., Demaine, E.D., Hajiaghayi, M., Leighton, T.: Basic network creation games. In: SPAA, pp. 106–113 (2010)
3. Bala, V., Goyal, S.: A noncooperative model of network formation. Econometrica 68(5), 1181–1230 (2000)
4. Barabasi, A.L., Albert, R.: Emergence of scaling in random networks. Science 286, 509–512 (1999)
5. Bollobás, B., Riordan, O., Spencer, J., Tusnády, G.: The degree sequence of a scale-free random graph process. Random Struct. Algorithms 18(3), 279–290 (2001)
6. Borgs, C., Chayes, J.T., Ding, J., Lucier, B.: The hitchhiker's guide to affiliation networks: A game-theoretic approach. In: ICS (2011)
7. Brautbar, M., Kearns, M.: A clustering coefficient network formation game, extended version, http://arxiv.org/abs/1010.1561
8. Easley, D., Kleinberg, J.: Networks Crowds and Markets: Reasoning about a Highly Connected World. Cambridge University Press, Cambridge (2010)
9. Even-Dar, E., Kearns, M., Suri, S.: A network formation game for bipartite exchange economies. In: SODA, pp. 697–706 (2007)
10. Even-Dar, E., Kearns, M.: A small world threshold for economic network formation. In: NIPS, pp. 385–392 (2006)
11. Fabrikant, A., Luthra, A., Maneva, E.N., Papadimitriou, C.H., Shenker, S.: On a network creation game. In: PODC, pp. 347–351 (2003)
12. Heider, F.: The Psychology of Interpersonal Relations. John Wiley & Sons, Chichester (1958)
13. Jackson, M.O.: Social and Economic Networks. Princeton University Press, Princeton (2008)
14. Jackson, M.O., Wolinsky, A.: A strategic model of social and economic networks. J. of Economic Theory 71, 44–74 (1996)
15. Johnson, C., Gilles, R.P.: Spatial social networks. Review of Economic Design 5, 273–299 (2000)
16. Koutsoupias, E., Papadimitriou, C.H.: Worst-case equilibria. In: Meinel, C., Tison, S. (eds.) STACS 1999. LNCS, vol. 1563, pp. 404–413. Springer, Heidelberg (1999)
17. Lattanzi, S., Sivakumar, D.: Affiliation networks. In: STOC, pp. 427–434 (2009)
18. Newman, M., Barabasi, A.L., Watts, D.J.: The Structure and Dynamics of Networks. Princeton University Press, Princeton (2006)
19. Watts, D.J.: Small worlds. Princeton University Press, Princeton (1999)
20. Watts, D.J., Strogatz, S.H.: Collective dynamics of 'small-world' networks. Nature 393(6684), 440–442 (1998)

On the Existence of Pure Strategy Nash Equilibria in Integer–Splittable Weighted Congestion Games

Long Tran-Thanh[1], Maria Polukarov[1], Archie Chapman[2],
Alex Rogers[1], and Nicholas R. Jennings[1]

[1] School of Electronics and Computer Science,
University of Southampton, UK
{ltt08r,mp3,acr,nrj}@ecs.soton.ac.uk
[2] The University of Sydney Business School, Sydney, Australia
a.chapman@econ.usyd.edu.au

Abstract. We study the existence of pure strategy Nash equilibria (PSNE) in *integer–splittable weighted congestion games* (ISWCGs), where agents can strategically assign different amounts of demand to different resources, but must distribute this demand in fixed-size parts. Such scenarios arise in a wide range of application domains, including job scheduling and network routing, where agents have to allocate multiple tasks and can assign a number of tasks to a particular selected resource. Specifically, in an ISWCG, an agent has a certain total demand (aka weight) that it needs to satisfy, and can do so by requesting one or more integer units of each resource from an element of a given collection of feasible subsets.[1] Each resource is associated with a unit–cost function of its level of congestion; as such, the cost to an agent for using a particular resource is the product of the resource unit–cost and the number of units the agent requests.

While general ISWCGs do not admit PSNE (Rosenthal, 1973b), the restricted subclass of these games with linear unit–cost functions has been shown to possess a potential function (Meyers, 2006), and hence, PSNE. However, the linearity of costs may not be necessary for the existence of equilibria in pure strategies. Thus, in this paper we prove that PSNE always exist for a larger class of convex and monotonically increasing unit–costs. On the other hand, our result is accompanied by a limiting asumption on the structure of agents' strategy sets: specifically, each agent is associated with its set of accessible resources, and can distribute its demand across any subset of these resources.

Importantly, we show that neither monotonicity nor convexity on its own guarantees this result. Moreover, we give a counterexample with monotone and semi–convex cost functions, thus distinguishing ISWCGs from the class of infinitely–splittable congestion games for which the conditions of monotonicity and semi–convexity have been shown to be sufficient for PSNE existence (Rosen, 1965). Furthermore, we demonstrate that the finite improvement path property (FIP) does not hold for convex increasing ISWCGs. Thus, in contrast to the case with linear costs, a potential function argument cannot be used to prove our result. Instead, we provide a procedure that converges to an equilibrium from an arbitrary initial strategy profile, and in doing so show that ISWCGs with convex increasing unit–cost functions are weakly acyclic.

---

[1] Additionally, strategy sets are restricted by certain domain–specific constraints—for instance, in network routing, an agent's strategy must define a feasible flow between its given pair of source and target nodes.

G. Persiano (Ed.): SAGT 2011, LNCS 6982, pp. 236–253, 2011.

1 Introduction

The study of interaction of multiple self–interested parties ("agents") sharing commonly–available facilities ("resources") is central to computational game theory. Such settings naturally arise in a wide range of typical application domains, from traffic routing in networks (e.g. roads, air traffic or information and communication networks (Rosenthal, 1973a; Roughgarden and Tardos, 2002)), to competition in job scheduling problems (e.g. for computational services or machine scheduling (Koutsoupias and Papadimitriou, 1999)).

In many real–world scenarios in these domains, agents may find it beneficial to assign different amounts of demand to different resources, but may have restrictions on the size of units in which this demand is distributed. For example, consider a job scheduling problem, comprised of n agents and m independent machines, where each agent has several indivisible jobs to be executed. To each selected machine, an agent pays a usage cost, which is equal to the number of jobs the agent allocates to that machine multiplied by the unit–cost per job, typically depending on the total level of demand on the machine (i.e., its congestion). A similar situation arises in communication networks (e.g. the Internet), where agents send packets (or, messages) and have to decide how many packets to route on each path in the network to minimise possible delays. Additional examples for a problem of this kind may include procuring factor inputs for manufacturing processes or purchasing transport capacity for logistics networks. Importantly, in all these situations, the agents cannot split their demands in arbitrary ways, but must do so in *integer* units.

Problems of this kind are addressed in the literature as *integer–splittable weighted congestion games* (ISWCGs), where agents strategically choose from a common set of resources, and are allowed to assign multi–unit requests to each of their selected resources; however, they are constrained to make their allocations in fixed-size parts (particularly, integer units). Each resource is equipped with a "unit–cost function" that indicates the cost that each agent pays per unit of request, depending on the aggregate level of congestion on that resource (i.e., the total number of units the users contribute to the resource). Since the agents may have different congestion impacts, the cost each agent has to pay for the use of a particular resource is the product of the amount of units it requests from that resource and the corresponding unit–cost. For example, in a computational services setting, if an agent were to purchase four units of processor time from a particular service provider, it would pay the same unit–cost for all four units, with the unit–cost determined by the total demand from all agents for that resource. The overall agent's cost is given by the sum of its costs for each resource it uses. In a ISWCG, each agent has a certain integer demand (or, weight) for resource units it needs to satisfy, and its aim is to minimise the total cost of the units by distributing its weight across the available resources. Unit–cost functions are resource–specific, but are the same for all agents (i.e., resource providers cannot discriminate between users), while demands for resource units can vary across the agents. Note that the above examples are captured in the ISWCG model by identifying the set of resources with the set of machines or network links, respectively, where differences in their technical parameters and performance factors, such as efficiency, or speed, are reflected by resource–dependent costs per unit (e.g. job, or data packet). An agent's demand represents the amount of resources (job, or data traffic) each agent has, and the set of feasible assignments (task allocations, traffic routes) corresponds to the set of feasible strategy profiles.

1.1 Related Work

Much of related work deals with a traditional *congestion game* model by Rosenthal (1973a), where agents have to choose from a given finite set of resources, and where the possible choices of an agent are given by the subsets of resources that satisfy its goals. The cost of a resource is determined by the total number of its users, and an agent's overall cost is given by the sum of resource costs over the set of the agent's selected resources. In a variant setting of network congestion models, agents have to choose subsets of edges on a graph forming a path from the agent's origin to destination, in order to route their demand (i.e. flow) through the network, and the cost (e.g. latency) of each edge varies with the number of agents traversing that edge.

The important property of congestion games shown by Rosenthal (1973a) is the existence of a Nash equilibrium in pure strategies (PSNE)—a profile where each agent plays a certain (non–randomised) strategy and no one has an incentive to unilaterally change it. Such solutions are highly desirable, since, from a system–wide perspective, they imply that a system has a deterministic stable state. This is necessary in a range of control problems where randomised strategies are not appropriate (e.g. in industrial processing or transport applications). Also, unlike mixed strategy and correlated equilibria, PSNE do not rely on the assumption that agents have the capacity to accurately randomise their actions according to an equilibrium prescription.

Moreover, congestion games are also known to possess a stronger charateristic, called the "finite improvement path property" (FIP), implying that any sequence of unilateral improvement deviations (i.e., strategy changes that decrease an agent's total cost) will converge to a PSNE in finite time. This is implied by the existence of a "potential function" that decreases along any such improvement path (Monderer and Shapley, 1996). Given this, the players can use a variety of simple potential–based search processes to find a PSNE in a distributed fashion, such as fictitious play or weighted regret monitoring (Leslie and Collins, 2006; Marden *et al.*, 2007).

The traditional model has been generalised to a variety of related situations. Such generalisations, for example, include *player–specific congestion games* (Milchtaich, 1996) where an agent's payoff depends on its identity, *weighted congestion games* (Milchtaich, 1996), in which agents may have different (although fixed) congestion impacts (weights), *local–effect games* (Leyton-Brown and Tennenholtz, 2003) with an agent's cost for a particular resource being also affected by a congestion on its neighbouring resources, *congestion games with failures* (Penn *et al.*, 2009a) and *random-order congestion games* (Penn *et al.*, 2009b) modelling faulty or asynchronous resources, and *congestion–averse games* (Byde *et al.*, 2009; Voice *et al.*, 2009) where the agents' utilities are determined by general real–valued functions of congestion vectors. Note that in all these settings, agents are restricted to request only a single or a fixed number of units from each particular chosen resource; that is, in terms of the network congestion model, they have to *unsplittably* route their flow within the network.

At the other extreme, *infinitely–splittable* congestion game models assume that agents have divisible demand, which can be fractionally split acroos an *arbitrary* number of resources (paths), in any proportion (Orda *et al.*, 1993; Cominetti *et al.*, 2009). For this setting, a result from Rosen (1965) implies that PSNE are guaranteed to exist if resource cost functions are semi–convex[2] and monotone increasing. As an intermediate concept between splittable and unsplittable games, the model of *k–splittable* network conges-tion models was introduced by Beier *et al.* (2004) to capture scenarios where agents are restricted to split their demand across at most *k* different paths. However, the portion of the demand that an agent allocates to a single path can be fractional. Beier *et al.* (2004) showed that it is NP–hard to decide whether a PSNE exists within such settings. In addition, Shachnai and Tamir (2002); Krysta *et al.* (2003) obtained similar results for *k*–splittable congestion games in the job scheduling domain.

More relevant to our work is the paper by Meyers (2006) where the *k*–splittable model is modified so that agents are only allowed to allocate integer amounts of de-mand to each chosen resource (or, path). The authors showed that the restricted sub-class of these games where the unit–cost functions are linear, possesses a potential function, and hence, the FIP holds and a PSNE is guaranteed to exist. For a gen-eral case, Rosenthal (1973b) gave an example of an asymmetric weighted network congestion game that does not have an equilibrium in pure strategies. More recently, Dunkel and Schulz (2008) strengthened this result by showing that the problem of de-ciding whether a weighted network congestion game with integer–splittable flows ad-mits a PSNE is strongly NP–hard.

1.2 Our Contribution

In this paper, we extend positive results on the existence of a pure strategy equilibrium in integer–splittable congestion games to a larger class of unit–cost functions which are monotonically increasing and convex. From a practical point of view, this class is important as convex increasing costs occur in a wide range of application domains. In-deed, in many real–world systems, marginal costs typically increase as total demand increases (e.g. energy cost in smart grids or delay in multi–server systems). Further-more, such systems are often regarded as overloaded, if the total demand exceeds a certain threshold. In this case, the users often have to pay extremely higher costs for using the resources (in smart grids, for example, each power plant has a finite produc-tion limit, and if the total demand exceeds the sum of these limits, additional expensive peaking plant must supply the excess). We note that our result is accompanied by a lim-iting asumption on the structure of agents' strategy sets. Specifically, we assume that each agent is associated with its set of accessible resources, wich is a part of a given superset, and can distribute its demand across any subset of these resources. For sake of brevity, in what follows we slightly abuse the notation and use the term ISWCG to define a game with such restricted strategy set structures.

The above assumption implies that negative results by Rosenthal (1973b) and Dunkel and Schulz (2008) do not apply to our setting. However, as we show, the exis-tence of PSNE is still violated. Moreover, PSNE are not guaranteed to exist in games with

---

[2] A function $f(x)$ is *semi-convex* if $x \cdot f(x)$ is convex.

either non–monotone or non–convex unit–costs, implying the necessity of these conditions for PSNE existence. Interestingly, our examples show that even functions which are monotone and semi–convex result in games with no pure strategy equilibria, thus distinguishing between the classes of ISWCGs and infinitely–splittable congestion games.

Following this, our main result proves that a pure strategy equilibrium is guaranteed for ISWCGs with monotonically increasing and convex unit–costs. Importantly, as we show, PSNE exist in these games despite of the non–existence of a potential function and the FIP. Consequently, in contrast to the case with linear costs (Meyers, 2006), potential–based methods cannot be used for proving PSNE existence and finding such equilibria. Based on this, we provide a search algorithm that returns a PSNE of a given game in finite time. Finally, we note that our algorithm shows convergence from an arbitrary initial strategy profile, thus showing that convex increasing ISWCGs possess the weak acyclicity property (Monderer and Shapley, 1996).

The remainder of the paper unfolds as follows. First, in Section 2 we formally define the model for ISWCGs. Then, in Section 3 we show that no guarantees on PSNE existence can be made if the unit–cost functions are not convex or monotone increasing. Following this, in Section 4 we study the case of ISWCGs with convex increasing costs. We show that these games do not generally possess a potential function by giving an example of an improvement cycle. Nonetheless, we prove that they are guaranteed to possess PSNE if the cost function is convex and monotone increasing, and devise an algorithm for computing them. Due to space limitations, some of the proofs are ommited from this version of the paper.

2   The Model

Consider a congestion domain with a set $N = \{1, \ldots, n\}$ of agents, where each agent $i \in N$ has a set R_i of $m_i \in \mathbb{N}$ accessible resources, which is a subset of a finite superset $R = \{r_1, \ldots, r_m\}$. An agent i needs to execute $X^i \in \mathbb{N}$ task units, and can distribute this demand (or, *weight*) arbitrarily among its resources. Note that each agent can use more than one *integer* unit from a single selected resource. An agent i's (pure) strategy is given by $x^i = \left(x_r^i\right)_{r \in R}$, where $x_r^i \in \mathbb{N}$ is the number of units that agent i demands from resource $r \in R$, such that $x_r^i = 0$ for all $r \notin R_i$ and

$$\sum_{r \in R} x_r^i = \sum_{r \in R_i} x_r^i = X^i \tag{1}$$

Every combination of strategies (a *strategy profile*) $x = \left(x^i\right)_{i \in N}$ corresponds to a *congestion vector* $h(x) = (h_r(x))_{r \in R}$, where

$$h_r(x) = \sum_{i \in N} x_r^i \tag{2}$$

indicates the *congestion*—the total number of assigned tasks (or, demanded units) on resource $r \in R$ in profile x.

From the perspective of agent i, a strategy profile x can be viewed as $\left(x^i, x^{-i}\right)$, where x^{-i} stands for the joint strategy of other agents. Similarly, for $r \in R$ we denote by

$$h_r^{-i}(x) = \sum_{j \neq i} x_r^j = h_r(x) - x_r^i \tag{3}$$

the congestion on resource r incurred by the collective demand of the agents, excluding agent i.

Each resource $r \in R$ is associated with a *unit–cost* (or simply, a *cost*) function $c_r : \mathbb{N} \to \mathbb{R}$ defining the cost for a unit of demand on resource r as a function of the total congestion on the resource. For simplicity, it is convenient to assume that cost functions are non–negative, although our results do not rely on this assumption.

Given this, the *payoff function* of an agent is defined as follows. The overall cost agent i has to pay in a strategy profile x is

$$C^i(x) = \sum_{r \in R} x_r^i c_r(h_r(x)) \tag{4}$$

Furthermore, the total cost of the system is given by

$$C(x) = \sum_{i \in N} C^i(x) = \sum_{r \in R} h_r(x) c_r(h_r(x)) \tag{5}$$

Definition 1. *An* integer–splittable weighted congestion game (ISWCG) $\Gamma = \left(N, R, (X^i)_{i \in N}, (c_r(\cdot))_{r \in R} \right)$ *consists of a set N of $n \in \mathbb{N}$ agents, a set R of $m \in \mathbb{N}$ resources, a unit–cost function c_r for each resource, and for each agent i a set of accessible resources $R_i \subseteq R$ and a total demand (aka weight) X^i. The strategy set for each agent $i \in N$ is the set of m-dimensional vectors $\left\{ (x_r^i)_{r \in R} \in \mathbb{N}^m \right\}$, such that $\sum_{r \in R} x_r^i = X^i$, $x_r^i = 0 \ \forall r \notin R_i$, and the cost to the agent for a combination of strategies x is $C^i(x) = \sum_{r \in R} x_r^i c_r(h_r(x))$, where $h_r(x)$ is the vector of congestion as determined by x.*

3 Non-existence of PSNE

In this section, we show that general ISWCGs do not necessarily admit pure strategy Nash equilibria (PSNE). We provide two examples, based on which we reason about conditions that would guarantee PSNE existence.

Example 1. Consider a two–player ISWCG with demands $X_1 = 2$ and $X_2 = 1$, and two resources with the following unit–cost functions:

$$c_{r_1}(1) = 12, \ c_{r_1}(2) = 5, \ c_{r_1}(3) = 7$$
$$c_{r_2}(1) = 10, \ c_{r_2}(2) = 6, \ c_{r_2}(3) = 10$$

The payoff matrix of the game is presented in Table 1. One can easily verify that there is no PSNE in this game.

Table 1. No PSNE in ISWCGs with non-monotone unit–costs

	$(0,2)$	$(1,1)$	$(2,0)$
$(0,1)$	10, 20	6, 18	10, 10
$(1,0)$	12, 12	5, 15	7, 14

Note that the cost functions in Example 1 are not monotone, but convex. That is, the convexity condition on its own is not sufficient for the existence of a pure strategy equilibrium. The next example demonstrates that neither is monotonicity sufficient.

Example 2. Consider a two–player ISWCG with demands $X_1 = 3$ and $X_2 = 1$, and two identical resources with a unit–cost function $c_{r_1}(\cdot) = c_{r_2}(\cdot) = c_r(\cdot)$ given by:

$$c_r(1) = 3, \ c_r(2) = 8, \ c_r(3) = 10 \, c_r(4) = 12$$

The payoff matrix of the game is presented in Table 2. Inspection shows that there is no PSNE in this game.

Table 2. No PSNE in ISWCGs with non-convex cost functions

	(0,3)	(1,2)	(2,1)	(3,0)
(1,0)	3, 30	8, 24	10, 23	12, 36
(0,1)	12, 36	10, 23	8, 24	3, 30

As mentioned above, Example 1 is convex, while Example 2 is monotone–increasing, implying that if either property of the cost functions is violated, a PSNE is not guaranteed. Furthermore, the cost function $c_r(x)$ in Example 2 is *semi–convex* (i.e., $x \cdot c_r(x)$ is convex). It implies that the conditions of monotonicity and semi-convexity, which have been shown to be sufficient for PSNE existence in infinitely–splittable congestion games, do not apply to the integer–splittable case! Based on this, in the following section we prove that a pure strategy equilibrium always exists in the ISWCGs whose resource unit–cost functions are monotone–increasing and convex.

4   Convex Increasing ISWCGs

In this section, we investigate the subclass of ISWCGs with convex and monotonically increasing cost functions (henceforth, *convex increasing* ISWCGs). Our main result proves that pure strategy Nash equilibria always exist in such games. Importantly, as we show in 4.1, an arbitrary sequence of myopic improving deviations may cycle even in this case; hence, the FIP property does not hold and a potential function argument is not applicable. Against this background, in 4.2 we propose a special dynamic procedure, that reaches an equilibrium from any starting point. This shows that convex increasing integer–splittable congestion games possess the weak–acyclicity property and implies an algorithm for finding PSNE in these games.

4.1   Violating the Finite Improvement Property

Given a pure strategy profile of a game, consider an arbitrary sequence of unilateral moves, where at each step a deviating agent improves its payoff with respect to the current one it gets from the game. If such a sequence of myopic improvement steps terminates, the resulting strategy profile is a Nash equilibrium. Now, if *every* such path leads to a PSNE, it is said that the game has the *finite improvement path property*

(FIP). Importantly, the FIP is equivalent to the existence of a *generalised ordinal potential* (Monderer and Shapley, 1996)—a real-valued function over the set of pure strategy profiles that strictly decreases along any improvement path. Thereby, if the FIP holds for a particular game, then the agents only need to search for a local minimum point of the potential, in order to find a pure strategy equilibrium. It is known that Rosenthal's congestion games always possess a potential function and the FIP and, in fact, are a central class of games with this property (see Monderer and Shapley (1996) for a detailed discussion).

Below, we demonstrate that convex increasing ISWCGs do not fall within the framework of congestion games, as these games generally violate the FIP property. Specifically, we provide an example of the convex increasing ISWCG that contains an improvement cycle, as follows.

Example 3. Consider a convex increasing ISWCG game with 2 agents $N = \{1, 2\}$ and 5 resources $R = \{r_1, r_2, \ldots, r_5\}$, where both agents have access to all of the resources. Agent 1 requires 14 units of resources, and agent 2's demand is 36. The unit–cost functions have the following particular values:

$$c_{r_1}(1) = 39$$
$$c_{r_2}(1) = 350$$
$$c_{r_3}(35) = 5, \ c_{r_3}(36) = 8, \ c_{r_3}(37) = 21$$
$$c_{r_4}(1) = 150$$
$$c_{r_5}(13) = 16, \ c_{r_5}(14) = 22, \ c_{r_5}(15) = 52$$

Consider profile $x = (x^1, x^2)$, where $x^1 = (1, 0, 10, 0, 3)$ and $x^2 = (0, 1, 25, 0, 10)$, with a corresponding congestion vector $h(x) = (1, 1, 35, 0, 13)$. Accordingly, the vector of unit–cost values as determined by x is $(39, 350, 5, 0, 16)$, and the agents' overall costs are $C^1(x) = 1 \cdot 39 + 10 \cdot 5 + 3 \cdot 16 = 137$ and $C^2(x) = 1 \cdot 350 + 25 \cdot 5 + 10 \cdot 16 = 635$. We construct an improvement cycle that starts at x and consists of simple improvement steps at which an agent moves a *single* task unit from one resource to another. First, agent 1 moves 1 unit from r_1 to r_3. The resulting cost to agent 1 is then given by $11 \cdot 8 + 3 \cdot 16 = 136$, which is less by 1 than what the agent paid before. Following this, agent 2 moves a unit from r_2 to r_3 and gets $26 \cdot 21 + 10 \cdot 16 = 706$, thus reducing the cost of $1 \cdot 350 + 25 \cdot 8 + 10 \cdot 16 = 710$ it paid after the first improvement step by agent 1. The whole sequence of moves and the corresponding cost reductions to deviating

Table 3. Improvement cycle in ISWCGs with convex increasing unit–cost functions

Step	Deviator	Move		Improvement
1	Agent 1	1 unit	$r_1 \to r_3$	137 - 136 = 1
2	Agent 2	1 unit	$r_2 \to r_3$	710 - 706 = 4
3	Agent 1	1 unit	$r_3 \to r_4$	279 - 278 = 1
4	Agent 2	1 unit	$r_3 \to r_5$	368 - 367 = 1
5	Agent 1	1 unit	$r_4 \to r_5$	266 - 258 = 8
6	Agent 2	1 unit	$r_5 \to r_2$	697 - 695 = 2
7	Agent 1	1 unit	$r_5 \to r_1$	138 - 137 = 1

agents is listed in Table 3. Note that after 7th step the system turns back to the initial strategy profile, and so the improvement path cycles.

However, the non-existence of the FIP and a potential function in a class of games does not generally contradict the existence of an equilibrium in pure strategies. Thus, in the following section, we prove that convex increasing integer–splittable congestion games do always possess such an equilibrium, despite of the non–existence of the FIP. Our proof is constructive and yields a natural procedure that achieves an equilibrium point in a finite number of steps. Importantly, the convergence is guaranteed, regardless of the initial strategy profile, and so convex increasing congestion games with multi–unit resource demands are weakly–acyclic.

4.2   Nash Equilibria

We start with the following Lemma 1, introducing a useful property of convex increasing functions that we will employ in proving results within this section.

Lemma 1. *Let $c : \mathbb{N} \to \mathbb{R}$ be a convex and monotonically increasing function. Then, for any $0 \le x \le y$ integer and $h \ge 0$, the following holds:*

- $yc(h + y) - xc(h + x) \ge (y - x)[(x + 1)c(h + x + 1) - xc(h + x)]$

- $yc(h + y) - xc(h + x) \le (y - x)[yc(h + y) - (y - 1)c(h + y - 1)]$

Moreover, the inequalities are strict if $y > x + 1$.

We now turn to prove our main result. In doing so, we first provide a useful characterisation of best response strategies in ISWCGs with convex increasing costs (Theorem 1). We then use this characterisation to prove PSNE existence (Theorem 2) and define a special type of improvement dynamics (Algorithm 1) that converges to a Nash equilibrium from an arbitrary starting point (Theorem 3).

Distances between Strategies

Definition 2. *The* modified Hamming distance *between agent i's strategies $x^i = \left(x_r^i\right)_{r \in R}$ and $y^i = \left(y_r^i\right)_{r \in R}$ is defined as*

$$H\left(x^i, y^i\right) = \sum_{r \in R} \left| x_r^i - y_r^i \right| \tag{6}$$

Now, since equation (1) must hold for any strategy of agent i, from Definition 2 we easily derive the following lemma.

Lemma 2. *In an integer–splittable congestion game, if $x^i \ne y^i$ are different strategies of agent i, then $H\left(x^i, y^i\right) \ge 2$.*

Based on this lemma, if the modified Hamming distance between two strategies x^i and y^i is exactly 2, we will refer to them as *neighbours*. The next lemma then states that an improving deviation from a particular strategy (if one exists) can always be found among its neighbours.

Lemma 3. *Let $x = \left(x^i, x^{-i}\right)$ be a strategy profile of a given ISWCG with convex increasing costs. If x^i is not agent i's best response against x^{-i}, then there exists a strategy y^i, such that $H\left(x^i, y^i\right) = 2$ and $C^i\left(y^i, x^{-i}\right) < C^i\left(x\right)$.*

Single Unit Moves. Given this, it will be useful to identify best improving deviations within the set of neighbouring strategies.

Definition 3. *Let $D^i_{\max}(x)$ denote the value of maximal improvement that agent i can achieve by deviating to a neighbouring strategy from profile x. That is,*

$$D^i_{\max}(x) = \max_{y^i:\, H(x^i, y^i) = 2} \left\{ C(x) - C\left(y^i, x^{-i}\right) \right\} \tag{7}$$

Obviously, if x is a Nash equilibrium profile, then for any $i \in N$ we have $D^i_{\max}(x) \leq 0$. Otherwise, if for some agent i its strategy x^i is not a best response against x^{-i}, then by Lemma 3, there exists a strategy y^i for agent i such that $H\left(x^i, y^i\right) = 2$ and $D^i_{\max}(x) \geq U(x) - U\left(y^i, x^{-i}\right) > 0$. This implies the following theorem.

Theorem 1. *Given a convex increasing ISWCG, a strategy x^i is a best response to agent $i \in N$ against s^{-i} if and only if $D^i_{\max}(x) \leq 0$.*

Thereby, a strategy profile x is a PSNE if and only if the condition in Theorem 1 holds for each agent $i \in N$. We seek such a profile by constructing an improvement path, where at each step an agent deviates to a best neighboring strategy. Let us now characterise these improving moves.

From Lemma 2, it is easy to see that x^i and y^i are neighboring strategies of agent $i \in N$ if and only if there are $p, q \in R_i$ such that $y^i_p = x^i_p - 1$ and $y^i_q = x^i_q + 1$. That is, agent i deviates from x^i to y^i by moving exactly one task unit from resource p to resource q. Hereafter, we refer to such deviations as *single unit moves*.

Let $D^i_{p \to q}(x)$ denote agent i's value of improvement by taking a single unit move $p \to q$ from profile x. That is,

$$D^i_{p \to q}(x) = C^i(x) - C^i\left(y^i, x^{-i}\right) \tag{8}$$

where y^i is such that $y^i_p = x^i_p - 1$, $y^i_q = x^i_q + 1$ and $y^i_r = x^i_r$ for all $r \in R \setminus \{p, q\}$. Given this, we can rewrite $D^i_{\max}(x)$ as follows:

$$D^i_{\max}(x) = \max_{p \neq q \in R_i} D^i_{p \to q}(x) \tag{9}$$

Now, let us calculate

$$D^i_{p \to q}(x) = \left[x^i_p c_p\left(h_p(x)\right) - \left(x^i_p - 1\right) c_p\left(h_p(x) - 1\right) \right]$$
$$+ \left[x^i_q c_q\left(h_q(x)\right) - \left(x^i_q + 1\right) c_q\left(h_q(x) + 1\right) \right] \tag{10}$$

and consider

$$p^{i*} \in \arg \max_{r \in R_i:\, x^i_r > 0} \left\{ x^i_r c_r\left(h_r(x)\right) - \left(x^i_r - 1\right) c_j\left(h_r(x) - 1\right) \right\} \tag{11}$$

That is, resource p^{i*} guarantees to agent i a maximal cost reduction if it removes one unit of demand from that resource. Similarly, resource

$$q^{i*} \in \arg \min_{r \in R_i} \left\{ \left(x^i_r + 1\right) c_j\left(h_r(x) + 1\right) - x^i_r c_j\left(h_r(x)\right) \right\} \tag{12}$$

guarantees a minimal increase in cost when i adds one unit of demand to q^{i*}.

Obviously, for any pair of resources p and q with $x_p^i > 0$ we have that $D_{p \to q}^i (x) \le D_{p^{i*} \to q^{i*}}^i (x)$. That is, if $p^{i*} \ne q^{i*}$ then $D_{\max}^i (x) = D_{p^{i*} \to q^{i*}}^i (x)$, and if $D_{\max}^i (x) > 0$ then $p^{i*} \to q^{i*}$ is a *best single unit move* to agent i from x. Otherwise, if $p^{i*} = q^{i*}$, then the following lemma implies that x^i is a best response strategy to agent i.

Lemma 4. *Given a convex increasing ISWCG and a strategy profile x, if for agent $i \in N$ there exist p^{i*} and q^{i*} (as defined in equations (11) and (12), respectively) such that $p^{i*} = q^{i*}$, then $D_{\max}^i (x) \le 0$.*

Best Response Dynamics. Let x be an arbitrary strategy profile of a given ISWCG with convex increasing costs. As we concluded before from Theorem 1, if $D_{\max}^i (x) \le 0$ holds for every agent $i \in N$ then x is a Nash equilibrium strategy profile. So assume otherwise, and let i be an agent with $D_{\max}^i (x) > 0$. By Lemma 4, we have that $p^{i*} \ne q^{i*}$, and let $B^i(x)$ denote the number of best single unit moves of i from x. We prove the following.

Theorem 2. *Given an ISWCG with convex increasing costs, let x be a strategy profile which is not in equilibrium. Then, there exists a profile y, such that for each agent $i \in N$, one of the following three conditions is satisfied:*

1. *$D_{\max}^i (x) > D_{\max}^i (y)$*
2. *$D_{\max}^i (x) = D_{\max}^i (y)$ and $B^i(x) > B^i(y)$*
3. *$D_{\max}^i (x) = D_{\max}^i (y)$ and $B^i(x) = B^i(y)$*

Moreover, for at least one agent either 1. or 2. holds.

Corollary 1. *Given an ISWCG with convex increasing costs and a strategy profile x, let*

$$P(x) = L \cdot \sum_{i \in N} D_{\max}^i(x) + \sum_{i \in N} B^i(x) \tag{13}$$

where L is a large number satisfying $L \ge \frac{nm(m-1)}{\min_{p,q,k,l} |c_p(k) - c_q(l)|}$. Then, if x is not a Nash equilibrium, then there exists a profile y, such that $P(x) > P(y)$.

Note that function $P(\cdot)$ in (13) does not decrease along *any* improvement path, and so the FIP does not follow. Nonetheless, Theorem 2 and Corollary 1 imply the existence of pure strategy Nash equilibria in convex increasing integer–splittable congestion games. To prove Theorem 2 we need the following auxiliary lemma.

Lemma 5. *Given a convex increasing ISWCG, assume there is a sequence $(x_1, x_2, \ldots x_T)$ of strategy profiles such that:*

- *x_1 is not a pure strategy Nash equilibrium, and x_2 is obtained from x_1 by a best single unit move of some agent i with $D_{\max}^i (x_1) > 0$*
- *$\forall 1 < t < T, \exists r_t^+, r_t^- \in R$, such that $h_{r_t^+} (x_t) = h_{r_t^+} (x_1)+1$, $h_{r_t^-} (x_t) = h_{r_t^-} (x_1) - 1$. Furthermore, $\forall r \in R \setminus \{r_t^+, r_t^-\}$ we have $h_r (x_t) = h_r (x_1)$*

- $\forall 1 < t < T,\ \exists j_t \in N$ with $D_{\max}^{j_t}(x_t) > D_{\max}^{j_t}(x_1)$ or $D_{\max}^{j_t}(x_t) = D_{\max}^{j_t}(x_1) \wedge$ $B^{j_t}(x_t) > B^{j_t}(x_1)$, and $\exists r \in R_{j_t}$, such that either $D_{\max}^{j_t}(x_t) = D_{r_t^+ \to r}^{j_t}(x_t)$ or $D_{\max}^{j_t}(x_t) = D_{r \to r_t^-}^{j_t}(x_t)$. Furthermore, x_{t+1} is obtained from x_t by the corresponding best single unit move by agent j, that removes a unit from r_t^+ (or adds one to r_t^-). Moreover, if $r = r_t^+$ or $r = r_t^-$ (i.e., $D_{\max}^j(x_t) = D_{r_t^+ \to r_t^-}^j(x_t)$), then $t + 1 = T$.

Then, for all $1 < t < T$ we have $D_{\max}^{j_t}(x_{t+1}) < D_{\max}^{j_t}(x_1)$ or $D_{\max}^{j_t}(x_{t+1}) = D_{\max}^{j_t}(x_1) \wedge$ $B^{j_t}(x_{t+1}) < B^{j_t}(x_1)$.

That is, at each step t in the sequence, we have an agent j_t, whose current maximal improvement is higher than the value it had in the initial strategy profile x_1 (or the number of best single unit moves available to i at step t is greater than that it had at the first step). Furthermore, the congestion levels in x_t differ from congestion levels in the initial profile x_1 for only two resources r_t^+ and r_t^-, plus/minus one unit each. A best move of agent j_t is to either move a unit from r_t^+ to some resource r (and so $r = r_{t+1}^+$, unless $r = r_t^-$), or to take one from some r and add to r_t^- (in which case, $r = r_{t+1}^-$, unless $r = r_t^+$). This best move by j_t then results in the subsequent strategy profile x_{t+1}, and if $r = r_t^-$ or $r = r_t^+$ (i.e., agent j_t's best move is from r_t^+ to r_t^-), then this is the last move in the sequence. Now, if such a sequence exists in a given game, then at each iteration, the value of maximal improvement for the corresponding deviator (or the number of its available best single unit moves) decreases comparing to what it had in the initial point of the sequence x_1.

Proof (of Theorem 2). We construct a finite sequence of best single unit moves that results in a strategy profile y for which the theorem holds. In particular, we first prove that during the sequence, if we reach a certain congestion level profile twice, then we can leave out the in between steps. Using this result, we then show that we cannot infinitely continue the sequence without reaching a strategy profile for which the theorem holds.

In doing so, we define a particular order of moves, as follows. Let $\{x_1, x_2, \ldots\}$ denote the sequence of strategy profiles resulted from a sequence of best single unit moves $x_t \to x_{t+1}$, $t \geq 1$, as defined in Lemma 5. We refer to the moves $r_t^+ \to r$ and $r_t^- \to r$ as *forward* and *backward* moves, respectively (moves $r_t^+ \to r_t^-$ can be both, but we will make it clear in the context). Note that by Lemma 5, if at step t some agent i violates the conditions of the theorem, then there is always either a forward or a backward move that it can apply. Given this, we start the sequence with a series of forward moves, and when no such move is available, we switch to backward moves if any exist. We prove that this construction leads to a desired strategy profile in any case. The steps involved within the proof are described below.

Step 1: By definition, we move from x_1 to x_2 with some agent i who applies its best single unit move. From x_2, we only allow agents to make forward moves (if exist); that is, for now, backward moves are out of consideration. Let $\{r_{f_1}, r_{f_2}, r_{f_3}, \ldots\}$ denote the sequence of such forward moves, where $r_{f_t} \to r_{f_{t+1}}$ denotes a forward move from resource r_{f_t} to $r_{f_{t+1}}$ at step t. For the sake of simplicity, we assume that the first move, that deviates x_1 to x_2, is also a forward move (i.e., that move is $r_{f_1} \to r_{f_2}$, and we start the sequence from the initial strategy profile).

Now, consider the case where $\exists u, v, u < v$, such that $r_{f_u} = r_{f_v}$, and none of them is equal to r_{f_1}; that is, the sequence of forward moves creates a loop by turning back to a previously visited resource, which is not the first resource. We show that if the sequence is $\{r_{f_1}, r_{f_2}, \ldots, r_{f_u}, \ldots, r_{f_v}, r_{f_v+1}, \ldots\}$, then if agent i is the one who makes the move $r_{f_v} \to r_{f_v+1}$, then it can make the move $r_{f_u} \to r_{f_u+1}$ as well; and thus, the sequence $\{r_{f_1}, r_{f_2}, \ldots, r_{f_u}, r_{f_v+1}, \ldots\}$ is also a feasible sequence of forward moves. That is, we can leave out the loop $\{r_{f_u+1}, \ldots, r_{f_v}\}$, without violating the conditions of Lemma 5.

Let x_{f_u} and x_{f_v} denote the profiles that result from subsequences $\{r_{f_1}, r_{f_2}, \ldots, r_{f_u}\}$ and $\{r_{f_1}, r_{f_2}, \ldots, r_{f_u}, \ldots, r_{f_v}\}$, respectively. Now, suppose that agent i makes the move $r_{f_v} \to r_{f_v+1}$ from x_{f_v}. We show that this move is available to i at x_{f_u} as well. Indeed, consider the congestion level and the demand of agent i on resource r_{f_u} in x_{f_u} and x_{f_v}. Since all the moves are forward moves, it is easy to see that the congestion level in both profiles is given by $h_{r_{f_u}}(x_1) + 1$ (this is since $r_{f_u} = r_{f_u}^+ = r_{f_v}^+$). Furthermore, it can be shown that the demand of agent i on r_{f_u} in x_{f_v} is at most as high as it was in x_{f_u}. One exceptional case is when agent i is the one who makes the move $r_{f_v-1} \to r_{f_v}$, and thus its demand on r_{f_u} may be increased by one. However, Lemma 5 implies that whenever an agent makes a best unit move (either forward or backward), at next step of the sequence it satisfies the conditions of the theorem. Hence, it cannot be the one who makes the subsequent move. Given this, if agent i is the one who makes the move $r_{f_v-1} \to r_{f_v}$, then it cannot make the move $r_{f_v} \to r_{f_v+1}$, which is a contradiction. This implies that the demand of agent i on r_{f_u} cannot be greater in x_{f_v} than that it has in x_{f_u}.

Now, if the demand of agent i on resource r_{f_v+1} in x_{f_v} is smaller than its demand on the same resource in x_{f_u}, then we show that this results in a contradiction. We prove this by indirection; that is, suppose that it is true. This implies that there exists $u < z < v$, such that $r_{f_z} = r_{f_v+1}$, and agent i moved a unit from r_{f_z} to r_{f_z+1} within the sequence. Furthermore, the demand of agent i on $r_{f_z} = r_{f_v+1}$ after the move is decreased by 1, compared to its demand in x_{f_u}. That is, since the demand of agent i on r_{f_v+1} in x_{f_v} is smaller than in x_{f_u}, agent i must move some units from that resource in between. Thus, we focus on the first move among these, which decreases agent i's demand by 1. Note that, by definition of the sequence, $r_{f_v+1} \to r_{f_z+1}$ is a best single unit move of agent i in x_{f_z}. Now, let a denote the amount of agent i's decreased cost by removing one unit from r_{f_v+1} in x_{f_z}, and b denote the agent's increased cost by adding one unit to r_{f_z+1}, also in x_{f_z}. Thus, the improvement that agent i gets by making $r_{f_v+1} \to r_{f_z+1}$ is $a - b > 0$. Similarly, let c denote the amount of agent i's decreased cost by removing one unit from r_{f_v} in x_{f_v}, and d denote the agent's increased cost by adding one unit to r_{f_v+1} (i.e. r_{f_z}), also in x_{f_v}. Since $r_{f_v} \to r_{f_v+1}$ is also a best single move, $c - d > 0$. It is easy to see that both the congestion level and the demand of agent i on r_{f_v+1} remain the same after the move $r_{f_v} \to r_{f_v+1}$, and before the move $r_{f_v+1} \to r_{f_z+1}$. Thus, we have $d = a$; and thus, $c - b > a - b > 0$. This implies that in x_{f_z}, the best single move is not moving from r_{f_v+1} to r_{f_z+1}, but from r_{f_v} (since both the congestion level and agent i's demand on r_{f_u} is not modified between x_{f_u} and x_{f_v}; that is, it stays unchanged within the loop). This, however, is a contradiction, since $r_{f_v+1} \to r_{f_z+1}$ is supposed to be the best single move in x_{f_z}.

Given this, the demand of agent i on resource $r_{f_{v+1}}$ in x_{f_v} is at least as its demand on the same resource in x_{f_u}. In this case, $r_{f_u} \rightarrow r_{f_{v+1}}$ is feasible for agent i from x_{f_u} as well. Indeed, since $r_{f_u} \rightarrow r_{f_{v+1}}$ is feasible for agent i in x_{f_v}, such that the demand of agent i on r_{f_u} in x_{f_v} is not higher than in x_{f_u}, and its demand on $r_{f_{v+1}}$ in x_{f_v} is not smaller than in x_{f_u}. That is, by choosing $r_{f_w} = r_{f_{v+1}}$, we get that $\{r_{f_1}, r_{f_2}, \ldots, r_{f_u}, r_{f_w}, \ldots\}$ is also a feasible sequence, without the $\{r_{f_{u+1}}, \ldots, r_{f_v}\}$ loop.

Thus, in summary, we can say that if there's a loop within the sequence, that does not return to r_{f_1}, then we can leave that loop out of the sequence.

Step 2: Now, we will show that if the sequence does not return to r_{f_1}, then it has to be finite. We prove this by contradiction as follows: Suppose that the sequence is infinite and never returns to r_{f_1}. Given this, there is an infinite subsequence of moves $r_{f_{u(t)}} \rightarrow r_{f_{u(t)+1}}$ applied by a particular agent i, such that $r_{f_{u(1)}} = r_{f_{u(2)}} = \ldots$ and $r_{f_{u(1)+1}} = r_{f_{u(2)+1}} = \ldots$. That is, agent i makes the same move $r_{f_{u(t)}} \rightarrow r_{f_{u(t)+1}}$ infinitely many times within the sequence. Furthermore, the demand of i on resources $r_{f_{u(t)}}$ and $r_{f_{u(t)+1}}$ are the same for every t. That is, if agent i's demands on $r_{f_{u(1)}}$ and $r_{f_{u(1)+1}}$ are a and b, respectively, then they are a and b for any t.

Now, consider the move $r_{f_{u(1)}} \rightarrow r_{f_{u(1)+1}}$ of agent i. After this move, agent i's demand on $r_{f_{u(1)}}$ and $r_{f_{u(1)+1}}$ becomes $a-1$ and $b+1$, respectively. However we know that when agent i makes the move $r_{f_{u(2)}} \rightarrow r_{f_{u(2)+1}}$, these values return to a and b again. That is, before applying $r_{f_{u(2)}} \rightarrow r_{f_{u(2)+1}}$, agent i had to make a move $r_{f_v} \rightarrow r_{f_{v+1}}$, where $r_{f_{v+1}} = r_{f_{u(1)}} = r_{f_{u(2)}}$, to increase its demand on $r_{f_{u(2)}}$ back to a. Now, note that $u(1) < v < u(2) - 1$. This implies that the subsequence $\{r_{f_{v+2}}, \ldots, r_{f_{u(2)}}\}$ forms a loop, and thus, according to the claim we stated in Step 1, we can leave this loop out from the sequence. That is, the moves $r_{f_v} \rightarrow r_{f_{v+1}}$ and $r_{f_{u(2)}} \rightarrow r_{f_{u(2)+1}}$ become subsequent moves within the sequence. However, as Lemma 5 implies, none of the agents can subsequently make more than one move within the sequence, and thus, this situation is not possible. This contradicts the initial assumption, and hence, sequence $\{r_{f_1}, r_{f_2}, \ldots\}$ either returns to r_{f_1}, or it is finite.

Step 3: Based on the results described in Step 2, if $\{r_{f_1}, r_{f_2}, \ldots\}$ (i.e. the sequence of forward moves) is not finite, then it has to return to r_{f_1}. That is, $\exists v$ such that in $\{r_{f_1}, r_{f_2}, \ldots, r_{f_v}\}$, $r_{f_v} = r_{f_1}$. If there is an inner loop within this sequence, then we can remove that loop (as proved in Step 1). Thus, we can assume that the sequence does not contain any inner loops (note that the sequence itself is also a loop). Let x_{f_v} denote the resulting strategy profile by making this sequence of forward moves. We show below that x_{f_v} satisfies the conditions of the theorem; that is, it is the strategy profile we are looking for.

Note that by returning to r_{f_1}, the congestion level on all the resources in x_{f_v} is the same as it is in x_1. Since the sequence does not contain any inner loops, it is easy to see that for any agent i, there is a set of disjoint pairs of resources $r_{f_{u(k)}}, r_{f_{u(k)+1}}$ such that agent i makes the move $r_{f_{u(k)}} \rightarrow r_{f_{u(k)+1}}$ within the sequence. This indicates that in x_{f_v}, agent i's demand on $r_{f_{u(k)}}$ is decreased by 1, compared to that it has on that resource in x_1 (since agent i removes one unit from that resource). On the other hand, agent i's demand on $r_{f_{u(k)+1}}$ is increased by 1, compared to that it has on that resource in x_1.

In order to prove the claim above, we show that the value of a best unit move of agent i in x_{f_v} is decreased, compared to that it has in x_1 (or the number of such moves is decreased). Since the congestion level is the same on all the resources in the two strategy profiles, we just need to consider the cases where agent i makes a move from $r_{f_{u(k)+1}}$ (where the demand is increased) to $r_{f_{u(l)}}$ (where the demand is decreased) for a particular pair of k, l.

If $k = l$, then $r_{f_{u(k)}} \rightarrow r_{f_{u(k)+1}}$ is a forward move of agent i. Let $x_{f_{u(k)}}$ and $x_{f_{u(k)+1}}$ denote the strategy profiles before and after the move. If $r_{f_{u(k)}} \rightarrow r_{f_{u(k)+1}}$ is not the first move in the sequence, then the congestion levels on resources $r_{f_{u(k)}}$ and $r_{f_{u(k)+1}}$ in $x_{f_{u(k)}}$ and $x_{f_{u(k)+1}}$ are: $h_{r_{f_{u(k)}}}\left(x_{f_{u(k)}}\right) = h_{r_{f_{u(k)}}}(x_1) + 1$, $h_{r_{f_{u(k)+1}}}\left(x_{f_{u(k)}}\right) = h_{r_{f_{u(k)}}}(x_1)$, and $h_{r_{f_{u(k)}}}\left(x_{f_{u(k)+1}}\right) = h_{r_{f_{u(k)}}}(x_1)$, $h_{r_{f_{u(k)+1}}}\left(x_{f_{u(k)+1}}\right) = h_{r_{f_{u(k)}}}(x_1) + 1$, respectively. Thus, after the move, the congestion level on $r_{f_{u(k)}}$ in $x_{f_{u(k)+1}}$ is the same as in x_{f_v}, while the congestion on $r_{f_{u(k)+1}}$ is greater by 1 than in x_{f_v}. Since $r_{f_{u(k)}} \rightarrow r_{f_{u(k)+1}}$ is a best unit move at $x_{f_{u(k)}}$, reversing this move (i.e., moving back from $x_{f_{u(k)+1}}$ to $r_{f_{u(k)}}$) in $x_{f_{u(k)+1}}$) is not possible. Given this, since the congestion on $r_{f_{u(k)+1}}$ in x_{f_v} is decreased, compared to that in $x_{f_{u(k)+1}}$, the move $x_{f_{u(k)+1}} \rightarrow r_{f_{u(k)}}$ is also not feasible. Note that the proof above also works for the case where $r_{f_{u(k)}} \rightarrow r_{f_{u(k)+1}}$ is the first move of the sequence (although the values of congestion levels are slightly different).

Now let $k \neq l$. Again, we first consider the case where none of the moves $r_{f_{u(k)}} \rightarrow r_{f_{u(k)+1}}$ and $r_{f_{u(l)}} \rightarrow r_{f_{u(l)+1}}$ is the first move of the sequence. If $k < l$ (i.e. the agent makes $r_{f_{u(k)}} \rightarrow r_{f_{u(k)+1}}$ earlier), then consider the move $r_{f_{u(l)}} \rightarrow r_{f_{u(l)+1}}$, and let $x_{f_{u(l)}}$ and $x_{f_{u(l)+1}}$ denote the strategy profiles before and after this move, respectively. Since agent i makes this move later, in $x_{f_{u(l)+1}}$, the congestion level of $r_{f_{u(k)+1}}$ and $r_{f_{u(l)}}$ is the same as they have in x_{f_v}. Given this, the improvement value of move $x_{f_{u(k)+1}} \rightarrow r_{f_{u(l)}}$ is exactly the same as it is in $x_{f_{u(l)+1}}$. Since $r_{f_{u(l)}} \rightarrow r_{f_{u(l)+1}}$ is a best unit move in $x_{f_{u(l)}}$, resource $r_{f_{u(l)}}$ belongs to the set defined in (11) (i.e. set of p^*); that is, reducing a unit from $r_{f_{u(l)}}$ guarantees a maximal cost reduction to agent i in strategy profile $x_{f_{u(l)}}$. This implies that the cost reduction by reducing a unit from $r_{f_{u(k)+1}}$ is not greater than the cost reduction by reducing a unit from $r_{f_{u(l)}}$. Given this, it is easy to see that the reverse move $x_{f_{u(k)+1}} \rightarrow r_{f_{u(l)}}$ in strategy profile $x_{f_{u(l)+1}}$ cannot be positive (i.e., it is not a feasible move). The proof for $k > l$ works in a similar way.

This implies that none of $x_{f_{u(k)+1}} \rightarrow r_{f_{u(l)}}$ is feasible in x_{f_v}. Thus, x_{f_v} satisfies the conditions of the theorem, where x_1 replaces x and x_{f_v} replaces y.

Step 4: Next, consider the case where the sequence of forward moves, $\{r_{f_1}, r_{f_2}, \ldots r_{f_K}\}$, is finite (i.e. $K < \infty$). At this point, we allow agents to make backward moves (i.e., moves that add a unit to r_t^- at each step t). Let $\{r_{b_1}, r_{b_2}, \ldots\}$ denote the sequence of backward moves, where $\forall t$, $r_{b_{t+1}} \rightarrow r_{b_t}$ is the backward move made by some agent i. Note that here $r_{b_1} = r_{f_1}$. Similarly to the case of forward moves, one can show that if there is a loop within $\{r_{b_1}, r_{b_2}, \ldots\}$, then we can leave that loop out from the sequence. Furthermore, one of the following must hold for $\{r_{b_1}, r_{b_2}, \ldots\}$: (i) apart from r_{b_1}, $\{r_{b_1}, r_{b_2}, \ldots\}$ also contains a resource r_{f_u} from the sequence of $\{r_{f_1}, r_{f_2}, \ldots r_{f_K}\}$; that is, $\exists v > 1, u > 0$ such that $r_{b_v} = r_{f_u}$; or (ii) it does not contain such resource, but then it must be finite. The proof is also based on contradiction, and is similar to the proof described in Step 2.

Now, if $\{r_{b_1}, r_{b_2}, \ldots\}$ contains a resource from the sequence of forward moves, then consider the following sequence of moves: $\{r_{f_1}, r_{f_2}, \ldots r_{f_u}, r_{b_1}, r_{b_2}, r_{b_v}\}$. That is, we leave out all the moves after r_{b_v} in the sequence of backward moves, and all the moves after r_{f_u} in the sequence of forward moves. It is easy to see that this sequence is also feasible, that is, all of the moves are best unit moves of some agent i who violates the conditions of the theorem. Now, if the sequence contains inner loops from the backward moves side (the subsequence of forward moves is loopless), we leave these loops out. This way, we obtain a loop similar to the loop described in Step 3. Let x_{b_v} denote the strategy profile resulted by the sequence of moves within $\{r_{f_1}, r_{f_2}, \ldots r_{f_v}, r_{b_1}, r_{b_2}, r_{b_v}\}$. We show that x_{b_v} satisfies the conditions of the theorem (the proof is similar to the one described in Step 3).

Finally, in the case where $\{r_{b_1}, r_{b_2}, \ldots, r_{b_L}\}$ is also finite (i.e. $L < \infty$), let x_T denote the resulting strategy profile by making the moves of the combined sequence $\{r_{f_1}, r_{f_2}, \ldots r_{f_K}\}$ and $\{r_{b_1}, r_{b_2}, \ldots, r_{b_L}\}$. One can easily see that the conditions of the theorem hold for x_T. This completes the proof.　　　　□

ISWCG Algorithm. The proof of Theorem 2 suggests a particular dynamic procedure that consists of best single unit moves (Algorithm 1) and arrives at a pure strategy Nash equilibrium from any starting point in finite time. This implies that convex increasing congestion games ISWCG are *weakly–acyclic* (Monderer and Shapley, 1996)—that is, possess an improvement dynamics whose convergence is guaranteed from an arbitrary initial strategy profile.

Theorem 3. *Algorithm 1 finds a pure strategy Nash equilibrium in a given convex increasing ISWCG.*

Proof. The algorithm constructs a sequence of strategy profiles, $\{x_1, x_2, \ldots\}$, such that $\forall t$, x_{t+1} satisfies Theorem 2 with respect to profile x_t (steps $4 - 19$). Then, Corollary 1 implies that $\forall t$, $P(x_t) > P(x_{t+1})$, where $P(x)$ is defined in equation (13). That is, sequence $\{P(x_1), P(x_2), \ldots\}$ is strictly decreasing. Hence, since the game is finite, the algorithm terminates in a PSNE after a finite number of steps.　　　　□

5　Conclusions

In this paper, we explore the conditions for PSNE existence in integer–splittalbe congestion games. Although these games do not necessarily admit such an equilibrium, we prove that it is guaranteed to exist in an important subclass of ISWCGs with convex increasing unit–cost functions. Furthermore, we demonstrate that although convex increasing ISWCGs do not have the FIP property, they do possess weak acyclicity, and we provide a natural procedure that achieves an equilibrium from an arbitrary initial strategy profile.

Our results suggest several directions for future research. Specifically, given PSNE existence and convergence, it is important to address further properties of integer–splittable congestion games, such as completeness of the model, quality of solutions

---

Algorithm 1. ISWCG Algorithm.

1: **Initialisation:** Let $t = 1$, $x_t = x$
2: If $\not\exists i : D^i_{\max} > 0 \rightarrow$ STOP
3: **while** PSNE not found **do**
4:     $x_t \leftarrow$ starting position
5:     $\{r_f\} \leftarrow$ sequence of forward moves, $k = 1$
6:     **while** forward move is feasible **do**
7:         make a forward move: $r_{f_k} \rightarrow r_{f_{k+1}}$, $k := k + 1$
8:         **if** there is an inner loop **then** leave out the loop
9:         **if** $r_{f_k} = r_{f_1}$ **then** $x_{t+1} \leftarrow$ resulting resource profile of $\{r_{f_1}, r_{f_2}, \ldots r_{f_k}\}$ from x_t,
         GOTO STEP 20
10:    **end while**
11:    $\{r_{f_1}, r_{f_2}, \ldots r_{f_K}\} \leftarrow$ resulting sequence of forward moves
12:    $\{r_b\} \leftarrow$ sequence of backward moves, $l = 1$
13:    **while** backward move is feasible **do**
14:        make a backward move: $r_{b_{l+1}} \rightarrow r_{b_l}$, $l := l + 1$
15:        **if** there is an inner loop **then** leave out the loop
16:        **if** $\exists r_{f_v} \in \{r_f\}$ such that $r_{b_l} = r_{f_v}$ **then** $x_{t+1} \leftarrow$ resulting resource profile of
         $\{r_{f_1}, r_{f_2}, \ldots r_{f_u}, r_{b_1}, r_{b_2}, r_{b_v}\}$ from x_t, GOTO STEP 20
17:    **end while**
18:    $\{r_{b_1}, r_{b_2}, \ldots, r_{b_L}\} \leftarrow$ resulting sequence of backward moves
19:    $x_{t+1} \leftarrow$ resulting resource profile of $\{r_{f_1}, r_{f_2}, \ldots r_{f_K}\}$ and $\{r_{b_1}, r_{b_2}, \ldots, r_{b_L}\}$ from x_t

20:    **if** $x_{t+1} =$ PSNE **then** STOP
21:    $t := t + 1$
22: **end while**

---

and computational complexity. To this end, we aim to (i) investigate how far the assumptions on the agents' strategy sets and payoff functions can be relaxed while still guaranteeing the existence of pure strategy equilibria, (ii) characterise the efficiency of PSNE in terms of prices of anarchy and stability, and (iii) provide a complexity analysis of the problem of finding equilibria and develop efficient algorithmic solutions, if applicable.

References

Beier, R., Czumaj, A., Krysta, P., Vöcking, B.: Computing equilibria for congestion games with (im)perfect information. In: Proc. of SODA 2004, pp. 746–755 (2004)

Byde, A., Polukarov, M., Jennings, N.R.: Games with congestion-averse utilities. In: Mavronicolas, M., Papadopoulou, V.G. (eds.) SAGT 2009. LNCS, vol. 5814, pp. 220–232. Springer, Heidelberg (2009)

Cominetti, R., Correa, J.R., Stier-Moses, N.E.: The impact of oligopolistic competition in networks. Operations Research 57(6), 1421–1437 (2009)

Dunkel, J., Schulz, A.: On the complexity of pure-strategy nash equilibria in congestion and local-effect games. Mathemathics of Operations Research 33(4), 851–868 (2008)

Koutsoupias, E., Papadimitriou, C.H.: Worst–case equilibria. In: Meinel, C., Tison, S. (eds.) STACS 1999. LNCS, vol. 1563, p. 404. Springer, Heidelberg (1999)

Krysta, P., Sanders, P., Vöcking, B.: Scheduling and traffic allocation for tasks with bounded splittability. In: Proc. of the 28th International Symposium on Mathematical Foundations of Computer Science, pp. 500–510 (2003)

Leslie, D.S., Collins, E.J.: Generalised weakened fictitious play. Games and Economic Behavior 56, 285–298 (2006)

Leyton-Brown, K., Tennenholtz, M.: Local-effect games. In: Proc. of IJCAI 2003, pp. 772–780 (2003)

Marden, J.R., Arslan, G., Shamma, J.S.: Regret based dynamics: Convergence in weakly acyclic games. In: Proc. of AAMAS 2007, pp. 194–201 (2007)

Meyers, C.: Network flow problems and congestion games: complexity and approximation results. Ph.D. thesis, Massachusetts Institute of Technology, Cambridge, MA, USA. AAI0809430 (2006)

Milchtaich, I.: Congestion games with player–specific payoff functions. Games and Economic Behavior 13(1), 111–124 (1996)

Monderer, D., Shapley, L.S.: Potential games. Games and Economic Behavior 14, 124–143 (1996)

Orda, A., Rom, R., Shimkin, N.: Competitive routing in multiuser communication networks. IEEE/ACM Trans. Networking 1(5), 510–521 (1993)

Penn, M., Polukarov, M., Tennenholtz, M.: Congestion games with load-dependent failures: Identical resources. Games and Economic Behavior 67(1), 156–173 (2009a)

Penn, M., Polukarov, M., Tennenholtz, M.: Random order congestion games. Mathematics of Op. Res. 34(3), 706–725 (2009b)

Rosen, J.: Existence and uniqueness of equilibrium points for concave n-person games. Econometrica 33, 520–534 (1965)

Rosenthal, R.W.: A class of games possessing pure-strategy Nash equilibria. Int. J. of Game Theory 2, 65–67 (1973a)

Rosenthal, R.W.: The network equilibrium problem in integers. Networks 3, 53–59 (1973b)

Roughgarden, T., Tardos, E.: How bad is selfish routing? J. ACM 49(2), 236–259 (2002)

Shachnai, H., Tamir, T.: Multiprocessor scheduling with machine allotment and parallelism constraints. Algorithmica 32, 651–678 (2002)

Voice, T.D., Polukarov, M., Jennings, N.R.: On the impact of strategy and utility structures on congestion-averse games. In: Leonardi, S. (ed.) WINE 2009. LNCS, vol. 5929, pp. 600–607. Springer, Heidelberg (2009)

On Dynamics in Basic Network Creation Games

Pascal Lenzner

Department of Computer Science,
Humboldt-Universität zu Berlin, Germany
lenzner@informatik.hu-berlin.de

Abstract. We initiate the study of game dynamics in the SUM BASIC
NETWORK CREATION GAME, which was recently introduced by Alon et
al.[SPAA'10]. In this game players are associated to vertices in a graph
and are allowed to "swap" edges, that is to remove an incident edge and
insert a new incident edge. By performing such moves, every player tries
to minimize her connection cost, which is the sum of distances to all
other vertices. When played on a tree, we prove that this game admits
an ordinal potential function, which implies guaranteed convergence to
a pure Nash Equilibrium. We show a cubic upper bound on the number
of steps needed for any improving response dynamic to converge to a
stable tree and propose and analyse a best response dynamic, where the
players having the highest cost are allowed to move. For this dynamic
we show an almost tight linear upper bound for the convergence speed.
Furthermore, we contrast these positive results by showing that, when
played on general graphs, this game allows best response cycles. This
implies that there cannot exist an ordinal potential function and that
fundamentally different techniques are required for analysing this case.
For computing a best response we show a similar contrast: On the one
hand we give a linear-time algorithm for computing a best response on
trees even if players are allowed to swap multiple edges at a time. On the
other hand we prove that this task is NP-hard even on simple general
graphs, if more than one edge can be swapped at a time. The latter
addresses a proposal by Alon et al..

1 Introduction

The importance of the Internet as well as other networks has inspired a huge body
of scientific work to provide models and analyses of the networks we interact with
every day. These models incorporate game theoretic notions to be able to express
and analyse selfish behavior within these networks. Such behavior by players can
be the creation or removal of links to influence the network structure to better
suit their needs. However, most of this work focused on *static* properties of such
networks, like structural properties of solution concepts. Prominent examples
are bounds on the Price of Anarchy or on the Price of Stability of (pure) Nash
Equilibria in games that model network creation. The problem is, that such
results do not explain how selfish and myopic players can actually *find* such
desired states.

G. Persiano (Ed.): SAGT 2011, LNCS 6982, pp. 254–265, 2011.

In this paper we focus on the process itself. That is, on the dynamic behavior of players which eventually leads to a state of the game having interesting properties like stability against unilateral deviations and low social cost. We initiate the study of myopic game dynamics in the SUM BASIC NETWORK CREATION GAME, which was introduced very recently by Alon et al.[2]. This elegant model incorporates important aspects of network design as well as network routing but is at the same time simple enough to provide insights into the induced dynamic process. The idea is to let players "swap" edges to resemble the natural process of weighing two decisions (possible edges) against each other. We investigate the convergence process of dynamics which allow players to myopically swap edges until a stable state of the game emerges. Furthermore, we take the mechanism design perspective and propose a specific dynamic, which yields near optimal convergence speed.

1.1   Related Work

The line of research which is closest to our work was initiated by Fabrikant et al.[5], who considered network creation with a fixed edge-cost of α. For some ranges of α they proved first bounds on the Price of Anarchy [8], which is the ratio of the social cost of the worst (pure) Nash Equilibrium and the minimum possible social cost achieved by central design. Subsequent work [9,1,4,10] has shown, that this ratio is constant for almost all values of α. Only for $\alpha \in \Theta(n)$ there remains a gap. However, there is a downside of this model: As already observed in [5], computing a best response is NP-hard, which implies, that players cannot efficiently decide if the game has reached a stable state. This computational hardness prevents myopic dynamics from being applied to finding a pure Nash Equilibrium.

Very recently, Alon et al.[2] proposed a slightly different model, which no longer depends on the parameter α but still captures important aspects of network creation. The authors consider two different cost-measures, namely the sum of distances to all other players and the maximum distance to all other players and give bounds on the price of anarchy. Here, we adopt the former measure. Alon et al. proved that in this case the star is the only equilibrium tree. Interestingly, as observed in [10], it is not true that the class of equilibria in the model without parameter is a super-class of the equilibria in the original model. Nevertheless, we believe that the model of Alon et al. is still interesting, because it models the natural process of locally weighing alternatives against each other. Furthermore, it has another striking feature: Best responses can be computed efficiently. Thus, applying myopic dynamics seems a natural choice for the task of finding stable states in the game. The authors of [2] also propose to analyse the case where players are allowed to swap more than one edge at a time.

The work of Baumann and Stiller [3] is very similar in spirit to our work. They provide deep insights into the dynamics of a related network creation game and show various structural properties, e.g. sufficient and necessary conditions for stability.

Due to space constraints we refer for further work on selfish network creation to Jackson's survey [6] and to the references in Nisan et al.[12, Chapter 19].

1.2 Model and Definitions

The SUM BASIC NETWORK CREATION GAME is defined as follows: Given an undirected, connected graph $G = (V, E)$, where each vertex corresponds to a player. Every player $v \in V$ selfishly aims to minimize her connection cost by performing moves in the game. A player's *connection cost* $c(v)$ is the sum of all shortest-path distances to all other players. If the graph is disconnected, then we define $c(v)$ to be infinite. At any time, a player can "swap" an incident existing edge with an incident non-edge at no cost. More formally, let u be a neighbor of v and w be a non-neighbor of v, then the edge swap (u, w) of player v removes the edge vu and creates the edge vw. Let $\Gamma_G(v)$ denote the closed neighborhood of v in G, which includes v and all neighbors of v. The set of pure strategies for player v in G is $S_G(v) = (\Gamma_G(v) \setminus \{v\} \times V \setminus \Gamma_G(v)) \cup \{\bot\}$, where $\bot$ denotes, that player v does not swap. Note, that this set depends on the current graph G and that moves of players in the game modify the graph. We allow only pure strategies and call a pure strategy $s \in S_G(v)$, which decreases player v's current connection cost most, a *best response*. Sometimes we say that a vertex x is a best response of a player v, which abbreviates, that v has a best response of the form (y, x), for some $y \neq x$.

We assume that players are lazy, in the sense that if for some player v the best possible edge-swap yields no decrease in connection cost, then player v prefers the strategy $\bot$, that is, not to swap. We say that G is *stable* or in *swap-equilibrium* if $\bot$ is a best response for every player.

Since the model does not include costs for edges, the utility of a player is simply the negative of her connection cost. Let $x \in G$ denote that G contains vertex x. The connection cost of player v in graph G is defined as $c_G(v) = \sum_{x \in G} d_G(v, x)$, where $d_G(v, x)$ is the number of edges on the shortest path from v to x in G. We omit the reference to G, if it is clear from the context. The *social cost* of a graph G is the sum of the connection costs of all players in G.

Furthermore, we use the convention, that for a graph G, we let $|G|$ denote the number of vertices in G and we define $G - x$ to be the graph G after the removal of vertex x.

1.3 Our Contribution

We provide a rigorous treatment of the induced game dynamics of the SUM BASIC NETWORK CREATION GAME on trees. For this case, Theorem 1 shows that the game dynamic has the desirable property that local improvements by players directly yield a global improvement in terms of the social cost. More formally, we show that the game on trees is an *ordinal potential game*[11], that is, there exists a function mapping states of the game to values with the property that pure Nash Equilibria of the game correspond to local minima of the function. A prominent feature of such games is, that a series of local improvements must eventually converge to a pure Nash Equilibrium – a stable state of the game in which no player wants to unilaterally change her strategy. Theorem 3 shows that this convergence is fast by providing a cubic upper bound on the number of steps any improving response dynamic needs to reach such a stable state. Furthermore we introduce and analyse a natural dynamic called *Max Cost Best Response Dynamic*. This dynamic is proven to be close to optimal in terms of convergence speed, since Theorem 4 shows that the number of steps needed by this dynamic almost matches the trivial lower bound. This implies, that the process of finding a pure Nash Equilibrium can be significantly sped up by introducing coordination and enforcing that best responses are played.

In contrast to these positive results on trees, Theorem 7 is a strong negative result for the SUM BASIC NETWORK CREATION GAME on general graphs. We show that in this case best response dynamics can cycle, which implies, that there cannot exist an ordinal potential function. Thus, any treatment of the game dynamics on general graphs requires fundamentally different techniques and is an interesting open problem for ongoing research.

Last, but not least, we use structural insights to obtain a linear-time algorithm for computing a best response on trees even for the case where players are allowed to swap multiple edges at a time. For the game on general graphs, we provide another sharp contrast by showing that computing a best response in the general case is NP-hard, if more than one edge can be swapped at a time. This is particularly interesting, since this addresses the proposal of Alon et al.[2] to analyse this case. Our results imply, that in this case best responses can be efficiently computed only if the game is played on trees or on very simple graphs.

Due to space constraints some proofs are omitted. They can be found in the full version of the paper.

2   Playing on a Tree

In this section we consider the special case where the given graph G is a tree. We show, that the SUM BASIC NETWORK CREATION GAME on trees belongs to the well-studied class of *ordinal potential games*[11]. This guarantees the desirable property that pure Nash Equilibria always exist and that such solutions can be found by myopic play.

Theorem 1. *The* SUM BASIC NETWORK CREATION GAME *on trees is an ordinal potential game.*

Before we prove the Theorem, we analyse the impact of an edge-swap on the connection cost of the swapping player and on the social cost.

Let $T = (V, E)$ be a tree having n vertices. Assume that player v performs the edge-swap vu to vw in the tree T. (Note, that this implies, that $vw \notin E$). Let T' be the tree obtained after this edge-swap. Let Φ and Φ' be the social cost of T and T', respectively. Let T_v and T_u be the tree T rooted at v and u, respectively. Let A be the subtree rooted at v in T_u and let B be the subtree rooted at u in T_v. See Fig. 1 for an illustration. Let $c_K(z) = \sum_{k \in K} d_K(z, k)$ denote the connection cost of player z within tree K.

Lemma 1. *The change in player v's connection cost induced by the edge-swap vu to vw is $\Delta(v) = c_B(u) - c_B(w)$.*

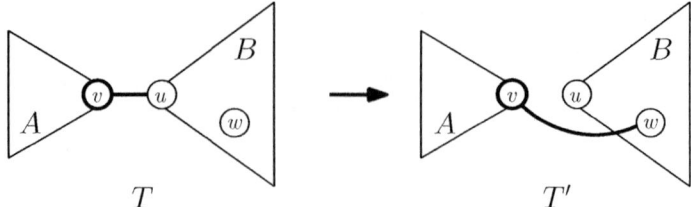

Fig. 1. Player v swaps edge vu to edge vw

The following Lemma implies the desired property, that local improvement of a player yields a global improvement in terms of social cost.

Lemma 2. *The change in social cost induced by the edge-swap vu to vw is*

$$\Delta(\Phi) = 2|A|\Delta(v) \ .$$

Proof (Theorem 1). By Lemma 2, we have that the social cost strictly decreases if and only if the connection cost of the swapping player strictly decreases. This implies, that the social cost Φ is an ordinal potential function for the SUM BASIC NETWORK CREATION GAME on trees.   □

Theorem 1 guarantees that a pure Nash Equilibrium of this game can be reached by myopic play, even if the players do not play in an optimal way. We only need one very natural ingredient for convergence: Whenever a player moves, this move must decrease this player's connection cost. We call every dynamic where a player strictly improves by making a move (or passing if no improving move is possible) an *improving response dynamic*(IRD). Such a dynamic stops if no player can strictly improve, which implies that any IRD stops if a stable graph is obtained.

2.1   Improving Response Dynamics on Trees

For trees it was shown by Alon et al.[2] that the star is the only stable tree. Using this observation and Theorem 1, we arrive at the following Corollary.

Corollary 1. *For every tree T, every IRD converges to a star.*

Having guaranteed convergence, the natural question to ask is how many steps are needed to reach the unique pure Nash Equilibrium by myopic play. The following Theorems provide a lower and an upper bound on that number.

Theorem 2. *Let P_n be a path having n vertices. Any IRD on P_n needs at least $\max\{0, n-3\}$ steps to converge.*

Lemma 3. *P_n is the tree on n vertices which has maximum social cost.*

Theorem 3. *Any IRD on trees having n vertices converges in $\mathcal{O}(n^3)$ steps.*

Proof. The idea is to start with the tree having the highest potential and to bound the number of steps any IRD needs by analysing the number of steps needed if this potential is decreased by the smallest possible amount per step. By Lemma 3, we have that P_n has the maximum social cost Φ_{P_n}. Observe, that $\Phi_{P_n} = \sum_{i=1}^{n-1} 2i(n-i) = \frac{n^3-n}{3}$. Let X_n be a star having n vertices. We have $\Phi_{X_n} = 2n^2 - 4n + 2$. To transform P_n into X_n any IRD has to decrease the social cost by $\Phi_{P_n} - \Phi_{X_n} = \frac{n^3}{3} - 2n^2 + \frac{11n}{3} - 2$. Since we have an IRD, every moving player decreases her connection cost by at least 1. By Lemma 2, we have that the minimum decrease in social cost by any move is 2. Hence, at most $\frac{n^3}{6} - n^2 + \frac{11n}{6} - 1 \in \mathcal{O}(n^3)$ steps are needed to transform P_n into X_n.   □

2.2   Best Response Dynamics on Trees

It is reasonable to assume, that players greedily try to decrease their connection cost most, whenever swapping an edge. In this section we analyse dynamics, where every move of a player is a best response move.

Since a best response is always an improving response, we have that every dynamic where every move is a best response must converge to a star for every tree T. We are left with the question of how fast best response dynamics converge. In the following, we analyse a specific best response dynamic, called *Max Cost Best Response Dynamic*(mcBRD), whose convergence speed almost matches the lower bound provided by Theorem 2. Hence, for best response dynamics we can significantly improve the upper bound of Theorem 3.

Definition 1. *The* Max Cost Best Response Dynamic *on a graph G is a dynamic, where in every step the player having the highest connection cost is allowed to play a best response. If two or more players have maximum connection cost, then one of them is chosen uniformly at random.*

In this section we show the following upper bound on the speed of convergence for the Max Cost Best Response Dynamic. Surprisingly, mcBRD behaves differently depending on whether the number of vertices in the tree is odd or even.

Theorem 4. *Let T be a tree having n vertices. The following holds:*

- *If n is even, then mcBRD(T) converges after at most $\max\{0, n-3\}$ steps and every player moves at most once.*
- *If n is odd, then at most $\max\{0, n + \lfloor n/2 \rfloor - 5\}$ steps are needed and every player moves at most twice.*

In order to prove Theorem 4, we first show some useful properties of the convergence process induced by the mcBRD-rule.

We begin with characterizing a player's best response on a tree. Here, the notion of a *center-vertex* is crucial.

Definition 2. *A center-vertex of a graph G is a vertex x, which satisfies*

$$x \in \arg\min_{v \in G} c(v) \ .$$

Lemma 4. *Let v be an arbitrary vertex of a tree T and let $F = T - v = \bigcup_{j=1}^{l} T_l$, where the trees T_j are connected components in the forest F. Let $u_1, \ldots, u_l$ be the neighbors of v in T, where u_j is a vertex of T_j for all $1 \leq j \leq l$. Let w_j be a center-vertex of the tree T_j. The best response of v in T is the edge-swap vu_j to vw_j, where $j \in \arg\max_j \{c_{T_j}(u_j) - c_{T_j}(w_j)\}$.*

The next Lemma provides a very useful property of neighbors in a tree.

Lemma 5. *Let u and w be neighbors in a tree T. Let T_u and T_w denote the tree T rooted at vertex u and w, respectively. Let U be the set of vertices in the subtree rooted at u in T_w. Analogously, let W be the set of vertices in the subtree rooted at w in T_u. Then we have $c(u) \leq c(w) \iff |U| \geq |W|$ and $c(u) < c(w) \iff |U| > |W|$.*

We can use Lemma 5, to show an important property of the mcBRD-process.

Lemma 6. *Let T be a tree. Every player who moves in a step of mcBRD(T) must be a leaf.*

The following Lemma provides the key to analysing mcBRD. It shows, that at some point in the dynamic a certain behavior is "triggered", which forces the dynamic to converge quickly.

Lemma 7 (First Trigger Lemma). *Let T be a tree. If the player who moves in step i of mcBRD(T) has a unique best response vertex w, then all players who move in a later step of mcBRD(T) will connect to vertex w.*

Lemma 8. *In any tree T on n vertices, there are at most two center-vertices. If this is the case, then they are neighbors and n must be even.*

Now we are ready, to prove the first part of Theorem 4.

Proof (Theorem 4, Part 1). We show, that if the number of vertices in a tree T is even, then mcBRD needs at most $\max\{0, n - 3\}$ to converge and every player moves at most once.

If T has two vertices, then it is already a star and no player will move in mcBRD(T). Thus, let T be a tree having at least $n \geq 4$ vertices, where n is even. By Lemma 6, we have that in every step of mcBRD(T) a leaf l of the current tree is allowed to move. By Lemma 4, we know that player l will connect to a center-vertex of $T' - l$, where T' is the tree before player l moves. Observe, that the tree $T' - l$ has an odd number of vertices. By Lemma 8, we have that any tree having an odd number of vertices must have a unique center vertex. It follows, that the leaf who moves in the first step of mcBRD(T) has a unique best response. Let this best response be the edge-swap towards vertex w. Lemma 7 implies, that all players who move in a later step of mcBRD(T), will connect to vertex w as well. Furthermore, again by Lemma 7, after the first step of mcBRD(T) it holds, that every vertex who is already connected to vertex w will never move again. Hence, every vertex moves at most once.

By Lemma 5, we have that w must be an inner vertex of T. Thus, w has at most $n - 3$ non-neighbors, which implies that the dynamic mcBRD(T) will need at most $n - 3$ steps to converge to a star having w as its center-vertex. $\square$

The next Theorem shows a lower bound on the speed of convergence for mcBRD on trees having an odd number of vertices. Surprisingly, the behavior of the dynamic on such instances is much more complex. The lower bound for odd n is roughly 50% greater than the upper bound for even n. Furthermore, the following Theorem together with Theorem 2 implies, that the analysis of mcBRD is tight.

Theorem 5. *There is a family of trees having an odd number of vertices greater than 5, where mcBRD can take $n + \lfloor n/2 \rfloor - 5$ steps to converge. Furthermore, every player moves at most twice.*

Figure 2 shows an example of a tree which belongs to the above mentioned family of trees and it sketches the convergence process induced by mcBRD.

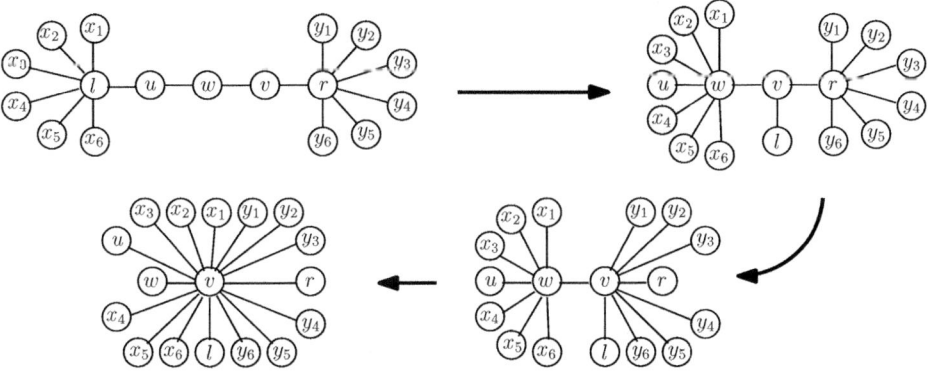

Fig. 2. Example of a tree T having 17 vertices, where mcBRD(T) takes $n + \lfloor n/2 \rfloor - 5 = 20$ steps to converge. The vertices $x_1, \ldots, x_6, u$ move twice.

Lemma 9 (Second Trigger Lemma). *Let T be an unstable tree having n vertices. If after any step i in mcBRD(T) a vertex w of T^i has degree $\lceil n/2 \rceil$, then this vertex will be the unique best response to connect to for all players moving in a later step of mcBRD(T).*

Lemma 10. *Let T be an unstable tree having an odd number of vertices. Only vertices which are best responses of the player who moves in the first step of mcBRD(T) will be best responses in any step of mcBRD(T).*

Finally, we have set the stage to prove the second part of Theorem 4.

Proof (Theorem 4, Part 2). We show that if a tree T has an odd number of vertices, then mcBRD(T) takes at most $\max\{0, n + \lfloor n/2 \rfloor - 5\}$ steps to converge and every player moves at most twice.

If $n = 5$, then the worst case instance is a path and thus the convergence takes at most 2 steps. Hence, we assume for the following that $n \geq 7$. Observe, that there are two events that force the dynamic to converge: Let E_1 be the event, where for the first time in the convergence process a vertex w becomes the unique best response of a moving player. Let E_2 be the event, where for the first time a vertex w has degree $\lceil n/2 \rceil$.

If event E_1 occurs in step j, then, by Lemma 7, all non-neighbors of the vertex w will connect to w in the subsequent steps of mcBRD(T). Thus, mcBRD(T) will converge in at most $j + |V \setminus \Gamma(w)|$ steps, where $\Gamma(w)$ is the closed neighborhood of w. If event E_2 occurs in step j, then, by Lemma 9, all non-neighbors of w will connect to w in the subsequent steps. Thus, in this case $j + \lfloor n/2 \rfloor - 1$ steps are needed for mcBRD(T) to converge.

Let T be any tree and v be the first player to move and assume that v has two best responses p and q, since otherwise the dynamic will converge in at most $n - 3$ steps. By Lemma 10, we have that in any step of mcBRD(T) a player will connect either to p or to q. Let $t_1(T)$ denote the number of steps until event E_1 is the first event to occur in mcBRD(T). Analogously, let $t_2(T)$ denote the number of steps until E_2 is the first occurring event. Let $r_1(T)$ denote the number of steps needed for convergence after event E_1. Hence, the maximum number of steps needed until mcBRD(T) converges is

$$t(T) = \max\{t_1(T) + r_1(T), t_2(T) + \lfloor n/2 \rfloor - 1\} \ .$$

We claim, that $t_1(T) + r_1(T) \leq n + \lfloor n/2 \rfloor - 5$. Observe, that $r_1(T) \leq n - 3$, since the vertex that becomes the center of the star must be an inner vertex of T and, thus, can have at most $n - 3$ non-neighbors. Furthermore, if $t_1(T) \leq \lfloor n/2 \rfloor - 2$, then the claim is true. Now let $t_1(T) > \lfloor n/2 \rfloor - 2$. Note, that both p and q must be inner vertices of T. Thus, they have at least degree 2. Since event E_2 did not occur in the first $t_1(T)$ steps of mcBRD(T) we have that not all players who moved within the first $t_1(T)$ steps can be connected to p. Thus, at least $x = t_1(T) - (\lfloor n/2 \rfloor - 2)$ players have connected to q. This yields $t_1(T) + r_1(T) \leq t_1(T) + n - 3 - x \leq n + \lfloor n/2 \rfloor - 5$. On the other hand, since all players move either to p or q and both p and q have degree at least 2, it follows that $t_2(T) \leq 2(\lfloor n/2 \rfloor - 2)$. Hence, $t_2(T) + \lfloor n/2 \rfloor - 1 \leq n + \lfloor n/2 \rfloor - 5$.

Observe, that any player x who is a neighbor of either p or q will not move again until event E_1 or E_2 happens. This holds because every leaf, which is not a neighbor of p or q must have higher connection cost than x and will therefore move before x. Thus, every player moves at most twice.   □

2.3   Computing a Best Response on Trees

Observe, that Lemma 4 directly yields an algorithm for computing a best response move of a player v: Compute the connection-costs of all other vertices in $T - v$ within their respective connected component to find a center-vertex for every component. Then choose the center-vertex, which yields the greatest cost decrease for player v. Clearly, the connection-cost of a player can be obtained using a BFS-computation. However the above naive approach of computing a center-vertex yields an algorithm with running time quadratic in n, since $\Omega(n)$ BFS-computations can occur. The following Lemma shows, that a center-vertex can be computed in linear time, which is clearly optimal. The algorithm crucially uses the structural property provided by Lemma 5.

Lemma 11. *Let T be a tree having n vertices. A center-vertex of T and its connection-cost can be computed in $\mathcal{O}(n)$ time.*

Proof. We give a linear time algorithm, which computes a center-vertex of T and its connection-cost. Let L be the set of leaves of T. Clearly, L can be computed in $\mathcal{O}(n)$ steps by inspecting every vertex.

Given T and L, the algorithm proceeds in two stages:

1. The algorithm computes for every vertex v of T two values n_v and c_v. This is done in reverse BFS-order: We define n_v to be the number of vertices in the already processed subtree T_v containing v and c_v to v's connection-cost to all vertices in T_v. For every leaf $l \in L$ we set $n_l := 1$ and $c_l := 0$. Let i be an inner vertex and assume that we have already processed all but one neighbor of i. Let $a_1, \ldots, a_s$ denote these neighbors. We set $n_i := 1 + n_{a_1} + \cdots + n_{a_s}$ and $c_i := n_i - 1 + c_{u_1} + \cdots + c_{a_s}$. By breaking ties arbitrarily, this computation terminates at a root-vertex r, for which all neighbors are already processed. Let $b_1, \ldots, b_q$ denote these neighbors. We set $n_r := n$ and $c_r := n - 1 + c_{b_1} + \cdots + c_{b_q}$.

2. Starting from vertex r, the algorithm performs a local search for the center-vertex with the help of Lemma 5. For all neighbors $b_i \in \{b_1, \ldots, b_q\}$ of r, the algorithm checks if $n_{b_i} > n_r - n_{b_i}$. Since T is a tree, this can hold for at most one neighbor x. In this case, x will be considered as new root-vertex. Let $c_1, \ldots, c_s, r$ be the neighbors of x. By setting $n_x := n$ and $c_x := n - 1 + c_1 + \cdots + c_s + c_r - c_x$ we arrive at the same situation as before and we now check for all neighbors $c_j \neq r$ if $n_{c_j} > n_x - n_{c_j}$ holds and proceed as above. Once no neighbor of the current root-vertex satisfies the above condition, the algorithm terminates and the current root-vertex is the desired center-vertex.

The correctness of the above algorithm follows by Lemma 5. Step 1 clearly takes time $\mathcal{O}(n)$. Step 2 takes linear time as well, since the condition is checked exactly once for every edge towards a neighbor and there are only $n - 1$ edges in T. □

Theorem 6. *If $p \geq 1$ edges can be swapped at a time, then the best response of a player v can be computed in linear time if G is a tree.*

Proofsketch. Let v be a degree d vertex in G and let $v_1, \ldots, v_d$ denote the neighbors of v in G. The $k \leq \min\{p, d\}$ edge swaps that decrease player v's connection cost most can be determined as follows.

Consider the forest $F = T_1 \cup T_2 \cup \cdots \cup T_d$ obtained by deleting v. By Lemma 4 we have that every swap in player v's best response is of the form (v_i, w_i), where w_i is a center-vertex of T_i. Thus, computing player v's best response reduces to finding a center-vertex in each tree T_i and to computing the corresponding cost decreases. By Lemma 11 we have that both tasks can be done in time linear to the number of vertices in each T_i.

3   Playing on General Graphs

3.1   Best Response Dynamics on General Graphs

Definition 3. *A cycle $x_1, \ldots, x_l$ is a best response cycle, if $x_1 = x_l$ and each x_i is a pure strategy profile in the* SUM BASIC NETWORK CREATION GAME *and for all $1 \leq k \leq l-1$ there is a player p_k whose best response move transforms the profile x_k into x_{k+1}.*

Theorem 7. *The* SUM BASIC NETWORK CREATION GAME *allows best response cycles.*

Proof. Consider the graph G depicted left in Figure 3 and let x_1 denote the corresponding strategy profile. Player a can decrease its connection cost and one of its best responses is to swap edge ab with edge ac. This leads to the second graph depicted in Figure 3. Call the corresponding strategy profile x_2. Now, player b has the swap bc to ba as its best response, which leads to the third graph depicted in the illustration, with x_3 as its strategy profile. Finally, player c can perform the swap ca to cb as its best response, which leads to profile $x_4 = x_1$. Thus, x_1, x_2, x_3, x_4 is a best response cycle in the SUM BASIC NETWORK CREATION GAME on graph G.   □

Voorneveld [13] introduced the class of *best-response potential games*, which is a super-class of ordinal potential games. Furthermore he proves, that if the strategy space is countable, then a strategic game is a best-response potential game if and only if there is no best response cycle. This implies the following Corollary.

Corollary 2. *There cannot exist an ordinal potential function for the* SUM BASIC NETWORK CREATION GAME *on graphs containing cycles.*

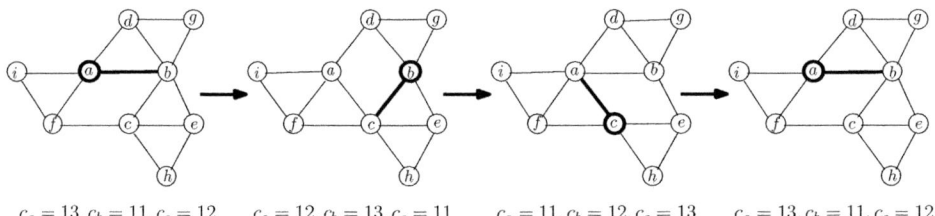

$c_a = 13, c_b = 11, c_c = 12$ $c_a = 12, c_b = 13, c_c = 11$ $c_a = 11, c_b = 12, c_c = 13$ $c_a = 13, c_b = 11, c_c = 12$

Fig. 3. Example of a graph, where the SUM BASIC NETWORK CREATION GAME allows a best response cycle. The steps of the cycle are shown.

3.2   Computing a Best Response in General Graphs

Given an undirected, connected graph G, then the best response for player v can be computed in $\mathcal{O}(n^2)$ time, since $|S_G(v)| < n^2$ and we can try all pure strategies to find the best one. Quite surprisingly, computing the best response is hard if we allow a player to swap $p > 1$ edges at a time.

Theorem 8. *If players are allowed to swap $p > 1$ edges at a time, then computing the best response is NP-hard even if G is planar and has maximum degree 3.*

Proofsketch. We reduce from the p-MEDIAN-PROBLEM [7].

References

1. Albers, S., Eilts, S., Even-Dar, E., Mansour, Y., Roditty, L.: On nash equilibria for a network creation game. In: Proceedings of the Seventeenth Annual ACM-SIAM Symposium on Discrete Algorithm, SODA 2006, pp. 89–98. ACM, New York (2006)
2. Alon, N., Demaine, E.D., Hajiaghayi, M., Leighton, T.: Basic network creation games. In: SPAA 2010: Proceedings of the 22nd ACM Symposium on Parallelism in Algorithms and Architectures, pp. 106–113. ACM, New York (2010)
3. Baumann, N., Stiller, S.: The price of anarchy of a network creation game with exponential payoff. In: Monien, B., Schroeder, U.-P. (eds.) SAGT 2008. LNCS, vol. 4997, pp. 218–229. Springer, Heidelberg (2008)
4. Demaine, E.D., Hajiaghayi, M., Mahini, H., Zadimoghaddam, M.: The price of anarchy in network creation games. In: Proceedings of the Twenty-Sixth Annual ACM Symposium on Principles of Distributed Computing, PODC 2007, pp. 292–298. ACM, New York (2007)
5. Fabrikant, A., Luthra, A., Maneva, E., Papadimitriou, C.H., Shenker, S.: On a network creation game. In: Proceedings of the Twenty-Second Annual Symposium on Principles of Distributed Computing, PODC 2003, pp. 347–351. ACM, New York (2003)
6. Jackson, M.O.: A survey of models of network formation: Stability and efficiency. Group Formation in Economics: Networks, Clubs and Coalitions (2003)
7. Kariv, O., Hakimi, S.L.: An algorithmic approach to network location problems. ii: The p-medians. SIAM Journal on Applied Mathematics 37(3), 539–560 (1979)
8. Koutsoupias, E., Papadimitriou, C.: Worst-case equilibria. In: Meinel, C., Tison, S. (eds.) STACS 1999. LNCS, vol. 1563, pp. 404–413. Springer, Heidelberg (1999)
9. Lin, H.: On the price of anarchy of a network creation game. Class final project (2003)
10. Mihalák, M., Schlegel, J.C.: The price of anarchy in network creation games is (mostly) constant. In: Kontogiannis, S., Koutsoupias, E., Spirakis, P.G. (eds.) Algorithmic Game Theory. LNCS, vol. 6386, pp. 276–287. Springer, Heidelberg (2010)
11. Monderer, D., Shapley, L.S.: Potential games. Games and Economic Behavior 14(1), 124–143 (1996)
12. Nisan, N., Roughgarden, T., Tardos, E., Vazirani, V.V.: Algorithmic Game Theory. Cambridge University Press, New York (2007)
13. Voorneveld, M.: Best-response potential games. Economics Letters 66(3), 289–295 (2000)

Pricing Exotic Derivatives Using Regret Minimization*

Eyal Gofer and Yishay Mansour

Tel Aviv University,
Tel Aviv, Israel
{eyalgofe,mansour}@post.tau.ac.il

Abstract. We price various financial instruments, which are classified as *exotic options*, using the regret bounds of an online algorithm. In addition, we derive a general result, which upper bounds the price of any derivative whose payoff is a convex function of the final asset price. The market model used is adversarial, making our price bounds robust. Our results extend the work of [9], which used regret minimization to price the standard European call option, and demonstrate the applicability of regret minimization to derivative pricing.

1 Introduction

Pricing derivatives is one of the most fundamental questions in finance. A *derivative* is a security whose price is dependent on the price of one or more underlying assets, for example, stocks. An important type of derivative is an *option*, which is a financial instrument that allows its holder to buy or sell a certain asset for a given price at a given time. For example, a *European call option* allows its holder, at its expiration time T, to buy the asset for a price K. In other words, the option pays $\max(S_T - K, 0)$ at time T, where S_T is the asset (stock) price at time T. Another standard option is the *European put option*, which allows its holder to *sell* the asset for a price K at time T, or, equivalently, to receive a payoff of $\max(K - S_T, 0)$. Buying such options enables stock holders to hedge their investments against future asset price changes. These options are also widely traded as securities in their own right. Apart from the standard call and put options, there are numerous other option types that are traded or developed to meet particular financial needs. Such options are collectively known as *exotic options*. One of the main contributions of this paper is pricing a variety of exotic options using regret minimization.

A pricing formula and model of great theoretic and practical influence are due to Black and Scholes [1] and Merton [14]. In their Nobel Prize-winning work,

---

* This research was supported in part by the Google Inter-university center for Electronic Markets and Auctions, by a grant from the Israel Science Foundation, by a grant from United States-Israel Binational Science Foundation (BSF), and by a grant from the Israeli Ministry of Science (MoS). This work is part of Ph.D. thesis research carried out by the first author at Tel Aviv University.

G. Persiano (Ed.): SAGT 2011, LNCS 6982, pp. 266–277, 2011.

they modeled the price of a stock as a geometric Brownian motion stochastic process. In addition, their model assumes an arbitrage-free market, namely, that market prices provide no opportunities for riskless profit. The Black-Scholes-Merton (BSM) model has several known drawbacks. First, the model is only an abstraction of price changes, while in reality prices are discrete and experience sharp jumps, and the daily returns are neither independent nor identically distributed. Second, the main parameter which is required, the stock volatility, is not observable, and has to be estimated.[1] In this work we investigate the pricing of derivatives in an adversarial online learning model which was introduced in [9]. A major advantage of such an adversarial online approach is that price jumps and discrete trading are inherently assumed in the model.

Many financial applications can be cast naturally in the framework of online learning. The basic actions are selling or buying of the basic assets, say stocks or bonds, and the payoff of the algorithm is compared to some benchmark. For example, in the problem of *portfolio selection*, an online algorithm distributes capital among n assets at each time period $1 \leq t \leq T$. The algorithm's returns are then measured against a reference class of investment strategies. By bounding the *regret* relative to that reference class, the return of the online algorithm can be guaranteed to exceed some proportion of the return of the best investment strategy in the reference class (see [5] for a comprehensive exposition of regret minimization, including portfolio selection).

Our work uses regret minimization to price derivatives. The general idea in pricing using regret minimization is that the regret bounds translate to a bound on the price of the financial instrument. More specifically, the regret bounds can guarantee that the payoff of a certain online algorithm will dominate the payoff of the financial instrument, and this implies that in an arbitrage-free market, the initial cost of the algorithm must be an upper bound on the cost of the financial instrument. This approach was pioneered in [9], where it was used to derive upper and lower bounds on the price of European call options. Our goal is to show that regret minimization based pricing is applicable to a wider range of financial instruments.

Our first contribution is extending the results of [9] by applying a unified regret minimization framework to pricing a variety of options. We give price bounds for various known exotic options, namely, the *exchange option*, the *shout option*, the *average strike call option*, and the *average price call option* (geometric and arithmetic averages). We derive these bounds by upper bounding the price of an option whose payoff is the maximum of several derivatives. We price this option based on regret bounds w.r.t. the underlying derivatives, and then express the above exotic options in terms of this option. In our analysis, we use the Polynomial Weights algorithm [6] as the regret minimization component.

---

[1] In fact, in many cases people compute the *implied volatility*, which is the volatility that under the BSM model would give the current option price. It is well documented that the implied volatility is not constant, even for a given expiration time, and depends on the strike price.

A second contribution is demonstrating that pricing using regret minimization is applicable to a broad class of securities, namely, *convex path-independent derivatives*. This class consists of all derivatives whose payoff is a convex function of the final asset price. Convex derivatives can be shown to have the same payoff as a portfolio of call options that contains no short positions. We can thus use any upper bound on the price of call options, e.g., the bound of [9], to derive an upper bound on the price of convex derivatives.

We conclude with an experimental demonstration of our bounds for the price of average strike call options, considering data from the S&P 500 index.

Related work. The main financial problem dealt with in the learning literature is portfolio selection, where the main goal is to maximize returns. For this purpose, the performance of algorithms is compared to benchmarks, which are expected in many cases to significantly outperform the best asset. It is important to note that competing against these benchmarks is, therefore, harder than competing against the underlying assets, which is our aim when pricing certain options.

The most widely used benchmark in works on portfolio selection is the *best constantly rebalanced portfolio* (BCRP). A key result by Cover [7] gives an algorithm, the *universal portfolio*, which achieves the same asymptotic growth rate as the BCRP, even under adversarial conditions. Subsequent work incorporated side information and transaction costs, proved optimal regret w.r.t. the BCRP, improved computational complexity, and considered short selling [8,15,2,13,17]. The work of [12] used a simple multiplicative update rule with a linear complexity in the number of assets, but gave worse regret bounds compared with the universal portfolio. The algorithms of [16] achieve bounded regret w.r.t. another benchmark, namely, the best switching regime between N fixed investment strategies, with and without transaction costs. Interestingly, the algorithms of both [12] and [16] were shown to outperform the universal portfolio on real data, with the latter occasionally outperforming the BCRP itself. Other approaches directly seek to exploit the underlying statistics of the market [3,10]. The authors of [10] show that their methods achieve the optimal asymptotic growth rate almost surely, assuming the markets are stationary and ergodic.

Work on portfolio selection has led to online trading algorithms with improved returns under statistical and adversarial assumptions, and provided regret bounds w.r.t. rich reference classes, in particular, constantly rebalanced portfolios. For pricing options, we also seek to prove adversarial regret bounds, but our reference class is simply a set of n assets. Working with this simple reference class, we can achieve better regret bounds than those given w.r.t. superior benchmarks such as the BCRP. Consider a naïve *buy and hold* algorithm which initially divides capital equally between assets, and performs no further action. This algorithm clearly has a regret bound of $\ln n$ w.r.t. the log returns of any of the assets. In comparison, the universal portfolio's (optimal) bound w.r.t. the BCRP is $\frac{n-1}{2} \ln 2T + \ln \frac{\Gamma(\frac{1}{2})^n}{\Gamma(\frac{n}{2})} + o(1)$ (see, e.g., [5]). If $n = T$, as in the case of some path-dependent options, we get an $\Omega(T)$ regret bound, which grows arbitrarily large as the trading frequency increases. The problematic dependence on T persists even if n is small, for example, in the case of call options, where

$n = 2$. Another algorithm by Hazan and Kale [11] has regret bounds that depend on the quadratic variability[2] of the single period returns. Their bound is a great improvement over the previous bound, under realistic conditions where the quadratic variability is much smaller than T. However, it is still lower bounded by the naïve bound of $\ln n$, for every n. Therefore, we do not use regret bounds w.r.t. the BCRP in the context of pricing options.

Outline. The outline of the paper is as follows. In Section 2 we provide notation and definitions. Section 3 presents an upper bound on the price of an option whose payoff is the maximum of several derivatives, given the regret bounds of a trading algorithm. In Section 4 we upper bound the price of a variety of options, based on the regret bounds of the Polynomial Weights algorithm. In Section 5 we show how any upper bound on the price of call options can be used to upper bound the price of any convex path-independent derivative. In Section 6 we give empirical results.

2   Preliminaries

We consider a discrete-time finite-horizon model, with a risky asset (stock) and a risk-free asset (bond or cash). The price of the stock at time $t \in \{0, 1, \ldots, T\}$ is S_t and the price of the risk-free asset is B_t. Initially, $B_0 = S_0 = 1$. We assume that the price of cash does not change, i.e., $B_t = 1$, which is equivalent to assuming a zero risk-free interest rate. We further assume that we can buy or sell any real quantity of stocks with no transaction costs. For the stock we denote by r_t the single period return between $t - 1$ and t, so $S_t = S_{t-1}(1 + r_t)$.

A realization of the prices is a *price path*, which is the vector $\mathbf{r}_t = (r_1, \ldots, r_t)$. We define a few parameters of a price path. We denote by M an upper bound on stock prices S_t, by R an upper bound on absolute single period returns $|r_t|$, and by Q an upper bound on the quadratic variation $\sum_{t=1}^{T} r_t^2$. We will assume that the bounds R, Q, and M are given, and $\Pi_{M,R,Q}$, or simply Π, will denote the set of all price paths satisfying these bounds. We note that, apart from M, stock prices are always upper bounded by $e^{\sqrt{QT}}$, as can be easily verified. Since $\max_{1 \le t \le T}(r_t^2) \le \sum_{t=1}^{T} r_t^2 < R^2 T$, we may assume that $R^2 \le Q \le R^2 T$. The number of time steps, T, is influenced by both the frequency of trading and the absolute time duration. For this reason it is instructive to consider M and Q as fixed, rather than as increasing in T.

A *trading algorithm* A starts with a total asset value V_0. At every time period $t \ge 1$, A sets weights $w_{s,t} \ge 0$ for stock and $w_{c,t} \ge 0$ for cash, and we define the fractions $p_{s,t} = w_{s,t}/W_t$ and $p_{c,t} = w_{c,t}/W_t$, where $W_t = w_{s,t} + w_{c,t}$. A fraction $p_{s,t}$ of the total asset value, V_{t-1}, is placed in stock, and likewise, $p_{c,t}V_{t-1}$ is placed in cash. Following that, the stock price S_t becomes known, and the asset value is updated to $V_t = V_{t-1}(1 + r_t p_{s,t})$, and time period $t + 1$ begins.

---

[2] Their definition is different from the quadratic variation Q, which we will use. In particular, their variability is centered around the average value of the daily returns, in a way similar to the variance of random variables.

We comment that since we assume that both weights are non-negative, the algorithms we consider use neither short selling of the stock, nor buying on margin (negative positions in cash). However, as part of the arbitrage-free assumption, we assume that short selling is, in general, allowed in the market.

We next define specific types of options referred to in this work:

C A *European call option* $\mathbf{C}(K,T)$ is a security paying $\max(S_T - K, 0)$ at time T, where K is the *strike price* and T is the *expiration time*.

EX An *exchange option* $\mathbf{EX}(\mathbf{X}_1, \mathbf{X}_2, T)$ allows the holder to exchange asset $\mathbf{X}_2$ for asset $\mathbf{X}_1$ at time T, making its payoff $\max(X_{1,T} - X_{2,T}, 0)$.

SH A *shout option* $\mathbf{SH}(K,T)$ allows its holder to "shout" and lock in a minimum value for the payoff at one time $0 \le \tau \le T$ during the lifetime of the option. Its payoff at time T is, therefore, $\max(S_T - K, S_\tau - K, 0)$. (If the holder does not shout, the payoff is $\max(S_T - K, 0)$.)

AP An *average price call option* $\mathbf{AP}(K,T)$ is a type of Asian option that pays $\max(\bar{S}_T - K, 0)$ at time T, where $\bar{S}_T$ may be either the arithmetic or the geometric mean of the stock's prices. To distinguish between these two possibilities, we will denote $\mathbf{AP}_A(K,T)$ for the option whose payoff is $\max(\bar{S}_T^A - K, 0)$, where $\bar{S}_T^A = \frac{1}{T+1}\sum_{t=0}^{T} S_t$ is the arithmetic mean, and $\mathbf{AP}_G(K,T)$ for the option whose payoff is $\max(\bar{S}_T^G - K, 0)$, where $\bar{S}_T^G = (\prod_{t=0}^{T} S_t)^{\frac{1}{T+1}}$ is the geometric mean.

AS An *average strike call option* $\mathbf{AS}(T)$ is a type of Asian option that allows its holder to get the difference between the final stock price and the average stock price, namely, a payoff of $\max(S_T - \bar{S}_T^A, 0)$.

We will use bold text for denoting securities and plain text for denoting their values at time 0. For example, we write $SH(K,T)$ for the value of the option $\mathbf{SH}(K,T)$ at time 0.

The above options are a special case of stock derivatives whose value at time T is some function of $\mathbf{r}_T$, the price path of the underlying stock.[3] A European call option is an example of a *path-independent* derivative, since its payoff of $\max(S_T - K, 0)$ depends only on the price at time T. An average strike option is an example of a *path-dependent* derivative, since its payoff depends on the entire price path of the stock.

Finally, we may assume that strike prices for European calls, average price calls, and shout options satisfy $K \in [0, M]$. The reason is that for $K > M$, the payoffs are always 0, and so are the option values. Working with $K < 0$ simply adds a constant to the payoff and value of the same option with $K = 0$.

3 Arbitrage-Free Bounds

We assume that the pricing is *arbitrage-free*, which is defined as follows. Trading algorithm $\mathbf{A}_1$ *dominates* trading algorithm $\mathbf{A}_2$ w.r.t. the set of price paths Π, if for every price path in Π, the final value of $\mathbf{A}_1$ is at least the final value of $\mathbf{A}_2$.

---

[3] Except for the shout option.

The *arbitrage-free assumption* says that if A_1 dominates A_2, then the initial value of A_1 is at least the initial value of A_2. This assumption is natural, because if it is violated, it becomes possible to make a riskless profit by buying into A_1 and selling A_2. The resulting flow of funds from A_2 to A_1 affects the stock price in a way that causes even a small arbitrage opportunity to quickly disappear.

For example, define trading algorithm A_{AS} which simply buys the average strike call option and holds it. Its initial value is $AS(T)$ and its final value is $\max(S_T - \bar{S}_T^A, 0)$. Assume we design a trading strategy A_1 whose initial value is V_0, and its value at time T always satisfies $V_T \geq \max(S_T - \bar{S}_T^A, 0)$. This implies that A_1 dominates A_{AS}. Therefore, by the arbitrage-free assumption, we have that $AS(T) \leq V_0$.

We now establish a connection between bounds on the multiplicative regret of a trading algorithm and arbitrage-free pricing, extending a result from [9].

Consider n derivatives, $\mathbf{X}_1, \ldots, \mathbf{X}_n$. Let $X_{1,t}, \ldots, X_{n,t}$ be their values at time t, and assume $X_{i,T} > 0$ for every $1 \leq i \leq n$. We now consider an option that pays the maximal value of a set of given derivatives.[4] More specifically, we will denote $\mathbf{\Psi}(\mathbf{X}_1, \ldots, \mathbf{X}_n, T)$ for an option that pays $\max(X_{1,T}, \ldots, X_{n,T})$ at time T, and $\Psi(\mathbf{X}_1, \ldots, \mathbf{X}_n, T)$ for its value at time 0.

Definition 1. *Let A be a trading algorithm with initial value $V_0 = 1$, and let $\beta_1, \ldots, \beta_n > 0$. A is said to have a $(\beta_1, \ldots, \beta_n)$ multiplicative regret w.r.t. derivatives $\mathbf{X}_1, \ldots, \mathbf{X}_n$ if for every price path and every $1 \leq i \leq n$, $V_T \geq \beta_i X_{i,T}$.*

Lemma 1. *If there exists a trading algorithm with a $(\beta_1, \ldots, \beta_n)$ multiplicative regret w.r.t. derivatives $\mathbf{X}_1, \ldots, \mathbf{X}_n$, then $\Psi(\mathbf{X}_1, \ldots, \mathbf{X}_n, T) \leq 1/\beta$, where $\beta = \min_{1 \leq i \leq n} \beta_i$.*

Proof. We have that $V_T \geq \beta \max_{1 \leq i \leq n} X_{i,T}$, therefore, the payoff of the algorithm dominates β units of the option $\mathbf{\Psi}(\mathbf{X}_1, \ldots, \mathbf{X}_n, T)$. By the arbitrage-free assumption, $\Psi(\mathbf{X}_1, \ldots, \mathbf{X}_n, T) \leq 1/\beta$. $\qquad\square$

Moreover, the lemma indicates exactly how improved regret bounds for a trading algorithm relate to tighter upper bounds on $\Psi(\mathbf{X}_1, \ldots, \mathbf{X}_n, T)$.

4 Price Bounds for a Variety of Options

In order to obtain concrete price bounds, we require a specific trading algorithm whose multiplicative regret bounds we can plug into Lemma 1. Following [9], we use an adaptation of the Polynomial Weights algorithm [6], called *Generic*.

It is important to note that in this section we consider derivatives that are *tradable*. We will define specific tradable derivatives later for pricing specific options. Let $\mathbf{X}_1, \ldots, \mathbf{X}_n$ be such derivatives, where $\mathbf{r}_{i,T} = (r_{i,1}, \ldots, r_{i,T})$ is the price path of $\mathbf{X}_i$, for $1 \leq i \leq n$. We will require that $|r_{i,t}| < R < 1 - 1/\sqrt{2} \approx 0.3$, and $\sum_{t=1}^T r_{i,t}^2 \leq Q_i$, for every $1 \leq i \leq n$. We will assume $X_{i,0} > 0$ for every $1 \leq i \leq n$, which implies that the derivatives have positive values at all times.

---

[4] Equivalently, this option is a call on the maximum with a zero strike price.

Theorem 1. ([9]) *Assume* $X_{1,0} = \ldots = X_{n,0} = 1$. *Let* V_T *be the final value of the Generic algorithm investing* 1 *unit of cash in* $\mathbf{X}_1, \ldots, \mathbf{X}_n$ *with initial fractions* $p_{1,1}, \ldots, p_{n,1}$, *and* $\eta \in [1, \frac{1-2R}{2R(1-R)}]$. *Then for every* $1 \le i \le n$,

$$V_T \ge p_{i,1}^{\frac{1}{\eta}} e^{-(\eta-1)Q_i} X_{i,T} .$$

In what follows, we will write $\eta_{max} = \frac{1-2R}{2R(1-R)}$ for short. We can now derive a bound on $\Psi(\mathbf{X}_1, \ldots, \mathbf{X}_n, T)$.

Theorem 2. *For every* $\eta \in [1, \eta_{max}]$,

$$\Psi(\mathbf{X}_1, \ldots, \mathbf{X}_n, T) \le \left(\sum_{i=1}^{n} e^{\eta(\eta-1)Q_i} X_{i,0}^{\eta} \right)^{\frac{1}{\eta}} .$$

Proof. For every $1 \le i \le n$, define $\mathbf{X}_i' = X_{i,0}^{-1} \mathbf{X}_i$, namely, a fraction of $\mathbf{X}_i$ with value 1 at time 0. Applying Theorem 1 to these new assets, we have that for every $1 \le i \le n$,

$$V_T \ge p_{i,1}^{\frac{1}{\eta}} e^{-(\eta-1)Q_i} X_{i,T}' = p_{i,1}^{\frac{1}{\eta}} e^{-(\eta-1)Q_i} X_{i,0}^{-1} X_{i,T} .$$

Denoting $\beta_i = p_{i,1}^{1/\eta} e^{-(\eta-1)Q_i} X_{i,0}^{-1}$ and $\beta = \min_{1 \le i \le n} \beta_i$, we have by Lemma 1 that $\Psi(\mathbf{X}_1, \ldots, \mathbf{X}_n, T) \le 1/\beta$. For any fixed η, we may optimize this bound by picking a probability vector $(p_{1,1}, \ldots, p_{n,1})$ that maximizes β. Clearly, β is maximized if $\beta_1 = \ldots = \beta_n = c$ for some constant $c > 0$. This is equivalent to having $p_{i,1} = c^{\eta} e^{\eta(\eta-1)Q_i} X_{i,0}^{\eta}$ for every $1 \le i \le n$. To ensure that $(p_{1,1}, \ldots, p_{n,1})$ is a probability vector, we must set $c = (\sum_{i=1}^{n} e^{\eta(\eta-1)Q_i} X_{i,0}^{\eta})^{-1/\eta}$. We thus have that $\Psi(\mathbf{X}_1, \ldots, \mathbf{X}_n, T) \le 1/\beta = 1/c = (\sum_{i=1}^{n} e^{\eta(\eta-1)Q_i} X_{i,0}^{\eta})^{1/\eta}$.   □

We next utilize the bound on $\Psi(\mathbf{X}_1, \ldots, \mathbf{X}_n, T)$ to bound the price of various exotic options, as well as the ordinary call option.

Theorem 3. *For every* $\eta \in [1, \eta_{max}]$, *the following bounds hold:*

- $EX(\mathbf{X}_1, \mathbf{X}_2, T) \le (e^{\eta(\eta-1)Q_1} X_{1,0}^{\eta} + e^{\eta(\eta-1)Q_2} X_{2,0}^{\eta})^{1/\eta} - X_{2,0}$
- $SH(K, T) \le (K^{\eta} + 2e^{\eta(\eta-1)Q} S_0^{\eta})^{1/\eta} - K$
- $AS(T) \le S_0(e^{(\eta-1)Q + (\ln 2)/\eta} - 1)$

Proof. Throughout this proof, we use the notation $\mathbf{X}_3 = \mathbf{X}_1 + \mathbf{X}_2$ to indicate that the payoff of the derivative $\mathbf{X}_3$ is always equal to the combined payoffs of the derivatives $\mathbf{X}_1$ and $\mathbf{X}_2$. Equivalently, we will write $\mathbf{X}_2 = \mathbf{X}_3 - \mathbf{X}_1$. We point out that by the arbitrage-free assumption, equal payoffs imply equal values at time 0. Therefore, we have that $X_{3,0} = X_{1,0} + X_{2,0}$.

- Since $\mathbf{EX}(\mathbf{X}_1, \mathbf{X}_2, T) = \Psi(\mathbf{X}_1, \mathbf{X}_2, T) - \mathbf{X}_2$, we have that $EX(\mathbf{X}_1, \mathbf{X}_2, T) = \Psi(\mathbf{X}_1, \mathbf{X}_2, T) - X_{2,0} \le (\sum_{i=1}^{2} e^{\eta(\eta-1)Q_i} X_{i,0}^{\eta})^{1/\eta} - X_{2,0}$, where the inequality is by Theorem 2.

– Let $\mathbf{X}_1$ be the stock. Let $\mathbf{X}_2$ be an algorithm that buys a single stock at time 0, and if the option holder shouts, sells it immediately. In addition, let $\mathbf{X}_3$ be K in cash (implying $Q_3 = 0$). Note that the quadratic variations of both $\mathbf{X}_1$ and $\mathbf{X}_2$ are upper bounded by Q. Since $\mathbf{SH}(K, T) = \mathbf{\Psi}(\mathbf{X}_1, \mathbf{X}_2, \mathbf{X}_3) - \mathbf{X}_3$, we have by Theorem 2 that $SH(K, T) \leq (K^\eta + \sum_{i=1}^{2} e^{\eta(\eta-1)Q_i} X_{i,0}^\eta)^{1/\eta} - K \leq (K^\eta + 2 e^{\eta(\eta-1)Q} S_0^\eta)^{1/\eta} - K$.

– For the bound on $AS(T)$, let $\mathbf{X}_1$ be the stock and let $\mathbf{X}_2$ be an algorithm that buys a single stock at time 0 and sells a fraction $\frac{1}{T+1}$ of the stock at each time $0 \leq t \leq T$. We thus have that $X_{2,T} = \frac{1}{T+1} \sum_{t=0}^{T} S_t$. Denote by Q_2 an upper bound on the quadratic variation of $\mathbf{X}_2$. Since $\mathbf{AS}(T) = \mathbf{EX}(\mathbf{X}_1, \mathbf{X}_2, T)$, then by our bound on $EX(\mathbf{X}_1, \mathbf{X}_2, T)$, we have that $AS(T) \leq (e^{\eta(\eta-1)Q} S_0^\eta + e^{\eta(\eta-1)Q_2} S_0^\eta)^{1/\eta} - S_0 = S_0((e^{\eta(\eta-1)Q} + e^{\eta(\eta-1)Q_2})^{1/\eta} - 1)$. For every t, $X_{2,t} = \frac{1}{T+1} \sum_{\tau=0}^{t} S_\tau + \frac{T-t}{T+1} S_t$, therefore,

$$|r_{2,t}| = \left| \frac{\frac{1}{T+1} \sum_{\tau=0}^{t} S_\tau + \frac{T-t}{T+1} S_t}{\frac{1}{T+1} \sum_{\tau=0}^{t} S_\tau + \frac{T+1-t}{T+1} S_{t-1}} - 1 \right| = \left| \frac{\frac{1}{T+1} S_t + \frac{T-t}{T+1} S_t - \frac{T+1-t}{T+1} S_{t-1}}{\frac{1}{T+1} \sum_{\tau=0}^{t-1} S_\tau + \frac{T+1-t}{T+1} S_{t-1}} \right|$$

$$= \frac{\frac{T+1-t}{T+1} |S_t - S_{t-1}|}{\frac{1}{T+1} \sum_{\tau=0}^{t-1} S_\tau + \frac{T+1-t}{T+1} S_{t-1}} \leq \frac{\frac{T+1-t}{T+1} |S_t - S_{t-1}|}{\frac{T+1-t}{T+1} S_{t-1}} = |r_t|.$$

Therefore, $\sum_{t=1}^{T} r_{2,t}^2 \leq \sum_{t=1}^{T} r_t^2 \leq Q$, and we may assume $Q_2 = Q$. We thus have that $AS(T) \leq S_0[(2 e^{\eta(\eta-1)Q})^{1/\eta} - 1]$, and the result follows. $\square$

Since an ordinary call is actually $\mathbf{EX}(\mathbf{X}_1, \mathbf{X}_2, T)$, where $\mathbf{X}_1$ is the stock and $\mathbf{X}_2$ is K in cash, we can derive the following bound from [9]:

Corollary 1. ([9]) *The price of a European call option satisfies $C(K, T) \leq \min_{1 \leq \eta \leq \eta_{max}} \left(K^\eta + S_0^\eta e^{\eta(\eta-1)Q} \right)^{1/\eta} - K$.*

For the average strike option we can optimize for η explicitly:

Corollary 2. *The price of an average strike call option satisfies $AS(T) \leq S_0 (e^{(\eta_{opt}-1)Q + (\ln 2)/\eta_{opt}} - 1)$, where $\eta_{opt} = \max\{1, \min(\sqrt{(\ln 2)/Q}, \eta_{max})\}$.*

We note that the above bound has different behaviors depending on the value of Q. The bound has a value of $S_0(e^{(\eta_{max}-1)Q + (\ln 2)/\eta_{max}} - 1)$ for $Q < (\ln 2)/\eta_{max}^2$, $S_0(e^{\sqrt{4Q \ln 2} - Q} - 1)$ for $(\ln 2)/\eta_{max}^2 \leq Q < \ln 2$, and (a trivial) S_0 for $Q \geq \ln 2$.

Average Price Call Options

Average price call options provide a smoothed version of European call options by averaging over the whole price path of the stock, and they are less expensive than European options. To allow a counterpart of this phenomenon in our model, we will allow the quadratic variation parameter Q to depend on time. More specifically, we will assume that Q_t is an upper bound on $\sum_{\tau=1}^{t} r_\tau^2$, where $Q_1 \leq \ldots \leq Q_T$.

The following simple relation is easily verified using the inequality of the arithmetic and geometric means:

Theorem 4. *The prices of average price call options satisfy* $AP_G(K,T) \leq AP_A(K,T) \leq \frac{1}{T+1} \sum_{t=0}^{T} C(K,t)$.

Using the bound of Corollary 1, we can obtain a concrete bound for both types of average price calls. However, its dependence on the sequence $Q_1, \ldots, Q_T$ is complicated. A simpler relation is given for $AP_G(K,T)$ by the following theorem (proof omitted):

Theorem 5. *Let* $\hat{Q}_T = \frac{1}{T+1} \sum_{t=1}^{T} Q_t$, $\hat{R} = \sqrt{\frac{T}{T+1}} R$, *and* $\hat{\eta}_{max} = \frac{1-2\hat{R}}{2\hat{R}(1-\hat{R})}$. *It holds that*

$$AP_G(K,T) \leq \min_{1 \leq \eta \leq \hat{\eta}_{max}} \left(K^\eta + S_0^\eta e^{\eta(\eta-1)\hat{Q}_T} \right)^{\frac{1}{\eta}} - K .$$

Since $\hat{\eta}_{max} \geq \eta_{max}$, the above expression does not exceed the bound of Corollary 1 with $\hat{Q}_T$ as the value of the quadratic variation. In other words, $AP_G(K,T)$ is upper bounded by the bound for a regular call with the *averaged* quadratic variation, which, depending on $Q_1, \ldots, Q_T$, may be significantly smaller than Q_T.

5   Convex Path-Independent Derivatives

In this section we move beyond specific options, and provide adversarial price bounds for a general class of derivatives. For this purpose, we add a new assumption to our model, namely, that the final stock price, S_T, has finite resolution. We thus assume that $S_T \in \mathcal{P} = \{j\Delta p : 0 \leq j \leq N\}$, where $\Delta p = M/N$, for some $N > 2$. This assumption mirrors the situation in reality.

It is a well-known result from finance, that exact pricing of European call options yields exact pricing for every path-independent derivative of a stock [4]. This result relies on the fact that every derivative is equivalent to a portfolio of call options with various strike prices.

Formally, for any $f: \mathcal{P} \to \mathbb{R}$, we define $f_{\Delta p}^{(1)}: \mathcal{P} \setminus \{M\} \to \mathbb{R}$ as $f_{\Delta p}^{(1)}(p) = \frac{f(p+\Delta p)-f(p)}{\Delta p}$ and $f_{\Delta p}^{(2)}: \mathcal{P} \setminus \{0, M\} \to \mathbb{R}$ as $f_{\Delta p}^{(2)}(p) = \frac{f(p+\Delta p)-2f(p)+f(p-\Delta p)}{\Delta p^2}$. The next lemma and theorem follow immediately from a result in [4], and essentially rephrase it:

Lemma 2. ([4]) *Define* $g_K: \mathcal{P} \to \mathbb{R}$ *as* $g_K(p) = \max(p-K, 0)$. *For every* $p \in \mathcal{P}$,

$$f(p) = f(0) + f_{\Delta p}^{(1)}(0) \cdot g_0(p) + \sum_{K=\Delta p}^{M-\Delta p} f_{\Delta p}^{(2)}(K) \cdot g_K(p) \cdot \Delta p .$$

Note that $g_K(p)$ is the payoff of $\mathbf{C}(K,T)$ and that, in addition, $C(0,T) = S_0$, because the payoff of $\mathbf{C}(0,T)$ is equivalent to a single stock. We thus get the following theorem:

Theorem 6. ([4]) *Let* **X** *be a path-independent derivative with payoff* $f\colon \mathcal{P} \to \mathbb{R}$. *The initial value of* **X** *is given by*

$$X = f(0) + f^{(1)}_{\Delta p}(0) \cdot S_0 + \sum_{K=\Delta p}^{M-\Delta p} f^{(2)}_{\Delta p}(K) \cdot C(K,T) \cdot \Delta p \;.$$

In the BSM model, where an exact pricing of call options is known, this amounts to pricing all path-independent derivatives exactly. In our model, however, where only *upper bounds* on call option prices are available, we cannot utilize this relation in every case. We must have that $f^{(2)}_{\Delta p}(K) \geq 0$ for every K in order to substitute those upper bounds for the terms $C(K,T)$. This requirement is fulfilled by convex derivatives.

Theorem 7. *Let* **X** *be a path-independent derivative with payoff* $f\colon \mathcal{P} \to \mathbb{R}$, *where* f *is the restriction to* $\mathcal{P}$ *of some convex function* $\bar{f}\colon [0,M] \to \mathbb{R}$. *Let* $C(K,T) \leq U(K)$ *for every* $K \in \mathcal{P}$, *where* $U\colon \mathcal{P} \to \mathbb{R}$. *Then*

$$X \leq f(0) + f^{(1)}_{\Delta p}(0) \cdot S_0 + \sum_{K=\Delta p}^{M-\Delta p} f^{(2)}_{\Delta p}(K) \cdot U(K) \cdot \Delta p \;.$$

Proof. By Theorem 6, it is enough to show that $f^{(2)}_{\Delta p}(K) \geq 0$ for every $K \in [\Delta p, M - \Delta p]$. By the convexity of $\bar{f}$, $f(K) = f(\frac{1}{2}(K - \Delta p) + \frac{1}{2}(K + \Delta p)) \leq \frac{1}{2}f(K - \Delta p) + \frac{1}{2}f(K + \Delta p)$, and thus $\frac{f(K+\Delta p)-2f(K)+f(K-\Delta p)}{\Delta p^2} \geq 0$. □

An Example: A Long Strangle Strategy

A long strangle investment strategy involves the purchase of a put option with a strike price of K_1 and a call option with a strike price of K_2, where $K_1 < K_2$. The payoff of this strategy is $\max(K_1 - S_T, 0) + \max(S_T - K_2, 0)$, which is a convex function of S_T. By Theorem 6, the value of a long strangle is $K_1 - S_0 + C(K_1, T) + C(K_2, T)$. Denoting $C_u(K, T)$ for the bound of Corollary 1, we can upper bound the price of a long strangle by $K_1 - S_0 + C_u(K_1, T) + C_u(K_2, T)$.

6   Empirical Results

In order to examine our results empirically, we consider the S&P 500 index data for the years 1950-2010. (The results are plotted in Figure 1.) We computed a price for a 1-year average strike call using our bound, with $R = 0.15$ and $Q = 0.1$. These R and Q values hold for all years but two in the test. In addition, for each year we computed the payoff of an average strike call option and also ran the Generic algorithm and computed the total profit. We used a single value of η, namely, the optimal value for the price bound. The stock prices for each year were normalized so that $S_0 = 1$ at the beginning of the year.

It is instructive to compare our upper bound on the option price, which was calculated to be 0.53, to the net payoff. The net payoff is the difference between

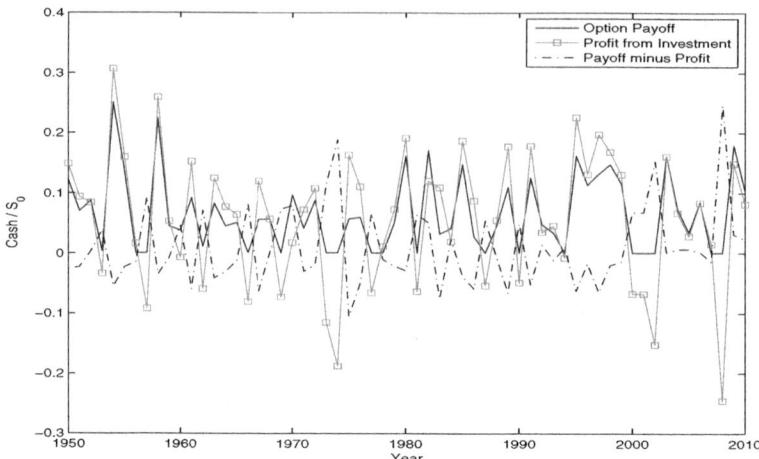

Fig. 1. A breakdown of the option writer's cash flow, for average strike call options, in terms of option payoff, option price, and total profit from the algorithm's trading. Data is calculated for 1-year average strike call options on the S&P 500 index for the years 1950-2010, with $R = 0.15$ and $Q = 0.1$. For the option writer to make a profit, the payoff minus profit line must be below the option price, which is 0.53 in this setting. Note that the "hindsight" empirical price is 0.25.

the payoff to the option holder (always non-negative) and the profit (or loss) the algorithm made in trading. It can be seen that our option price dominates the net payoff for every year, with the maximal net payoff at 0.25.

We point out that our results are influenced by the fact that we assume a zero risk-free interest rate, while in reality the interest rate can be substantial.

References

1. Black, F., Scholes, M.: The pricing of options and corporate liabilities. Journal of Political Economy 81(3), 637–654 (1973)
2. Blum, A., Kalai, A.: Universal portfolios with and without transaction costs. Machine Learning 35(3), 193–205 (1999); special issue for COLT 1997
3. Borodin, A., El-Yaniv, R., Gogan, V.: Can we learn to beat the best stock. Journal of Artificial Intelligence Research 21, 579–594 (2004)
4. Breeden, D.T., Litzenberger, R.H.: Prices of state-contingent claims implicit in option prices. Journal of Business 51(4), 621–651 (1978)
5. Cesa-Bianchi, N., Lugosi, G.: Prediction, Learning, and Games. Cambridge University Press, New York (2006)
6. Cesa-Bianchi, N., Mansour, Y., Stoltz, G.: Improved second-order bounds for prediction with expert advice. Machine Learning 66(2-3), 321–352 (2007)
7. Cover, T.M.: Universal portfolios. Mathematical Finance 1(1), 1–29 (1991)
8. Cover, T.M., Ordentlich, E.: Universal portfolios with side information. IEEE Transactions on Information Theory 42(2), 348–363 (1996)

9. DeMarzo, P., Kremer, I., Mansour, Y.: Online trading algorithms and robust option pricing. In: Proceedings of the Thirty-Eighth Annual ACM Symposium on Theory of Computing, pp. 477–486. ACM, New York (2006)
10. Györfi, L., Lugosi, G., Udina, F.: Nonparametric kernel-based sequential investment strategies. Mathematical Finance 16(2), 337–357 (2006)
11. Hazan, E., Kale, S.: On stochastic and worst-case models for investing. In: Bengio, Y., Schuurmans, D., Lafferty, J., Williams, C.K.I., Culotta, A. (eds.) Advances in Neural Information Processing Systems, vol. 22, pp. 709–717 (2009)
12. Helmbold, D.P., Schapire, R.E., Singer, Y., Warmuth, M.K.: On-line portfolio selection using multiplicative updates. In: Proc. 13th International Conference on Machine Learning, pp. 243–251. Morgan Kaufmann, San Francisco (1996)
13. Kalai, A., Vempala, S.: Efficient algorithms for universal portfolios. Journal of Machine Learning Research 3, 423–440 (2002)
14. Merton, R.C.: Theory of rational option pricing. Bell Journal of Economics and Management Science 4(1), 141–183 (1973)
15. Ordentlich, E., Cover, T.M.: The cost of achieving the best portfolio in hindsight. Mathematics of Operations Research 23(4), 960–982 (1998)
16. Singer, Y.: Switching portfolios. In: UAI, pp. 488–495 (1998)
17. Vovk, V., Watkins, C.: Universal portfolio selection. In: Proceedings of the Eleventh Annual Conference on Computational Learning Theory, pp. 12–23. ACM Press, New York (1998)

Strategic Pricing in Next-Hop Routing with Elastic Demands

Elliot Anshelevich[1], Ameya Hate[1], and Koushik Kar[2]

[1] Department of Computer Science, Rensselaer Polytechnic Institute, Troy, NY
[2] Department of Electrical, Computer & Systems Engineering, Rensselaer
Polytechnic Institute, Troy, NY
{eanshel,hatea}@cs.rpi.edu, koushik@ecse.rpi.edu

Abstract. We consider a model of next-hop routing by self-interested
agents. In this model, nodes in a graph (representing ISPs, Autonomous
Systems, etc.) make pricing decisions of how much to charge for forwarding traffic from each of their upstream neighbors, and routing decisions
of which downstream neighbors to forward traffic to (i.e., choosing the
next hop). Traffic originates at a subset of these nodes that derive a utility when the traffic is routed to its destination node; the traffic demand
is elastic and the utility derived from it can be different for different
source nodes. Our next-hop routing and pricing model is in sharp contrast with the more common source routing and pricing models, in which
the source of traffic determines the entire route from source to destination. For our model, we begin by showing sufficient conditions for prices
to result in a Nash equilibrium, and in fact give an efficient algorithm
to compute a Nash equilibrium which is as good as the centralized optimum, thus proving that the price of stability is 1. When only a single
source node exists, then the price of anarchy is 1 as well, as long as some
minor assumptions on player behavior is made. The above results hold
for arbitrary convex pricing functions, but with the assumption that the
utilities derived from getting traffic to its destination are linear. When
utilities can be non-linear functions, we show that Nash equilibrium may
not exist, even with simple discrete pricing models.

Keywords: Network Pricing, Selfish Routing, Elastic Demand, Price of
Stability, Nash Equilibrium.

1 Introduction

The ubiquitous impact of the Internet on modern life is a testimony to its growth
in the past two decades. One of the principal factors behind this growth has been
the decentralization of control, which also allows it to be modeled naturally as a
system of interacting but independent, self-interested agents. More specifically,
the Internet can be viewed as a collection of ISPs or ASes (Autonomous Systems)
that are interested in routing and pricing traffic to maximize their individual
revenues [3, 8, 16]. Similar frameworks have also been applied to the study of
relaying/routing of traffic in wireless ad-hoc networks [20, 21]. The study of large

G. Persiano (Ed.): SAGT 2011, LNCS 6982, pp. 278–289, 2011.
© Springer-Verlag Berlin Heidelberg 2011

decentralized networks of self-interested agents, with regard to their efficiency, has sparked an enormous amount of interest, as the insight thus earned can be used to extract maximum utility from existing infrastructure, as well as to make good policy decisions.

We consider the interactions of self-interested agents in a network at a very abstract level, where each agent is modeled as a self-interested node in a graph. Traffic originates at a subset of these nodes that derive a utility when the traffic is routed to its destination node, which may be many hops away. We consider next-hop routing, where each node on the path of the traffic individually determines which node(s) the traffic should be forwarded to (i.e., chooses the "next hop"). Nodes are allowed to *charge* their upstream neighbors for the traffic that they are asked to forward, as is typically done in contracts formed by neighboring Autonomous Systems in the Internet. Thus, a node obtains payments from its upstream neighbors for accepting their traffic for forwarding, and must in turn pay its downstream neighbors for receiving its traffic (i.e., the traffic it has accepted to forward, plus the traffic that it originates). Moreover, the nodes, being self-interested, will always choose to send traffic to the downstream neighbors with the cheapest price. As remarked in [4], for example, there is an interplay between setting the price to forward traffic, and choosing the routing policies of a node, since both decisions can change the profit/cost of a node (Autonomous System). In this paper, we consider both decisions to be under the control of each node, and study the properties of the equilibrium solutions of this game, which we believe captures the fundamental aspects of next-hop routing by self-interested agents.

Our next-hop routing and pricing model is in sharp contrast with the more common "source routing" and pricing models (see e.g., [7, 11, 15, 18]). In the latter models, the source node of the traffic determines its entire route from the source to the destination. Next-hop models provide a better representation of the routing protocols and pricing practices in the current Internet, as well as those that are likely to dominate the future multi-hop wireless networks [16]. In the Internet, traffic flow and service pricing negotiations occur at the inter-domain level, between an ISP and its neighboring ISPs (i.e., ISPs with which it shares a POP (Point-Of-Presence) and has a customer/provider/peering relationship) [12]. Inter-domain routing follows the BGP protocol, where hops at the AS level are determined one at a time [17]. Even though BGP determines this next hop based on information on the entire AS-level path, the benefits of making it strictly next-hop has been argued recently [19]. Our next-hop routing and pricing model also closely captures the Path Vector Contract Switching framework proposed for the future Internet [22], where neighboring ISPs establish contracts (on the amount of flow and its pricing) towards forwarding traffic for a specific destination. Source routing requires knowledge of the entire path at the source node, and this practical limitation has restricted the use of source routing in the Internet, while next-hop routing involves decision making by agents that is much more local and distributed.

In addition to its focus on next-hop routing, our model differs from most existing models in several other aspects as well. We assume that links in our network have fixed capacities, which represents the constraints associated with routing somewhat better than having linear cost functions that depend on how much traffic is being routed. An important feature of our model is the existence of multiple sources that have elastic demands with non-uniform utilities. See the Related Work section on further contrast with existing models.

Model Summary. We now give a brief outline of our model. A more detailed description is given in Section 2.

We are given a directed acyclic graph $G = (V, E)$ containing a special sink node t, and edge capacities c_e. As commonly done when analyzing competition in networks [5, 7, 16], we assume that all edges of this DAG, except the ones that are incident on the sink node, have a special non-monopolistic property. For our model, this property essentially ensures that enough capacity exists that no node could charge an infinitely high price for forwarding traffic, and yet have other nodes pay this price because they have no alternative.

The players of our game are all vertices of G except the sink node. An edge $e = (u, v) \in E$ with capacity c_e denotes that player u has the capacity to send a flow of size c_e to player v. Additionally, every player v has an associated *source utility* λ_v. This means that if a player v sends f_v amount of its own flow (flow originating form vertex v) to the sink t, then the player will obtain a utility of $\lambda_v f_v$. Thus, the player demands are *elastic*, since each player can choose an amount of traffic to send in order to maximize its utility. We consider an extension of this model in Section 5 where source utilities are allowed to be non-linear.

Players choose prices on their incoming edges. For every edge $e = (u, v)$ player v chooses a price p_e such that if u sends a flow of size f_e on edge e then u pays an amount $p_e f_e$ to v. Players route flow on outgoing edges such that this minimizes their cost, but are obligated to forward all flow that they receive. Finally, the utility of a player is the total amount of money it receives from upstream players and the utility obtained by sending its own flow minus the amount of money paid to the downstream nodes.

Example.   To illustrate some key consequences of our model, consider the example in Figure 1. The utility values of all nodes except the ones mentioned are 0. Edge (j, k), (k, l), (l, t) have capacity 2 and all other edges have capacity 1. Not all the edges in the graph are pictured: for every edge (u, v) shown in the figure that does not satisfy the non-monopolistic property, there also exist (non-pictured) edges (u, w), (w, t) of capacities 1 such that w has a high source utility (say 1000). In any optimal solution edge (u, w) will not have any flow on it whereas (w, t) will be saturated. Now it is not difficult to see that any optimal solution will consist of the flow indicated in Figure 1, where the double walled edges are saturated with flow, except edge (j, k) has a flow of size 1 whereas its capacity is 2.

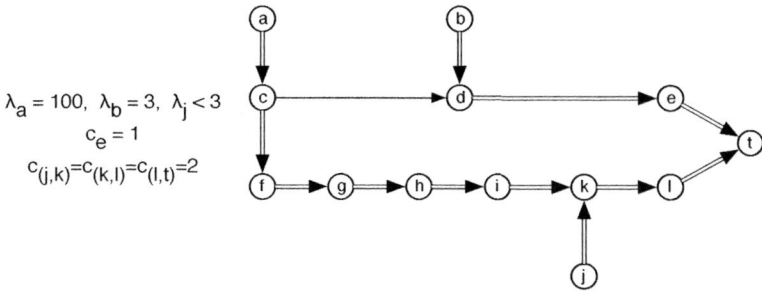

Fig. 1. An example illustrating our model, and the "long-range" effects that node utilities can have on edge prices

A more careful analysis shows that any Nash equilibrium strategy that yields this optimal flow will have the following edge prices: $p_{(j,k)} = p_{(k,l)} = \lambda_j$, $p_{(b,d)} = 3$. Also $p_{(c,d)} \geq p_{(c,f)} > p_{(f,g)} \geq p_{(g,h)} \geq p_{(h,i)} \geq p_{(i,k)} \geq \lambda_j$. This means that the price of edge (c, d) depends on the source utility value of node j, and thus there are "long-range" interactions between source utilities and edge prices.

Results. Our main goal involves understanding the properties of stable solutions in this pricing game: specifically we focus on pure Nash equilibrium. In Section 3 we give an efficient algorithm that constructs a Nash equilibrium strategy that is as good as the optimal solution. In other words, we show that the price of stability is 1, and thus it is always possible to implement traffic pricing that maximizes social welfare. This holds for an arbitrary number of sources with elastic demands of heterogenous value. We also show in Section 4 that in case of a single source, under some reasonable behavioral assumptions for the players, the price of anarchy is 1, and in fact player prices at equilibrium are unique.

Until this point, the source utilities λ_v and the allowed price functions p_e were considered to be linear. In Section 5, we instead consider the more general case where the source utility can be an arbitrary concave function $\Lambda_v(f_v)$, and the prices can be arbitrary convex functions $\Pi_e(f_e)$. We show that the above results still hold if arbitrary convex prices are allowed, and thus allowing non-linear prices does not impact the quality of equilibrium solutions. On the other hand, if source utilities can be non-linear functions, then we show that pure Nash equilibrium may no longer exist, even for discrete pricing models.

Related work. Selfish routing and pricing games have been studied in many contexts (see e.g., [2–5, 10, 11, 13, 14, 18], and the many references in [9, 15]). As mentioned before, most of the work in this area has been done using source routing, where a source of traffic chooses the entire path that the traffic takes. In one such model, [6, 7] consider a game where there are two sets of players. One set of players own edges of the network (edges have finite capacities) and sell capacity to players of the second set. The second set of players obtain utility for routing a unit amount of flow from its source to destination and hence buys capacity on edges along the route if it is profitable to do so. The essential difference between source routing models like the one in [6, 7] and ours is that in our model when a node changes its price on an incoming edge, only its upstream neighbors are immediately affected and may change their routes; whereas in the source routing model, any change in prices by the first set of players is seen by all players of the second set, and can immediately result in a globally different routing. Thus next hop routing operates much more on local knowledge [19].

Perhaps the most relevant paper to ours is by Papadimitriou and Valiant [16], in which they define a next hop routing model where players are edges of a network, and argue for the importance of next-hop routing models. The strategy of players in their game is similar to our model: players charge neighbors for processing and forwarding flow. Unlike our model where edges have capacities, their model has edges with (linear) latencies, and deals only with a single source with a fixed demand. Although their results can most likely be extended to networks with multiple sources, the crucial complications in our model arise from the fact that we consider sources with elastic demands and non-uniform utilities. This leads to complex interaction of prices as illustrated by the example in Figure 1, and prevents us from using the methods from [16] for analysis. To further illustrate the differences between our model and the one from [16], notice that in our model, the price of anarchy for single source games is 1, while in [16], the price of anarchy can be large.

Another recent next-hop routing model is discussed by Xi and Yeh [21]. As in [16], their model only considers a single source with a fixed amount of traffic demand. The links in [21] have latency functions instead of capacities. For these reasons, just as with [16], equilibrium solutions in [21] are very different from the ones in our model, and have a very different structure. In essence, the complexity in our game arises from the interplay between different source utilities and edge capacities, while in [16] and [21] it arises due to the presence of latency functions. Finally, [4] considers a somewhat general routing game that can also include next-hop routing as a special case, but uses a very different pricing mechanism from the one considered here.

2   Model

We are given a directed acyclic graph $G = (V, E)$ containing a special sink node t, and edge capacities c_e. To arrive at a meaningful model of price competition, we assume that all edges of this DAG, except the ones that are incident on the sink node, have a special non-monopolistic property. This property says that for any edge $e = (u, v)$, even if this edge is removed from the graph, the total capacity of outgoing edges of node u will be greater than the total capacity of its incoming edges. The rationale behind this property will be explained once the model has been illustrated in more detail.

The set of players in our game consists of all vertices of G, except the sink node. An edge $e = (u, v) \in E$ with capacity c_e denotes that player u has the capacity to send a flow of size c_e to player v. Additionally, every player v has an associated *source utility* λ_v. This means that if a player v sends f_v amount of its own flow (flow originating form vertex v) to the sink t, then the player will obtain a utility of $\lambda_v \cdot f_v$. We consider an extension of this model in Section 5 where source utilities are allowed to be non-linear.

Player Strategy. The core behavior of our model is that a vertex can charge a price for the flow being sent through it. For example, since edge $e = (u, v)$ is incident on v, node v can set a price p_e on this edge. If vertex u sends a flow f_e on this edge then vertex u has to pay an amount equalling $f_e \cdot p_e$ to vertex v. (We will consider a model where the price per packet changes with the amount of flow being sent on the edge in Section 5.)

So apart from gaining utility by sending their own flow, vertices also gain utility for receiving flows and lose utility by paying the next-hop vertices that receive their flow. Also, observe that for the flow to reach the sink, it is essential that intermediate nodes forward the flow reliably. Hence every player is required to forward all incoming flow (alternatively, we can think of there being a very large penalty for accepting payment for incoming flow that the player has no intention of forwarding). The forwarding of all flow is always possible since the non-monopolistic property ensures that for every node the total outgoing capacity is always greater that the total incoming capacity.

Imagine a situation where in order to route all its incoming flow, node u always has to forward some flow to vertex v. Since it is obligatory for u to forward all its incoming flow, v can charge an exorbitantly high price on edge (u, v), and u would have to pay it. In other words, v can act as a monopoly. Similarly to [16], existence of such a structure in G may lead to no meaningful equilibrium and the non-monopolistic property for edge (u, v) obviates precisely this situation. Since the sink node is not a player, it does not set prices for edges that are incident on it (we assume that price is fixed at 0, and that wlog these edges are always saturated); neither do these edges need to satisfy the non-monopolistic property, as long as the outgoing capacity of each node is at least as large as the incoming capacity.

In order to provide a more formal definition of the player strategies and the resulting flow, we define the following terminology. Let E_v^{in} and E_v^{out} be the set of incoming and outgoing edges for node v respectively. The vector of flows on the incoming edges of vertex v is denoted by f_v^{in} and on the outgoing edges is denoted by f_v^{out}. Similarly, the vector of prices on the incoming edges is denoted by p_v^{in} and on the outgoing edges is denoted by p_v^{out}. Let f_v be the amount of own flow (flow originating at vertex v) sent by vertex v to sink t.

We assume that a vertex always sends or forwards flow by choosing the outgoing edges that have the lowest price and have free capacity. Also, if there exists an outgoing edge with free capacity and has $p_e < \lambda_v$ then the node will always send its own flow on such an edge, and will never send its own flow on edges with $p_e > \lambda_v$. To make this precise, we define a notion of valid flows, which are flows where every vertex forwards flow in order to maximize its utility. Specifically, given the prices p_v^{out} and flows f_v^{in}, we define the set of valid resulting flows f_v^{out} to be $\mathcal{F}_v(f_v^{in}, p_v^{out})$, which are all flows satisfying the following conditions:

- $\forall e \in E_v^{out} : f_e \le c_e$ (usual capacity constraint);
  $f_v = \sum_{e \in E_v^{in}} f_e - \sum_{e \in E_v^{out}} f_e \ge 0$ (usual flow conservation)
- $\forall e \in E_v^{out} : f_e > 0$ only if for every $e' \in E_v^{out} \backslash e$ with $p_{e'} < p_e$, e' is saturated (send on cheapest edges first),
- $\forall e \in E_v^{out} : p_e < \lambda_v$ implies that e is saturated (send own flow if profitable), and
  $p_e > \lambda_v$ and $f_e > 0$ imply that $f_v = 0$ (don't send own flow if unprofitable).

Any way of forwarding flow to maximize v's utility obeys these conditions. The last condition holds since if v is sending its own flow, but $f_e > 0$ for some edge $e \in E_v^{out}$ with $p_e > \lambda_v$, then v could re-distribute its flow so that it is sending its own flow on edge e, and then improve its utility by sending less of its own flow.

When all prices in p_v^{out} are distinct from each other and λ_v, then $\mathcal{F}_v(f_v^{in}, p_v^{out})$ is a unique flow resulting from forwarding flow from all from all incoming edges, beginning with the edges of least cost, and then sending its own flow on remaining edges with free capacity and cost $p_e < \lambda_v$. Now consider instead a situation where two or more outgoing edges have the same price. So long as they have the same price, the utility of player v is not affected by the choice of edge on which it sends a flow. Similarly, when there exists an outgoing edge with free capacity and $p_e = \lambda_v$, the vertex is indifferent towards the choice of sending its own flow on the edge. In this model we assume that both these tie-breaking choices are left up to the players and are part of their strategy. More formally, since each valid flow in $\mathcal{F}_v(f_v^{in}, p_v^{out})$ corresponds to a tie-breaking rule selected by player v, we associate these tie-breaking rules with a flow generation function γ_v which, given the incoming flows and out going prices, produces an outgoing flow. The set of these flow generation functions is denoted by Γ_v:

$$\Gamma_v = \{\gamma_v | \forall f_v^{in}, p_v^{out} : \gamma_v(f_v^{in}, p_v^{out}) = f_v^{out} \in \mathcal{F}_v(f_v^{in}, p_v^{out})\}$$

In other words, Γ_v contains all functions that generate only valid out-flows. Hence the strategy set of each player v is $\mathcal{R}_+^{|E_v^{in}|} \times \Gamma_v$, and a strategy of the player is given by the tuple $\{p_v^{in}, \gamma_v\}$ where $p_v^{in} \in \mathcal{R}_+^{|E_v^{in}|}$ and $\gamma_v \in \Gamma_v$. We denote the collective strategy of all players by $\{P, \gamma\}$.

Outcome. Each flow generating function γ_v needs incoming flow and prices on outgoing edges in order to compute the resulting flow. Given a strategy $\{P, \gamma\}$, the prices are already known. The algorithm to produce the resulting flow is then simply: Iterate over $v \in V$ in topologically sorted order (recall that our graph is a DAG), and set $f_v^{out} = \gamma_v(f_v^{in}, p_v^{out})$. We denote the resulting flow by $f(P, \gamma)$: this is the outcome of the strategy $\{P, \gamma\}$.

Utility and Best Response. Given the output flow $f(P, \gamma)$, the utility of player v is given by the following expression:

$$utility_v(P, \gamma) = \sum_{e \in E_v^{in}} f_e \cdot p_e - \sum_{e \in E_v^{out}} f_e \cdot p_e + \lambda_v \cdot f_v$$

Consider a player v who is computing its best response to a strategy $\{P, \gamma\}$. Notice that by changing its prices p_v^{in}, the resulting flow f_v^{in} may become completely different from $f(P, \gamma)$. If this were not the case, then v could always raise its incoming price, knowing that this would increase its utility since the flow would remain the same. In essence, players in this game anticipate changes in flow that result from price changes, but myopically assume that the prices of all other nodes remain the same when computing their own best response. Such behavior is reasonable in ISP routing settings, for example, since price setting takes place on a much slower time scale than routing.

3   Uniform Nash Equilibrium and Price of Stability

We first show useful sufficient conditions for a strategy to be a Nash equilibrium.

Theorem 1. *If flow $f(P, \gamma)$ and prices P satisfy the following conditions, then strategy $\{P, \gamma\}$ is a Nash equilibrium:*

(a) *For every node u, the price on all edges of E_u^{out}, except edge (u, t) if it exists, is the same. Let this price be denoted by y_u.*
(b) *If $f_u > 0$ then $y_u = \lambda_u$; if $f_u = 0$ then $y_u \geq \lambda_u$.*
(c) *For $\forall v \neq t$, if edge (u, v) has a positive flow on it, then $y_u \geq y_v$.*
(d) *For $\forall v \neq t$, if edge $e = (u, v)$ is unsaturated ($f_e < c_e$), then $y_u \leq y_v$.*

Proof of the theorem can be found in full version of the paper [1].

Theorem 1 gives sufficient conditions for a strategy to be a Nash equilibrium. We will call such strategies *uniform*, since all the outgoing prices are the same for every node in such a solution. As we will show below, good uniform Nash equilibria always exist, and can be efficiently computed.

Definition 1. *Uniform Nash equilibrium: Any Nash equilibrium strategy that satisfies the conditions of Theorem 1 is a **uniform** Nash equilibrium.*

3.1   Computing a Good Nash Equilibrium

By an optimal solution to this game, we will mean one in which the sum of the utilities of all players is maximized. Since the price paid by players to each other cancels out in the sum, optimal solutions are ones in which $\sum_v \lambda_v f_v$ is maximized. We will call a flow f^* *socially optimal* if $\sum_v \lambda_v f_v^*$ is maximum over all flows that obey capacity constraints and where f_v^* flow originates at node v, with all flow ending at the sink t (it is easy to see that, without loss of generality, all edges incident on t are saturated). Clearly, $\sum_v \lambda_v f_v^*$ is the social welfare in an optimal solution, since if all prices are set to 0, and γ is such that f^* is the resulting flow, then this results in social welfare of $\sum_v \lambda_v f_v^*$. The following theorem states that flow f^* can also be achieved by a Nash equilibrium solution, i.e., that the price of stability of this game is 1.

Theorem 2. *Given a socially optimal flow f^*, there exists a collective strategy $\{P, \gamma\}$ such that $f(P, \gamma) = f^*$ and $\{P, \gamma\}$ is a uniform Nash Equilibrium. In other words, the price of stability is 1.*

Full proof of this theorem appears in the full version of the paper [1]. However, a gist of the proof is given here.

Given an optimal solution f^*, we give an algorithm that assigns prices to edges. The assignment of prices is such that for a given vertex, the price of all outgoing edges is the same. The price of outgoing edges of node v is set to the smallest λ_w with $f_w^* > 0$ that is reachable from v in the residual flow graph for optimal flow f^*. Finally we show that such an assignment of prices satisfies all conditions of Theorem 1.

4   Reasonable Assumptions and Price of Anarchy

Consider a game where all players with non-zero λ_v value are not neighbors of the sink. Now consider a strategy for this game where every vertex charges a very high price (say bigger than the highest λ_v value). Given this pricing strategy, no vertex will send its own flow and still every vertex will have no incentive to deviate, i.e., the strategy will be in Nash equilibrium. This is because no vertex would unilaterally reduce the prices of its incoming edges, given that they will have to pay a large amount to forward any flow sent to them. Nodes that have edges incident to the sink will not change their prices as there is no hope of obtaining any flow and hence, any profit. In this Nash equilibrium strategy the total utility of players is 0 and hence the price of anarchy is unbounded. These "bad equilibria" cannot be eliminated even after introducing pairwise deviations.

In order to eliminate such unrealistic solutions from consideration, work dealing with similar scenarios made some reasonable assumptions about player behavior. For example, [16] assumes that if a player does not receive any flow on its incoming edge, then she never charge an unnecessarily large price for this edge. In this section, we make the same assumption on the players' pricing strategy:

Property 1. If a vertex v does not receive any flow on edge (u, v), then it sets $p_{(u,v)}$ to be the price of the cheapest unsaturated outgoing edge of v, if one exists.

We call pricing strategies that satisfy this property *reasonable*. This property simply says that given an edge (u, v) that has no flow on it, node v will charge the minimum price such that potential flow on this edge will not result in loss of utility for v. Below we show that, at least for single-source games (i.e., games where only one node has a non-zero λ value), this additional assumption on player behavior causes all equilibria to become as good as the optimum solution. Proof of the following theorem appears in full version of the paper [1].

Theorem 3. *For a single source game where players form reasonable pricing strategies, the price of anarchy is 1.*

5   Non-linear Utility and Price Functions

In previous sections we analyzed the case where the utility of sending one unit of own flow (will also be referred to as 'per packet') was constant for the player. We will now study the case where the utility is a concave function of the amount of flow sent. This mirrors the fact that sending more flow usually has diminishing returns for the player. We denote this utility function as $\Lambda_v(f_v)$ where f_v is the total amount of own flow sent by node v and Λ_v is continuously differentiable, concave, and non-decreasing. Additionally, denote the derivative of Λ_v by λ_v: in the old model this was a constant, but now it is a non-increasing function.

Similarly when players receive flow on an incoming edge, the processing cost for each unit of flow generally increases with the total amount of flow. Hence we look at the case where the price charged by each player for incoming flow is a convex function. We denote this by $\Pi_e(f_e)$ where f_e is the flow on edge e and Π_e is a continuously differentiable convex non-decreasing function. Let π_e be the derivative of Π_e: in the old model this was called p_e and was a constant; now it is a non-decreasing function.

This more general model is formally defined in full version of the paper [1]: it is a strict generalization of the model in Section 2.

If utility functions Λ_v are linear, then we show that all our results for linear price functions also hold for arbitrary convex price functions, thus showing that allowing players to set non-linear prices does not make the system any worse. To do this, we prove analogues of Theorems 1 and 3. By proving that the conditions from Theorem 1 imply that a solution is a Nash equilibrium, even when changing your strategy to an arbitrary price function is allowed, we immediately get the consequence that the price of stability is 1, since we already showed how to create an optimal solution satisfying these conditions in Theorem 2. The proofs of all following theorems can be found in full version of the paper [1].

Theorem 4. *For instances with linear utility functions and non-decreasing, convex price functions, the price of stability is 1.*

In section 4 we showed that when prices have to be linear, the price of anarchy is 1 for networks with a single source under the mild assumption that players do not set large prices without a good reason (Property 1). We also show that the same result holds if prices are allowed to be convex non-decreasing functions. Note that utilities are still linear. Since edge prices are allowed to be functions, we call pricing strategies that satisfy the following property as reasonable:

Property 2. If a vertex v does not receive any flow on edge (u, v), then it sets $\Pi_{(u,v)}(x) = p_{(u,v)}x$ where $p_{(u,v)}$ is the cheapest marginal price of all unsaturated outgoing edges of v, if one exists.

Proof of the following theorem can be found in full version of the paper [1].

Theorem 5. *If node utilities are linear and edge prices are allowed to be convex non-decreasing functions then, for a single source game where players form reasonable pricing strategies, the price of anarchy is 1.*

If utility functions Λ_v can be non-linear, then we give an example (Figure 2) which does not admit a pure Nash equilibrium. We show that in any Nash equilibrium strategy for example in Figure 2, there exists another equilibrium strategy where the price functions for edges (c, a) and (c, b) are linear and the edges will have the same price. We also show that in any Nash equilibrium strategy, edges (c, a) and (c, b) will be saturated. These observations cannot be satisfied simultaneously given the utility function Λ_v and consequently, a pure Nash equilibrium strategy does not exist.

Again the details can found in full version of the paper [1].

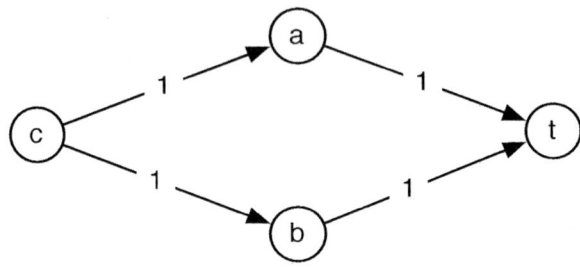

Fig. 2. $\Lambda_c(f) = -9f^2 + 37f$ for $f \le 2$. $\Lambda_a = \Lambda_b = 0$. All edges have capacity 1

We also show that this example does not admit a Nash equilibrium even in the (nicer) case when all prices must be discrete, and thus the non-existence of equilibrium does not stem from the fact that prices can be changed by an infinitesimal amount.

Theorem 6. *If the player utilities Λ_v are concave non-decreasing functions, then pure Nash equilibrium may not exist.*

References

1. Full version can be found at, http://www.cs.rpi.edu/~eanshel/pubs.html
2. Acemoglu, D., Johari, R., Ozdaglar, A.E.: Partially optimal routing. IEEE Journal on Selected Areas in Communications 25(6), 1148–1160 (2007)
3. Anshelevich, E., Shepherd, B., Wilfong, G.: Strategic Network Formation through Peering and Service Agreements. Games and Economic Behavior (2011), doi:10.1016/j.geb, 01.002
4. Anshelevich, E., Wilfong, G.: Network Formation and Routing by Strategic Agents using Local Contracts. In: Papadimitriou, C., Zhang, S. (eds.) WINE 2008. LNCS, vol. 5385, pp. 386–393. Springer, Heidelberg (2008)
5. Archer, A., Tardos, É.: Frugal path mechanisms. ACM Transactions on Algorithms, TALG (2007)
6. Chawla, S., Niu, F.: The Price of Anarchy in Bertrand Games. In: EC 2009 (2009)
7. Chawla, S., Roughgarden, T.: Bertrand Competition in Networks. In: Monien, B., Schroeder, U.-P. (eds.) SAGT 2008. LNCS, vol. 4997, pp. 70–82. Springer, Heidelberg (2008)

8. Feigenbaum, J., Papadimitriou, C.H., Sami, R., Shenker, S.: A BGP-based mechanism for lowest-cost routing. Distributed Computing 18(1), 61–72 (2005)
9. Feigenbaum, J., Schapira, M., Shenker, S.: Distributed Algorithmic Mechanism Design. In: Nisan, N., Roughgarden, T., Tardos, É., Vazirani, V.V. (eds.) Algorithmic Game Theory, ch. 14, Cambridge University Press, Cambridge
10. Hall, A., Nikolova, E., Papadimitriou, C.: Incentive-Compatible Interdomain Routing with Linear Utilities. Internet Mathematics 5(4), 395–410 (2008) (Special Issue for Selected papers from WINE 2007)
11. Hayrapetyan, A., Tardos, É., Wexler, T.: A network pricing game for selfish traffic. In: Distributed Computing (March 2007)
12. Huston, G.: Interconnection, peering, and settlements. In: Proceedings of the Internet Global Summit. The Internet Society, San Jose (1999)
13. Johari, R., Mannor, S., Tsitsiklis, J.N.: A contract-based model for directed network formation. Games and Economic Behavior 56(2), 201–224 (2005)
14. Levin, H., Schapira, M., Zohar, A.: Interdomain Routing and Games. SIAM Journal on Computing (SICOMP); Special Issue on Selected Papers from STOC 2008
15. Ozdaglar, A., Srikant, R.: Incentives and Pricing in Communication Networks. In: Nisan, N., Roughgarden, T., Tardos, É., Vazirani, V.V. (eds.) Algorithmic Game Theory, ch. 22, Cambridge University Press, Cambridge
16. Papadimitriou, C., Valiant, G.: A New Look at Selfish Routing. Innovations in Computer Science (2010)
17. Rekhter, Y., Li, T.: A Border Gateway Protocol 4 (BGP-4). RFC 4271 (January 2006)
18. Roughgarden, T., Tardos, É.: How Bad is Selfish Routing? Journal of the ACM (2002)
19. Schapira, M., Zhu, Y., Rexford, J.: Putting BGP on the right path: A case for next-hop routing. In: Proceedings of HotNets-IX, Monterey, CA (October 2010)
20. Srinivasan, V., Nuggehalli, P., Chiasserini, C.F., Rao, R.R.: An Analytical Approach to the Study of Cooperation in Wireless Ad Hoc Networks. IEEE Transactions on Wireless Communications 4(2), 722–733 (2005)
21. Xi, Y., Yeh, E.M.: Pricing, competition, and routing in relay networks. In: Proceedings of the 47th Annual Allerton Conference on Communication, Control, and Computing, Allerton 2009 (2009)
22. Yuksel, M., Gupta, A., Kar, K., Kalyanaraman, S.: Contract-Switching for Managing Inter-Domain Dynamics. In: Ramamurthy, B., Rouskas, G.N., Sivalingam, K.M. (eds.) Next-Generation Internet Architectures and Protocols, pp. 136–153. Cambridge University Press, Cambridge (2010)

Weakly-Acyclic (Internet) Routing Games

Roee Engelberg[1,*] and Michael Schapira[2]

[1] Computer Science Department, Technion, Haifa 32000, Israel
`roee@cs.technion.ac.il`
[2] Department of Computer Science, Princeton University, NJ, USA
`ms7@cs.princeton.edu`

Abstract. Weakly-acyclic games – a superclass of potential games – capture distributed environments where simple, globally-asynchronous interactions between strategic agents are guaranteed to converge to an equilibrium. We explore the class of routing games in [4, 12], which models important aspects of routing on the Internet. We show that, in interesting contexts, such routing games are weakly acyclic and, moreover, that pure Nash equilibria in such games can be found in a computationally efficient manner.

Keywords: Weakly-acyclic games, routing games, convergence to Nash equilibrium, best-response dynamics.

1 Introduction

1.1 Weakly-Acyclic Games

Convergence to a pure Nash equilibrium (PNE) is an important objective in a large variety of application domains – both computerized and economic. Ideally, this can be achieved via simple and natural dynamics, e.g., better-response or best-response dynamics. Under better-response dynamics, players start at some initial strategy profile and take turns selecting strategies. At each (discrete) time step, a single player selects a strategy that *increases* his utility (given the others' current strategies). Under best-response dynamics, at every time step the "active" player chooses a strategy that *maximizes* his utility. Better-response and best-response dynamics are simple, low-cost behaviors to build into distributed systems, as evidenced by today's protocol for routing on the Internet [4, 12].

Convergence of better-/best-response dynamics to PNE is the subject of much research in game theory. Clearly, a *necessary* condition for better-/best-response dynamics to converge to a PNE regardless of the initial state is that, for every such state, there exist *some* better-/best-response improvement path to a PNE, i.e., a sequence of players' better-/best-response strategies which lead to a PNE.[1] Games for which this holds (e.g., potential games [15]) are called "*weakly*

---

* Current affiliation: Google Inc. This work was done while the author was a doctoral student at the Technion.
[1] Observe that this is equivalent to requiring that the game have no "non-trivial" sink equilibria [8, 4] under better-response dynamics (i.e., that it have no sink equilibrium of size greater than 1).

G. Persiano (Ed.): SAGT 2011, LNCS 6982, pp. 290–301, 2011.

acyclic" [18, 14]. Weak acyclicity has also been shown to imply that simple dynamics (e.g., *randomized* better-/best-response dynamics, no-regret dynamics) are guaranteed to reach a PNE [13, 14, 18]. Thus, weak acyclicity captures distributed environments where a PNE can be reached via simple, globally-asynchronous interactions between strategic agents, regardless of the starting state of the system.

While the class of potential games – a subclass of weakly-acyclic games – is the subject of extensive research, relatively little attention has been given to the much broader class of weakly-acyclic games (see, e.g., [3, 13]). As a result, very few concrete examples of weakly-acyclic games that do not fall in the category of potential games are known. One famous result along these lines is that of Milchtaich [14]. [14] studies Rosenthal's congestion games [17] and proves that, in interesting cases where the payoff functions (utilities) are player-specific, such games are weakly acyclic (but not necessarily potential games).

Our focus in this work is on another extensively studied environment: routing on the Internet. We show that weak acyclicity is important for analyzing such environments. Our work, alongside its implication for Internet routing, provides concrete examples of weakly acyclic games that lie beyond the space of potential games, as well as technical insights into the structure of such games.

1.2   (Internet) Routing Games

The Border Gateway Protocol (BGP) establishes routes between the smaller, independently administered, often competing networks that make up the Internet. Hence, BGP can be regarded as the glue that holds today's Internet together. Over the past decade there has been extensive research on the computational and strategic facets of routing with BGP. Recent advances along these lines were obtained via game-theoretic analyses (see, e.g., [4, 9, 12, 16]), which rely on the simple, yet important, observation that BGP can be regarded as best-response dynamics in a specific class of routing games [4, 12]. We now provide an intuitive exposition of the class of routing games in [4, 12]. We refer the reader to Sect. 2 for a formal presentation.

In the game-theoretic framework of [4, 12], the players are source nodes residing on a network graph, which aim to send traffic to a unique destination in the network. Each source node has a (private) ranking of all simple (loop-free) routes between itself and the destination. We stress that, in practice, different source nodes can have very different, often conflicting, rankings of routes, reflecting, e.g., local business interests [7] (in particular, source nodes do not always prefer shorter routes to longer ones). Every source node's strategy space is the set of its neighboring nodes in the network; a choice of strategy represents a choice of a single neighbor to forward traffic to. Observe that every combination of source nodes' strategies thus captures how traffic is forwarded (hop-by-hop) towards the destination. A source node's utility from every such combination of strategies reflects how highly its ranks its induced route to the destination.

Fabrikant and Papadimitriou [4] and, independently, Levin et al. [12], observed that BGP can be regarded as best-response dynamics in this class of routing games and that PNEs in such games translate to the notion of stable routing states, which has been extensively studied in communication networks literature. These observations laid the foundations for recent results regarding the dynamics and incentive compatibility of routing on the Internet (see [9, 11, 16]).

1.3   Our Contributions: Weakly-Acyclic Routing Games

We present two interesting subclasses of routing games – (1) routing games with Byzantine players and (2) backup routing games – which capture important aspects of routing on the Internet. Routing games with Byzantine players intuitively capture scenarios where all but a few players are "well behaved", and the remaining players behave in an arbitrary manner. Such erratic "Byzantine" (in distributed computing terminology) misbehavior can, for instance, be the consequence of router configuration errors. Backup routing games model the common practice of backup routing with BGP [6].

We prove that games in both these classes are weakly acyclic, even under best-response (i.e., from every initial state there exists a best-response improvement path to a PNE). Our results thus establish that, in these two contexts, a PNE is guaranteed to exist and can be reached via simple, globally-asynchronous interactions between strategic agents regardless of the initial state of the system. Moreover, we prove that not only is a PNE reachable from every initial state via a best-response improvement path, but that this path is "short" (of polynomial length). Hence, in these subclasses of routing games, a PNE can be found in a computationally-efficient manner; simply start at an arbitrary initial state and follow the short best-response improvement path – whose construction we give explicitly – until a PNE is reached.

Routing games with Byzantine players. To illustrate this subclass of games, consider the scenario that all source nodes but a single source node m have a "shortest-path ranking" of routes, i.e., they always prefer shorter routes to longer routes. Unlike the other source nodes, m's ranking of routes need not necessarily be a shortest-path ranking and is not restricted in any way, e.g., m might even always prioritize longer routes over shorter routes. We aim to answer the following question: "Can m's erratic behavior render the network unstable?".

We prove a surprising positive result: every routing game of the above form (i.e., with a single "misbehaving" source node) is weakly acyclic under best-response. Hence, in particular, routing games where each player has a shortest-path ranking are guaranteed to posses a PNE even in the presence of an arbitrary change in a single source node's behavior! We generalize this result to a broader class of routing policies. We point out that our work is one of few to explore the impact of "irrational" behavior in game-theoretic settings (see [1, 2, 10]).

Backup routing games. In this subclass of routing games each edge in the network graph is either categorized as a "primary" edge or as a "backup" edge. A source node with multiple outgoing edges prefers forwarding traffic to neighboring nodes to which it is connected via primary edges over forwarding traffic to neighbors to which it is connected via backup edges. Such "backup relationships" are often established in practice to provide connectivity in the event of network failures via redundancy; the intent is that backup edges be used for carrying traffic only in case of failures in the primary edges [6]. We consider natural restrictions on source nodes' routing policies which capture this notion of "backup routing". We prove that the resulting routing games are weakly acyclic.

1.4   Organization

We present the class of weakly-acyclic games and the class of weakly-acyclic under best-response games in Sect. 2, where we also present the class of routing games of [4, 12]. In Sect. 3, we illustrate the type of results we obtain via two simple families of weakly-acyclic routing games. We present our results for routing games with Byzantine players, and for backup-routing games, in Sects. 4 and 5, respectively. Due to space constraints, all proofs appear in the full version of the paper.

2   Model

2.1   Weakly-Acyclic Games

We use standard game-theoretic notation. Consider a normal-form game with n players $1, \ldots, n$, where each player i has strategy space S_i and utility function u_i (which specifies player i's utility for every combination of players' strategies). Let $S = S_1 \times \ldots \times S_n$ and $S_{-i} = S_1 \times \ldots \times S_{i-1} \times S_{i+1} \times \ldots \times S_n$. For every $s_i \in S_i$ and $s_{-i} \in S_{-i}$, (s_i, s_{-i}) denotes the combination of players' strategies where player i's strategy is s_i and the other players' strategies are as in s_{-i}.

Definition 2.1. (better-response strategies) *We call a strategy* $s_i^* \in S_i$ *a "better-response" of player i to a strategy vector* $s = (s_i, s_{-i}) \in S$ *if* $u_i(s_i^*, s_{-i}) > u_i(s_i, s_{-i})$.

Definition 2.2. (best-response strategies) *We call a strategy* $s_i^* \in S_i$ *a "best response" of player i to a combination of other players' strategies* $s_{-i} \in S_{-i}$ *if* $s_i^* \in argmax_{s_i \in S_i} u_i(s_i, s_{-i})$.

Definition 2.3. (pure Nash equilibria) *A strategy vector* $s = (s_1, \ldots, s_n) \in S$ *is a* pure Nash equilibrium (PNE) *if s_i is a best response to s_{-i} for every player i.*

Definition 2.4. (better- and best-response improvement paths) *A* better-response (best-response) improvement path *in a game Γ is a sequence of strategy vectors* $s^{(1)}, \ldots, s^{(k)} \in S$, *each reachable from the previous via a better response (best response) of a single player.*

We are now ready to present the class of weakly-acyclic games and the class of weakly-acyclic under best-response games.

Definition 2.5. (weak acyclicity and weak acyclicity under best response) *A game Γ is* weakly acyclic (weakly acyclic under best response) *if, from every $s \in S$ there exists a better-response (best-response) improvement path to a pure Nash equilibrium of Γ.*

2.2 (Internet) Routing Games

In the class of routing games in [4, 12] the players are n source nodes $1, \ldots, n$ residing on a network $G = (V, E)$ who wish to send traffic to a *unique* destination node d. Let P_i be the set consisting of all simple (loop-free) routes from source node i to d in G and of the "empty route" $\bot$. Each source node i's strategy is its choice of an outgoing edge $e_i \in E(G)$ (intuitively, a neighboring node to forward traffic to), or the empty set $\emptyset$ (intuitively, not forwarding traffic). Observe that every combination of source nodes' strategies $s \in S$ thus specifies a (directed) subgraph G_s of G in which each source node has outdegree at most 1. Given a combination of nodes' strategies $s \in S$ we define i's *induced route* R_i^s to be i's unique simple route to d in G_s if such a route exists, and $\bot$ otherwise.

We now define source nodes' utility functions. Each source node i has a *routing policy* with two components: (1) a *ranking function* π_i that maps elements in P_i to the integers, such that $\pi_i(\bot) < \pi_i(R)$ for all $R \in P_i \setminus \{\bot\}$; and (2) an *export policy* that, for each neighboring node $j \in V(G)$, specifies a set of routes $R_{ij} \subseteq P_i$ that i is willing to make available to j. To simplify notation, when $\pi_i(R) < \pi_i(Q)$ ($\pi_i(R) \leq \pi_i(Q)$) for some routes $R, Q \in P_i$, we write $R <_i Q$ ($R \leq_i Q$). We say that a route $R_i \in P_i$ is "*permitted*" if each node on R is willing to export its (sub)route to its predecessor on R. Given a combination of nodes' strategies $s \in S$, i's utility is $\pi(R_i^s)$ if R_i^s is permitted and 0 otherwise.

3   Illustration: Simple Weakly-Acyclic Routing Games

We now illustrate the kind of results we obtain via two simple families of weakly-acyclic games.

Shortest-path routing with Byzantine players. Consider the scenario that all source nodes have shortest-path rankings (where shorter routes are always preferred to longer ones) and "export-all policies", i.e., each node i is willing to make all routes in P_i available to all neighboring nodes. We call games of this form "shortest-path routing games". We make the simple observation that shortest-path routing games are potential games. Now, consider the case that there is a single Byzantine player, i.e., that the routing policy of a single source node in a shortest-path routing game is changed arbitrarily. We now present the following corollary of a more general result proved in Sect. 4.

Corollary 3.1. *Every shortest-path routing game with a single Byzantine player is weakly-acyclic under best response and, moreover, a PNE in such a game can be found in a computationally-efficient manner.*

In contrast, we show (Appendix A) that shortest-path routing games with a single Byzantine player are no longer necessarily potential games.

Theorem 3.1. *There exists a shortest-path routing game with a single Byzantine player which is not a potential game.*

Shortest-path backup routing. In this setting, every edge in the network graph is either "primary" or "backup". A source node always prefers a route through a neighbor to which it is connected via a primary edge ("primary route") over a route through a neighbor to which it is connected via a backup edge ("backup route"). When faced with a choice between two (or more) primary routes, or two (or more) backup routes, nodes always prioritize shorter routes. Every source node i has an export-all policy (i.e., i is willing to make all routes in P_i available to all neighboring nodes). We prove the following.

Theorem 3.2. *Every shortest-path backup routing game is a potential game and, moreover, a PNE in such a game can be found in a computationally-efficient manner.*

In Sect. 5 we examine a more complex class of backup routing games and show that games in that class are guaranteed to be weakly acyclic under best-response yet are not necessarily potential games.

4 Routing Games with Byzantine Players

We now present our results for the class of routing games with Byzantine players.

4.1 Routing Policies

[5] introduces the notions of policy consistency and of consistent export, which generalize natural classes of routing policies, e.g., shortest-path routing and next-hop routing. We now present these two concepts.

Policy-Consistent Ranking. Two well-studied classes of ranking functions are shortest-path rankings and next-hop rankings. Shortest-path rankings always prioritize shorter routes. Next-hop rankings, in contrast, rank routes based solely on the identity of the "next-hop" – the immediate neighbor – en route to the destination, i.e., a next-hop ranking assigns the same preference to all routes that share the same next-hop node. [5] generalizes these two classes of rankings as follows.

Definition 4.1. (policy consistency) *[5] Let i and j be two adjacent source nodes in G. We say that i is policy consistent with j iff for every two routes $Q, R \in P_j$ such that $i \notin Q, R$, if $R <_j Q$, then $(i,j)R \leq_i (i,j)Q$. We say that policy consistency holds if each source node is policy consistent with each of its neighboring source nodes.*

Observe that in the scenario that all source nodes have shortest-path rankings, and also in the scenario that all nodes have next-hop rankings, policy consistency indeed holds. (See [5] for more details.)

Consistent Export. The simplest export policy is the export-all policy, where a source node i is willing to make all routes in P_i available to all neighboring nodes. [5] presents the following generalization of export-all.

Definition 4.2. (consistent export) *[5] Let i and j be two adjacent nodes in G. We say that i consistently exports with respect to j iff there is some route $R \in P_i$ such that $R_{ij} = \{Q|\ Q \in P_i \text{ and } R \leq_i Q\}$. We say that a node i consistently exports if it consistently exports with respect to each neighboring node j. We say that consistent export holds if all nodes consistently export.*

Observe that when all source nodes have all-export policies then consistent export indeed trivially holds.

4.2   Positive Result

Consider games for which policy consistency and consistent export hold. These games include, among others, routing games with shortest-path rankings and export-all policies, as well as routing games with next-hop rankings and export-all policies, and can easily be shown to be potential games. We turn our attention to the scenario that there exists a single Byzantine player, i.e., that the routing policy of a single player can be changed in an arbitrary manner. We prove the following surprising positive result.

Theorem 4.1. *If policy consistency and consistent export hold for a routing game then the game is weakly acyclic under best response even in the presence of a single Byzantine player. Moreover, a PNE in such a game (with a single Byzantine player) can be found in a computationally-efficient manner.*

Can this result be extended to more than a single Byzantine player? Simple examples show that the answer to this question is, in general, No. We believe, however, that under certain reasonable conditions (e.g., that the number of Byzantine nodes not exceed a certain threshold, and that the Byzantine nodes not be "too concentrated" in a single part of the network) our result can be made to hold more generally. We leave this as an interesting direction for future research.

5   Backup Routing Games

In Sect. 3, we considered the following simple setting. Every edge in the network graph is either categorized as a "primary" edge or as a "backup" edge. A source node with multiple outgoing edges ranks routes in which it is connected to the next-hop node via a primary edge ("primary routes") more highly than routes in which it is connected to the next-hop node via a backup edge ("backup routes"). When faced with a choice between multiple routes in the same category (primary/backup) a source node always prioritizes shorter routes over longer routes. In addition, every source node has an export-all policy. We have shown that games that fall within this subclass of routing games are weakly-acyclic (and, in fact, even potential games). Next, we present a more realistic model, inspired by today's commercial Internet.

5.1   Commercial Backup-Routing Games

As before, each edge in the network graph is either "primary" or "backup". In addition, neighboring nodes in the network graph have one of two business relationships: either one node is a *customer* of the other (which is its *provider*) or the two nodes are *peers*. We make the standard assumption that no node is an indirect customer of itself, i.e., that there are no customer-provider cycles in this business hierarchy [7]. We now present constraints on source nodes' ranking functions and export policies that are naturally induced by this business hierarchy and extend the famous economic Gao-Rexford constraints [7] to handle backup routing. See [7] for a detailed explanation of this economic framework.

Ranking. A source node with multiple outgoing edges ranks primary routes more highly than backup routes. When faced with a choice between multiple routes in the same category (primary/backup), a source node always prioritizes (revenue-generating) routes in which its next-hop is its customer ("customer routes") over routes in which its next-hop is its peer/provider ("peer/provider routes"). Consider the network in Fig. 1(a). In the event that node 3 has a primary edge to its peer, node 2, and a backup edge to its customer, node 1, node 3 should prefer routes through 2 over routes through 1. However, if 3's edges to nodes 1 and 2 are both primary or both backup, node 3 should prefer routes through 1 over routes through 2.

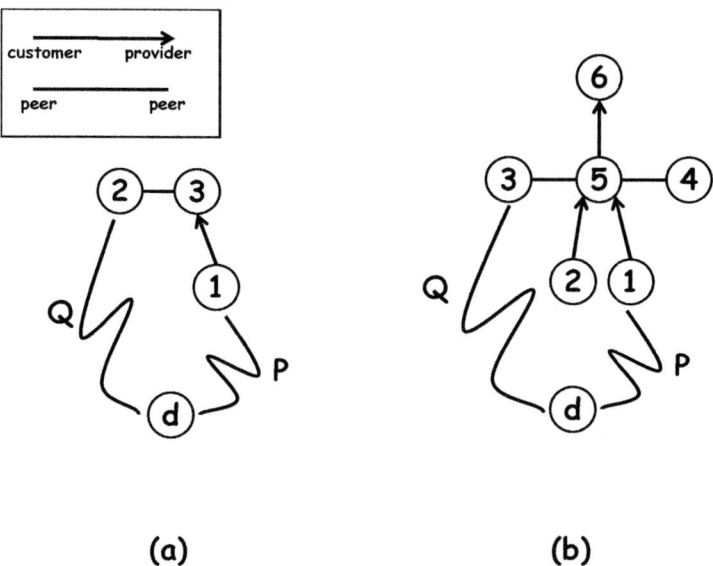

(a)   (b)

Fig. 1. Commercial Backup Routing

Export. A source node is willing to export all routes through it to its customers, but is only willing to export (all of) its customer routes to neighbors its peers and providers. Intuitively, this captures a source node's willingness to carry transit traffic for its customers, but not for its peers and providers (by whom it is not paid). Consider the network in Fig. 1(b), and suppose that all edges are primary. Node 5 should announce routes through node 1, its customer, to all neighboring nodes. However, node 5 should only announce routes through node 3, its peer, to its customers (nodes 1 and 2), and not to its other peer (node 4) and provider (node 6).

We call routing games where each node has a ranking function and export policy as above "commercial backup-routing games".

5.2   Positive Result

We prove the following positive result for the class of commercial backup-routing games.

Theorem 5.1. *Every commercial backup-routing game is weakly acyclic under best-response and, moreover, a PNE in such a game can be found in a computationally-efficient manner.*

We show (Appendix B) that commercial backup-routing games are, in fact, not contained in the class of potential games.

Theorem 5.2. *There exists a commercial backup-routing game which is not a potential game.*

References

[1] Babaioff, M., Kleinberg, R., Papadimitriou, C.H.: Congestion games with malicious players. In: MacKie-Mason, J.K., Parkes, D.C., Resnick, P. (eds.) ACM Conference on Electronic Commerce, pp. 103–112. ACM, New York (2007)

[2] Eliaz, K.: Fault tolerant implementation. Review of Economic Studies 69, 589–610 (2002)

[3] Fabrikant, A., Jaggard, A.D., Schapira, M.: On the structure of weakly acyclic games. In: Kontogiannis, S., Koutsoupias, E., Spirakis, P.G. (eds.) Algorithmic Game Theory. LNCS, vol. 6386, pp. 126–137. Springer, Heidelberg (2010)

[4] Fabrikant, A., Papadimitriou, C.H.: The complexity of game dynamics: BGP oscillations, sink equilibria, and beyond. In: Teng, S.-H. (ed.) SODA, pp. 844–853. SIAM, Philadelphia (2008)

[5] Feigenbaum, J., Ramachandran, V., Schapira, M.: Incentive-compatible interdomain routing. In: Feigenbaum, J., Chuang, J.C.-I., Pennock, D.M. (eds.) ACM Conference on Electronic Commerce, pp. 130–139. ACM, New York (2006)

[6] Gao, L., Griffin, T., Rexford, J.: Inherently safe backup routing with BGP. In: INFOCOM, vol. 1, pp. 547–556. IEEE, Los Alamitos (2001)

[7] Gao, L., Rexford, J.: Stable internet routing without global coordination. IEEE/ACM Trans. Netw. 9(6), 681–692 (2001)

[8] Goemans, M., Mirrokni, V., Vetta, A.: Sink equilibria and convergence. In: FOCS, pp. 142–151. IEEE Computer Society, Los Alamitos (2005)

[9] Goldberg, S., Halevi, S., Jaggard, A.D., Ramachandran, V., Wright, R.N.: Rationality and traffic attraction: Incentives for honest path announcements in BGP. In: Bahl, V., Wetherall, D., Savage, S., Stoica, I. (eds.) SIGCOMM, pp. 267–278. ACM, New York (2008)

[10] Gradwohl, R.: Fault tolerance in distributed mechanism design. In: Papadimitriou, C., Zhang, S. (eds.) WINE 2008. LNCS, vol. 5385, pp. 539–547. Springer, Heidelberg (2008)

[11] Jaggard, A.D., Schapira, M., Wright, R.N.: Distributed computing with adaptive heuristics. In: ICS, pp. 417–443. Tsinghua University Press, Beijing (2011)

[12] Levin, H., Schapira, M., Zohar, A.: Interdomain routing and games. In: Dwork, C. (ed.) STOC, pp. 57–66. ACM, New York (2008)

[13] Marden, J.R., Young, H.P., Arslan, G., Shamma, J.S.: Payoff-based dynamics in multi-player weakly acyclic games. SIAM J. on Control and Optimization 48, 373–396 (2009)

[14] Milchtaich, I.: Congestion games with player-specific payoff functions. Games and Economic Behavior 13, 111–124 (1996)

[15] Monderer, D., Shapley, L.S.: Potential games. Games and Economic Behavior 14(1), 124–143 (1996)

[16] Nisan, N., Schapira, M., Valiant, G., Zohar, A.: Best-response mechanisms. In: ICS, pp. 155–165. Tsinghua University Press, Beijing (2011)

[17] Rosenthal, R.W.: A class of games possessing pure-strategy nash equilibria. International Journal of Game Theory 2, 65–67 (1973)

[18] Young, H.P.: The evolution of conventions. Econometrica 61(1), 57–84 (1993)

A Shortest-Path Routing Games with a Single Byzantine Player are not Necessarily Potential Games

Consider the network G partially described by Fig. 2. We are interested in three nodes: x, y and m, where m is the Byzantine node. The paths P_1, P_2, P_3, P_4 and P_5 are all disjoint, and their lengths are $8, 6, 2, 2$ and 10 respectively.

Hence, x prefers the path $(x, m)P_4$ (length 3) over P_1 (8) which is preferred over $(x, m)P_5$ (11).

Also, y prefers the path $(y, x)(x, m)P_4$ (length 4) over P_2 (6) which is preferred over $(y, x)P_1$ (9).

The Byzantine node m has the following preferences:

$$P_3P_2 <_m P_4 <_m P_5 <_m P_3(y, x)P_1 \ .$$

We now present a better-response improvement cycle. In this cycle, all nodes but x, y and m are fixed, and the paths P_1, P_2, P_3, P_4 and P_5 are all valid routes resulting from the fixed strategies of these nodes. Now, consider the following sequence of transitions.

- x chooses P_1, y chooses x and m chooses P_3.
- x chooses P_1, y chooses P_2 and m chooses P_3.
- x chooses P_1, y chooses P_2 and m chooses P_4.

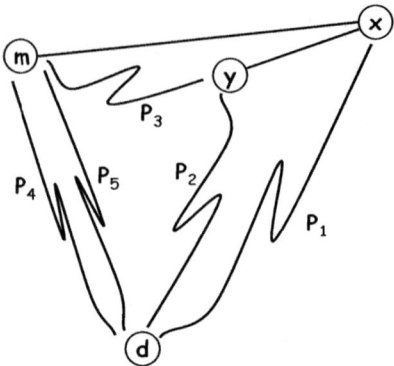

Fig. 2. A shortest-path routing game with a single Byzantine player that is not a potential game

- x chooses m, y chooses P_2 and m chooses P_4.
- x chooses m, y chooses x and m chooses P_4.
- x chooses m, y chooses x and m chooses P_5.
- x chooses P_1, y chooses x and m chooses P_5.
- x chooses P_1, y chooses x and m chooses P_3.

Observe that every strategy profile is reachable from the strategy profile that comes before it via the better response of a single node in $\{x, y, m\}$, and that the first strategy and the last strategy in this sequence are identical. Hence, there exists a better-response improvement cycle and so this game is not a potential game.

B   Commercial Backup-Routing Games Are Not Necessarily Potential Games

Consider the network in Fig. 3. There are 6 source nodes $1, 2, 3, a, b, c$ and a unique destination node d. The business relationships between nodes, and the classification of edges into primary edges and backup edges are described in the figure. Each of the source nodes $1, 2$, and 3 has a next-hop ranking, and its preferences over next-hops are as in the figure (e.g., node 1 prefers all routes through its peer a over the direct route to its customer d, over all routes through its provider c). Each of the source nodes a, b, and c prefers routes through its peer (to which it is connected via a primary edge) over routes through its customer (to which it is connected via a backup edge). Observe that these rankings are indeed backup/primary commercial rankings (each node prefers primary routes over backup routes and, within each category of routes, prefers customer routes to peer/provider routes). Each source node has a commercial export-all policy, that is, it exports all routes to its customers and all customer routes to its peers/providers.

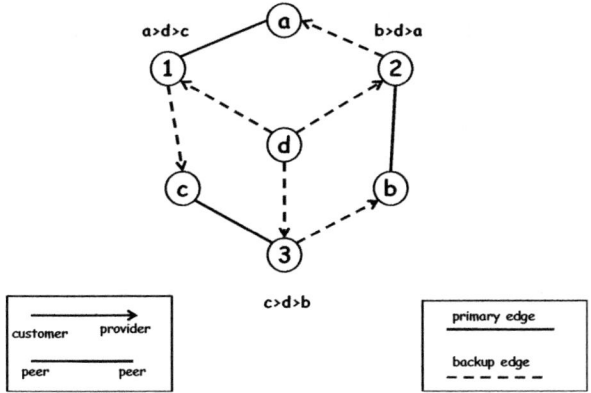

Fig. 3. A commercial backup routing game that is not a potential game

Observe that this routing game possesses (multiple) PNE, e.g., the routing state in which nodes 1 and 2 forward traffic directly to d, and node 3 forward traffic to node 1. We now show that the game is not a potential game by presenting a better-response improvement cycle. Consider the case that each of the source nodes a, b, and c's strategy is fixed to be the outgoing link to its customer (e.g., c sends traffic to 1). Now, consider the following sequence of transitions between 3-tuples of source nodes 1, 2, and 3's strategies (listed in that order): $((1d), (2d), (3c)) \rightarrow ((1a), (2d), (3c)) \rightarrow ((1a), (2d), (3d)) \rightarrow ((1a), (2b), (3d)) \rightarrow ((1d), (2b), (3d)) \rightarrow ((1d), (2b), (3c)) \rightarrow ((1d), (2d), (3c))$. Observe that every 3-tuple of strategies is reachable from the 3-tuple that comes before it via the best-response of a single node in $\{1, 2, 3\}$, and that the first 3-tuple and last 3-tuple in this sequence are identical. Hence, there exists a best-response improvement cycle and so this game is not a potential game.

Efficiency of Restricted Tolls in Non-atomic Network Routing Games

Vincenzo Bonifaci[1,*], Mahyar Salek[2,**], and Guido Schäfer[3]

[1] Università Telematica Internazionale Uninettuno, Rome, Italy
vincenzo.bonifaci@uninettunouniversity.net
[2] Department of Computer Science, University of Southern California, USA
salek@usc.edu
[3] CWI and VU University Amsterdam, The Netherlands
g.schaefer@cwi.nl

Abstract. An effective means to reduce the inefficiency of Nash flows in non-atomic network routing games is to impose tolls on the arcs of the network. It is a well-known fact that marginal cost tolls induce a Nash flow that corresponds to a minimum cost flow. However, despite their effectiveness, marginal cost tolls suffer from two major drawbacks, namely (i) that potentially every arc of the network is tolled, and (ii) that the imposed tolls can be arbitrarily large.

In this paper, we study the *restricted network toll problem* in which tolls can be imposed on the arcs of the network but are restricted to not exceed a predefined threshold for every arc. We show that optimal restricted tolls can be computed efficiently for parallel-arc networks and affine latency functions. This generalizes a previous work on taxing subnetworks to arbitrary restrictions. Our algorithm is quite simple, but relies on solving several convex programs. The key to our approach is a characterization of the flows that are inducible by restricted tolls for single-commodity networks. We also derive bounds on the efficiency of restricted tolls for multi-commodity networks and polynomial latency functions. These bounds are tight even for parallel-arc networks. Our bounds show that restricted tolls can significantly reduce the price of anarchy if the restrictions imposed on arcs with high-degree polynomials are not too severe. Our proof is constructive. We define tolls respecting the given thresholds and show that these tolls lead to a reduced price of anarchy by using a (λ, μ)-smoothness approach.

1 Introduction

Congestion in traffic networks has several negative effects as it causes, e.g., environmental pollution, waste of natural resources and time, stress on the traffic participants, etc. With the increase in traffic in recent years, it becomes an increasingly important issue to implement regulation means that efficiently reduce congestion in networks. In this context, road pricing has long been recognized as being one of the most effective regulation means. The idea is to charge traffic participants for the usage of roads by

---

* Part of this work has been done while the author was at MPI Informatics, Saarbrücken, Germany.

** Part of this work has been done while the author spent a summer internship at CWI.

G. Persiano (Ed.): SAGT 2011, LNCS 6982, pp. 302–313, 2011.
© Springer-Verlag Berlin Heidelberg 2011

imposing tolls. Such negative incentives usually lead to a change in behavior in that traffic participants, for example, travel along longer (but less congested) routes, avoid certain parts of the network (at certain times), or do not travel at all during peak-times, etc. Recent technological advances, in particular, in satellite technology, facilitate the realization of such pricing schemes, e.g., by enabling to collect tolls electronically. Furthermore, they open up the possibility to implement *dynamic* pricing schemes, in which tolls may vary over time or depend on congestion.

In this paper, we study the problem of computing efficient pricing schemes to reduce congestion in network applications caused by selfish behavior. In our studies we use a well-established model of traffic routing in networks, also known as the *Wardrop model*. In this model, we are given a directed graph $G = (V, A)$ with latency functions $\ell := (\ell_a)_{a \in A}$ on the arcs, k commodities $(s_1, t_1), \ldots, (s_k, t_k) \in V \times V$, and a non-negative demand r_i for every commodity $i \in [k]$. The latency functions are used to model the (flow-dependent) congestion on the arcs and are assumed to be non-negative and non-decrasing. The demand r_i of commodity $i \in [k]$ specifies the amount of flow that needs to be routed from s_i to t_i. A common interpretation is that the r_i units of flow represent an (infinitely) large population of players, each controlling an infinitesimal amount of the r_i flow units. The goal of every player is to send his flow along a shortest latency path from its source s_i to its destination t_i. The resulting game is also called a *non-atomic network routing game*. A flow f in which no player has an incentive to unilaterally deviate from its path is called a *Nash flow* (or *Wardrop flow*).

In general, a Nash flow can be inefficient in the sense that it does not correspond to an optimal flow that minimizes *social cost*, i.e., the total average latency. The *price of anarchy* [14] is a measure to quantify the efficiency loss caused by selfish behavior. In the context of network routing games, it is defined as the worst-case ratio over all instances between the cost of a Nash flow and the cost of an optimal flow. In a seminal work, Roughgarden and Tardos [19] show that the price of anarchy of non-atomic network routing games is unbounded in general and provide bounds for specific classes of latency functions, e.g., polynomial latency functions.

An effective means to reduce the price of anarchy in network routing games is to impose non-negative tolls $\tau := (\tau_a)_{a \in A}$ on the arcs. We consider both *dynamic* and *static* tolls in this paper. In the dynamic case, the toll that is imposed on arc $a \in A$ is defined by a (flow-dependent) toll function τ_a which maps every flow value x to a non-negative toll $\tau_a(x)$. In the static case, the toll on arc $a \in A$ is specified by a non-negative constant τ_a. By traversing an arc $a \in A$ with flow value x, a player now experiences a delay of $\ell_a(x)$ and additionally has to pay a toll of $\tau_a(x)$. We let $\alpha > 0$ be a parameter that specifies how players value time over money. That is, the combined cost of an arc $a \in A$ with flow value x is defined as $\phi_a(x) := \ell_a(x) + \alpha \tau_a(x)$. We assume that every players' goal is to choose a path that minimizes his total combined cost. A stable outcome of this game is a Nash flow with respect to the combined cost functions $\phi := (\phi_a)_{a \in A}$.

A fundamental result due to Beckman, McGuire and Winsten [2] states that *marginal cost tolls* induce a Nash flow that is socially optimal. That is, if we define $\tau_a(x) := \frac{1}{\alpha} x \cdot \ell_a'(x)$ for every arc $a \in A$ then a Nash flow with respect to ϕ is an optimal flow with respect to ℓ. Even though marginal cost tolls are theoretically appealing, they have two major drawbacks: (i) It is assumed that tolls can be imposed on *every* arc of the network.

(ii) The tolls imposed on the arcs can be arbitrarily large. These are severe drawbacks that rule out the applicability of marginal cost tolls in several situations.

In this paper, we overcome these drawbacks by restricting the set of feasible tolls. These restrictions are assumed to be given exogenously by means of threshold functions on the arcs. That is, in a *restricted network toll problem* we are given an instance of the network routing game together with some threshold functions $\theta := (\theta_a)_{a \in A}$ that specify an upper bound on the maximum toll chargeable on each arc. As for tolls, we call threshold functions θ *dynamic* if they are flow-dependent and *static* otherwise. We call the tolls $\tau = (\tau_a)_{a \in A}$ θ-*restricted* if for every $a \in A$, $0 \le \tau_a(x) \le \theta_a(x)$ for all flow values $x \ge 0$. Given θ-restricted tolls τ, let f^τ denote a Nash flow that is induced by τ, i.e., f^τ is a Nash flow with respect to $\phi = \ell + \alpha\tau$.

Our model incorporates several interesting special cases. For example, we can enforce that tolls are only imposed on a subnetwork induced by a subset $T \subseteq A$ of the arcs by setting $\theta_a = \infty$ for every $a \in T$ and $\theta_a = 0$ otherwise. Another example is that we can restrict the toll on each arc $a \in A$ by a (flow-independent) threshold value θ_a. Yet another example is that we can require that the toll on each arc $a \in A$ does not exceed a certain fraction of the latency of that arc, e.g., $\theta_a(x) = \varepsilon\ell_a(x)$ for some $\varepsilon > 0$.

Given the restrictions imposed on the set of feasible tolls, the following two natural questions arise and will be studied in this paper:

1. Can one quantify the efficiency of θ-restricted tolls?
   We are interested in studying the efficiency of θ-restricted tolls in relation to the cost of a socially optimal flow. To this aim, we define the *efficiency* of θ-restricted tolls as the minimum ratio of the cost of a Nash flow f^τ inducible by θ-restricted tolls τ and the cost of an optimal flow. We also address the problem of computing θ-restricted tolls that guarantee a certain efficiency.

2. Can one compute (approximately) optimal θ-restricted tolls?
   We consider the problem of computing (approximately) optimal θ-restricted tolls. We call θ-restricted tolls τ *optimal* if the Nash flow f^τ induced by τ has cost less than or equal to any other Nash flow that is inducible by θ-restricted tolls. Similarly, θ-restricted tolls τ are said to be λ-*approximate* for some $\lambda \ge 1$ if the cost of f^τ is at most λ times the cost of any other Nash flow inducible by θ-restricted tolls.

Clearly, from the discussion above it follows that we obtain an efficiency of one if $\theta_a = \infty$ for every $a \in A$. On the other hand, the efficiency coincides with the price of anarchy if $\theta_a = 0$ for every $a \in A$.

The special case that tolls can only be imposed on a subset $T \subseteq A$ of the arcs has recently been studied by Hoefer, Olbrich and Skopalik [12]. For this case, the authors derive an algorithm to compute optimal T-restricted tolls for parallel-arc networks with affine latency functions. They also prove that the problem of computing optimal tolls is NP-hard, even for two-commodity networks and affine latency functions. Note that the restricted network toll problem that we consider here is more general, and thus this hardness result extends to our setting.

Our Results. The main contributions presented in this paper are as follows:

In Section 3 we show that optimal θ-restricted tolls can be computed efficiently in parallel-arc networks with affine latency functions. This extends the result of

Hoefer et al. [12] to arbitrary dynamic threshold functions on the arcs. Our approach is different from the one described in [12]. Despite its generality, our algorithm is quite simple. The key to our approach is a characterization of the flows that are inducible by θ-restricted tolls. Our characterization applies to single-commodity networks in general. It allows us to determine whether a given flow is inducible by θ-restricted tolls by verifying whether there is a negative cycle in a properly constructed graph (which can be done in polynomial time). Based on this characterization, we derive an algorithm to compute optimal θ-restricted tolls for parallel-arc networks. Our algorithm works for general latency functions; however, we can only guarantee polynomial running time if all latency and threshold functions are affine (in which case we need to solve a series of convex quadratic programs).

In Section 4 we derive upper bounds on the efficiency of dynamic θ-restricted tolls for multi-commodity networks and polynomial latency functions of degree p. Our pricing scheme is a simple and natural adaptation of marginal cost tolls to the restricted setting: for every arc $a \in A$ we charge marginal cost tolls if this does not exceed the threshold θ_a, and we charge θ_a otherwise. Essentially, we show that these tolls achieve an efficiency that depends on the degree of the polynomial and the smallest ratio between the threshold value and the latency of an arc (see Section 4 for precise statements). The technique that we use to prove these bounds rests on a (λ, μ)-*smoothness* approach [18] that was used (implicitly) in previous works to bound the price of anarchy of routing games (see, e.g., [1,3,4,10]) and put into a more general context in [18]. We also prove that our bounds are tight, even for parallel-arc networks. Our pricing scheme also provides a way to compute θ-restricted tolls for multi-commodity networks and polynomial latency functions that are λ-approximate, where λ is equal to the established efficiency.

Our findings support the intuition that, in order to achieve a good efficiency, it is more important to be able to impose tolls on the arcs that are sensitive to flow changes (high degree polynomials) than on the arcs that are relatively insensitive to flow changes (low degree polynomials).

For the special case that all restrictions are of the form $\theta_a(x) = \varepsilon \ell_a(x)$, our bound matches exactly the *price of stability* of ε-Nash flows shown by Christodoulou, Koutsoupias and Spirakis [5]. Our result therefore shows that such tolls allow us to reduce the (generally large) inefficiency of Nash flows to at least the price of stability of ε-Nash flows; the actual instance-dependent efficiency of such tolls might be better than that.

All our results mentioned above hold for dynamic threshold functions (and thus also for static ones).

Related Work. As mentioned above, most related to our work is the recent article [12] by Hoefer et al. who study the problem of taxing subnetworks, a special case of the restricted toll problem that we consider here. The authors focus on the problem of computing optimal tolls. They show that this problem is NP-hard for two-commodity networks and affine latency functions by a reduction from partition. They also derive an algorithm to compute optimal tolls for parallel-arc networks and affine latency functions. Their algorithm is sophisticated and crucially exploits that the restrictions are of the form $\theta_a \in \{0, \infty\}$ for every arc $a \in A$.

The classic result that marginal cost pricing induces optimal flows is due to Beckmann, McGuire and Winsten [2]. More recently, it has been shown that optimal-inducing tolls exist even when users are heterogeneous, i.e., have different latency/toll trade-offs: this was first shown for single-commodity networks by Cole, Dodis and Roughgarden [6] and then extended to the multi-commodity case by Fleischer, Jain and Mahdian [9] and independently by Karakostas and Kolliopoulos [13].

Cole et al. [7] study the setting in which the cost of each user is defined as the latency plus the taxes paid by the user. For heterogeneous users, Fleischer [8] shows that if there is a single commodity, then tolls that are linear in the maximum latency of the optimal flow are sufficient to force the users to the system optimum. The question of computing tolls that enforce particular flows has been studied in [11]. The above papers all study the non-atomic model; tolls for heterogeneous users in the context of atomic routing games have been considered by Swamy [20].

Bounds on the price of anarchy and the price of stability of ε-Nash flows in non-atomic and atomic congestion games, including network congestion games, have been derived recently by Christodoulou et al. [5].

2   Preliminaries

We provide formal definitions of the concepts introduced in the Introduction. Suppose we are given an instance $\mathscr{I} = (G, (\ell_a)_{a \in A}, (s_i, t_i)_{i \in [k]}, (r_i)_{i \in [k]})$ of the non-atomic network routing game. Let $\mathscr{P}_i$ denote the set of all simple directed s_i, t_i-paths in G and define $\mathscr{P} := \cup_{i \in [k]} \mathscr{P}_i$. An outcome of the game is a flow $f : \mathscr{P} \to \mathbb{R}_+$ that is feasible, i.e., $\sum_{P \in \mathscr{P}_i} f_P = r_i$ for every $i \in [k]$. Given a flow f, the total flow on arc $a \in A$ is defined as $f_a := \sum_{P \in \mathscr{P}: a \in P} f_P$. We define the latency of a path $P \in \mathscr{P}$ with respect to f as $\ell_P(f) := \sum_{a \in P} \ell_a(f_a)$. The total cost $C(f)$ of f is given by its average latency, i.e., $C(f) := \sum_{P \in \mathscr{P}} f_P \ell_P(f)$. A flow that minimizes $C(\cdot)$ is called optimal and denoted by f^*. A feasible flow f is called a *Nash flow* (or *Wardrop flow*) with respect to $\ell := (\ell_a)_{a \in A}$ if and only if

$$\forall i \in [k], \forall P \in \mathscr{P}_i, f_P > 0 : \qquad \ell_P(f) \leq \ell_{P'}(f) \quad \forall P' \in \mathscr{P}_i. \tag{1}$$

Throughout this paper, we assume that the latency functions are non-negative, non-decreasing, differentiable and semi-convex, i.e., $x \cdot \ell_a(x)$ is convex for every arc $a \in A$; such latency functions are also called *standard* [16]. The cost of a Nash flow is unique if the latency functions are standard.

In a *restricted network toll problem* we are given an instance $\mathscr{I}$ of the network routing game and threshold functions $\theta := (\theta_a)_{a \in A}$ on the arcs. In this setting, non-negative tolls $\tau := (\tau_a)_{a \in A}$ can be imposed on the arcs that have to obey the bounds defined by the threshold functions $(\theta_a)_{a \in A}$. In the most general setting, both tolls and threshold functions are flow-dependent. Unless stated otherwise, we assume that both tolls and threshold functions are non-decreasing and continuous. Given a feasible flow f, we define the combined cost that a player experiences by traversing arc $a \in A$ as $\phi_a(f_a) = \ell_a(f_a) + \alpha \tau_a(f_a)$. We assume that every players' goal is to choose a path P that minimizes the combined cost $\ell_P(f) + \alpha \tau_P(f)$, where $\tau_P(f) := \sum_{a \in P} \tau_a(f_a)$.

For notational convenience, we assume that α is normalized to 1. This is without loss of generality because we can always divide all toll functions by α.

The tolls $\tau = (\tau_a)_{a \in A}$ are called θ-*restricted* if for every arc $a \in A$, $0 \leq \tau_a(x) \leq \theta_a(x)$ for all flow values $x \geq 0$. We define $\mathscr{T}(\theta)$ as the set of all θ-restricted tolls, i.e.,

$$\mathscr{T}(\theta) := \{(\tau_a)_{a \in A} \mid \forall a \in A : 0 \leq \tau_a(x) \leq \theta_a(x) \ \forall x \geq 0\}.$$

Given θ-restricted tolls τ, let f^τ denote a Nash flow that is induced by τ, i.e., f^τ is a Nash flow with respect to $\phi = \ell + \tau$. The *efficiency* of θ-restricted tolls for a given instance of the restricted network toll problem is defined as $\min_{\tau \in \mathscr{T}(\theta)} C(f^\tau)/C(f^*)$. That is, we relate the cost of the best Nash flow f^τ that is inducible by θ-restricted tolls τ to the cost of an optimal flow. Note that we account for the average latency of the network here rather than the total disutility (latency plus toll) of the players. The reason for that is that we are interested in characterizing the effect of tolls on the performance (measured in terms of average latency) of the network.

Given the restrictions $\theta = (\theta_a)_{a \in A}$ on the arcs, θ-restricted tolls τ are *optimal* if the Nash flow f^τ induced by τ satisfies $C(f^\tau) \leq C(f^{\bar{\tau}})$ for all Nash flows $f^{\bar{\tau}}$ induced by θ-restricted tolls $\bar{\tau}$. Similarly, θ-restricted tolls τ are ρ-*approximate* for some $\rho \geq 1$ if $C(f^\tau) \leq \rho C(f^{\bar{\tau}})$ for all Nash flows $f^{\bar{\tau}}$ induced by θ-restricted tolls $\bar{\tau}$.

3 Computing Optimal θ-Restricted Tolls

We first give a characterization of the flows that are inducible by θ-restricted tolls for single-commodity networks. This characterization will be the key to derive an algorithm that computes optimal θ-restricted tolls for parallel-arc networks. All results presented in this section hold for flow-dependent threshold functions θ.

3.1 Characterization of Inducible Flows for Single-Commodity Networks

We consider the problem of determining whether a given flow f is inducible by θ-restricted tolls. We focus on the single-commodity case. As we will see, this problem reduces to verifying whether there is a negative cycle in a properly constructed graph.

Suppose we are given a flow f. Recall that f is a Nash flow with respect to $\ell + \tau$ iff for every two s,t-paths $P, P' \in \mathscr{P}$ with $f_P > 0$ it holds $\ell_P(f) + \tau_P \leq \ell_{P'}(f) + \tau_{P'}$. Said differently, every flow-carrying path must be a shortest path with respect to the combined cost $\phi := \ell + \tau$. Subsequently, let ℓ_a, τ_a and θ_a refer to $\ell_a(f_a)$, $\tau_a(f_a)$ and $\theta_a(f_a)$, respectively. (In the discussion below, several definitions will depend on the flow f; however, for notational convenience we often do not state this dependence explicitly.)

We use the following alternative characterization of Nash flows (see, e.g., [17]). For every vertex $u \in V$, let δ_u be the length of a shortest path from s to u with respect to $\ell + \tau$. Define A^+ as the set of arcs with positive flow, i.e., $A^+ := \{a \in A : f_a > 0\}$. Then f is a Nash flow with respect to $\phi = \ell + \tau$ if and only if (i) $\delta_v \leq \delta_u + \ell_a + \tau_a$ for every arc $a = (u,v) \in A$, and (ii) $\delta_v = \delta_u + \ell_a + \tau_a$ for every arc $a = (u,v) \in A^+$.

We can thus express the set $\mathscr{F}(\theta)$ of θ-restricted tolls that induce f as follows:

$$\mathscr{F}(\theta) := \{(\tau_a)_{a \in A} \mid \begin{array}{ll} \delta_v - \delta_u \leq \ell_a + \tau_a & \forall a = (u,v) \in A \setminus A^+ \\ \delta_v - \delta_u = \ell_a + \tau_a & \forall a = (u,v) \in A^+ \\ \delta_u \text{ free} & \forall u \in V \\ 0 \leq \tau_a \leq \theta_a & \forall a \in A \}. \end{array} \tag{2}$$

Note that the $(\delta_u)_{u \in V}$ are unrestricted in this formulation. Alternatively, we could have required that $\delta_s = 0$ and $\delta_u \geq 0$ for every $u \in V$. However, this is equivalent to the formulation (2) stated above.

We define a graph $\hat{G} = \hat{G}(f) = (V, \hat{A})$ with arc-costs $c : \hat{A} \rightarrow \mathbb{R}$ as follows: $\hat{G}$ contains all arcs $a = (u,v) \in A$ and, additionally, for every arc $a = (u,v) \in A^+$ the reversed arc (v,u). We call the former type of arcs *forward arcs* and the latter type of arcs *backward arcs*. The cost of each forward arc $a = (u,v) \in \hat{A}$ is equal to $c_a := \ell_a + \theta_a$. Every backward arc $a = (v,u) \in \hat{A}$ has a cost equal to the negative of the latency of its reversed arc $(u,v) \in A$, i.e., $c_a := -\ell_{(u,v)}$.

Given some subset X of arcs and functions $(g_a)_{a \in X}$, we define $g(X)$ as a short for $\sum_{a \in X} g_a$.

Theorem 1. *Let f be an arbitrary feasible flow. Then f is inducible by θ-restricted tolls if and only if $\hat{G}(f)$ does not contain a cycle of negative cost.*

Proof. Suppose $\hat{G} = \hat{G}(f)$ contains a cycle $C \subseteq \hat{A}$ of negative cost. Since only backward arcs have negative cost, at least one backward arc is part of C. Partition C into the set F of forward arcs and the set B of backward arcs, respectively. Let $\bar{B}$ denote the set of reversed arcs in B. Note that $\bar{B} \subseteq A^+$. We have $c(C) = c(F) + c(B) = \ell(F) + \theta(F) - \ell(\bar{B}) < 0$.

Suppose for the sake of contradiction that $\tau = (\tau_a)_{a \in A} \in \mathscr{F}(\theta)$ are feasible tolls that induce f. By the feasibility of τ, we have for every forward arc $a = (u,v) \in F$, $\delta_v - \delta_u \leq \ell_{(u,v)} + \tau_{(u,v)}$ and for every backward arc $a = (u,v) \in B$, $\delta_u - \delta_v = \ell_{(v,u)} + \tau_{(v,u)}$, or equivalently, $\delta_v - \delta_u = -\ell_{(v,u)} - \tau_{(v,u)}$. Summing over all arcs in C, we obtain

$$0 = \sum_{(u,v) \in C} \delta_v - \delta_u \leq \sum_{(u,v) \in F} \ell_{(u,v)} + \tau_{(u,v)} - \sum_{(u,v) \in B} \ell_{(v,u)} + \tau_{(v,u)}$$

$$\leq \ell(F) + \theta(F) - \ell(\bar{B}) - \tau(\bar{B}) < -\tau(\bar{B}),$$

where the last inequality follows from the observation above. Thus $\tau(\bar{B}) < 0$ which is a contradiction since $\tau_a \geq 0$ for every arc $a \in A$.

Next suppose that $\hat{G}$ does not contain a negative cycle. We can then determine the shortest path distance δ_u from s to every node $u \in V$ in $\hat{G}$ with respect to c. (These distances are well-defined because $\hat{G}$ does not contain a negative cycle.) Note that for every arc $a = (u,v) \in \hat{A}$ we have $\delta_v \leq \delta_u + c_{(u,v)}$. Based on these distances, we extract tolls $\tau := (\tau_a)_{a \in A}$ as follows: For every arc $a = (u,v) \in A$, we define $\tau_a := \max\{0, \delta_v - \delta_u - \ell_a\}$. We show that τ induces f. By definition, we have for every arc $a = (u,v) \in A$: $\delta_v - \delta_u - \tau_a \leq \ell_a$. Consider an arc $a = (u,v) \in A^+$. Then $\delta_u - \delta_v \leq -\ell_a$, or equivalently, $\delta_v - \delta_u - \ell_a \geq 0$. Thus, $\delta_v - \delta_u - \tau_a = \ell_a$. Clearly, $\tau_a \geq 0$ for every $a \in A$. Moreover, for every arc $a = (u,v) \in A$ we have $\delta_v - \delta_u \leq \ell_a + \theta_a$ and thus $\delta_v - \delta_u - \ell_a \leq \theta_a$. We can infer that $\tau_a \leq \theta_a$ for every $a = (u,v) \in A$. $\qquad \square$

Note that the proof of the theorem also provides a way to extract the respective tolls if f is inducible by θ-restricted tolls: Given f, we compute the shortest path distance δ_u with respect to c from s to u for every $u \in V$ and define the toll τ_a for every arc $a = (u,v) \in A$ as in the proof of Theorem 1.

The following corollary is an immediate consequence of the above theorem and the fact that negative cycles can be detected efficiently (e.g., by the Bellman-Ford algorithm).

Corollary 1. *Given a flow f, we can determine in polynomial time whether f is inducible by θ-restricted tolls.*

3.2 Computing Optimal Tolls in Parallel-Arc Networks

In light of the above characterization, the problem of computing θ-restricted tolls such that the cost $C(f^\tau)$ of the induced Nash flow f^τ is minimized is equivalent to the problem of computing a minimum cost flow f such that $\hat{G}(f)$ does not contain a negative cost cycle. Once we have determined f, we can extract the optimal θ restricted tolls τ as defined in the proof of Theorem 1. This equivalence constitutes the basis of our algorithm to compute optimal tolls in parallel-arc networks.

Let $G = (V,A)$ be a parallel-arc network and let f be a feasible flow. The condition of Theorem 1 then reduces to the following property: f is inducible by θ-restricted tolls if and only if

$$\forall a \in A, \ f_a > 0: \qquad \ell_a(f_a) \leq \ell_{a'}(f_{a'}) + \theta_{a'}(f_{a'}) \quad \forall a' \in A. \tag{3}$$

Note that these conditions are similar to the Nash flow conditions in (1) (specialized to parallel-arc networks) with the difference that we allow some additional slack $\theta_{a'}(f_{a'})$ on the right-hand side. Thus, our goal is to determine a minimum cost flow f among all flows that satisfy (3).

Corollary 2. *The problem of computing optimal θ-restricted tolls for the parallel-arc restricted network toll problem is equivalent to computing a minimum cost flow f satisfying (3).*

Computing a minimum cost flow can be done efficiently by solving a convex program. However, here we need to ensure (3) additionally and it is a-priori not clear how to encode these constraints. Note that for Nash flows the corresponding conditions are ensured by applying the Karush-Kuhn-Tucker conditions to a convex program with an appropriately chosen objective function. A similar approach does not work here because we cannot deliberately choose an objective function and because of the asymmetry in (3) (due to the slack).

Our approach exploits the following key insight. Fix some minimum cost flow f^* satisfying (3) and suppose we knew the minimum value $z = \min\{\ell_a(0) + \theta_a(0) \mid a \in A, \ f_a^* = 0\}$ among all zero-flow arcs in f^*. Let $z = \infty$ if all arcs have positive flow in f^*. We can then compute an minimum cost flow $f^z = (f_a^z)_{a \in A}$ satisfying (3) as follows. From (3) we infer that $f_a^z = 0$ for every arc $a \in A$ with $\ell_a(0) > z$. Let $A^z = \{a \in A \mid \ell_a(0) \leq z\}$ be the remaining arcs. On the arcs in A^z, we compute a

---

Algorithm 1. Algorithm to compute a minimum cost flow satisfying (3)

---
1   Let $Z = \{\ell_a(0) + \theta_a(0) \mid a \in A\}$.
2   **for** *every* $z \in Z \cup \{\infty\}$ **do**
3   Define $A^z := \{a \in A \mid \ell_a(0) \leq z\}$.
4   Set $f_a^z = 0$ for every $a \notin A^z$.
5   Let $(f_a^z)_{a \in A^z}$ be an optimal solution of cost C^z to the program in (4).
       (Remark: $(f_a^z)_{a \in A^z}$ is undefined and $C^z = \infty$ if (4) is infeasible.)
6   **end**
7   Return $f = f^z$ with C^z minimum among all $z \in Z \cup \{\infty\}$.

---

feasible flow $(f_a^z)_{a \in A^z}$ of minimum cost satisfying $\ell_a(f_a^z) \leq z$ for every $a \in A^z$ and $\ell_a(f_a^z) \leq \ell_{a'}(f_{a'}^z) + \theta_{a'}(f_{a'}^z)$ for every $a, a' \in A^z$. The latter can be done by solving the program:

$$
\begin{aligned}
C^z = \min \ & \sum_{a \in A^z} f_a^z \ell_a(f_a^z) \\
\text{s.t. } & \sum_{a \in A^z} f_a^z = r \\
& f_a^z \geq 0 && \forall a \in A^z \\
& \ell_a(f_a^z) \leq z && \forall a \in A^z \\
& \ell_a(f_a^z) \leq \ell_{a'}(f_{a'}^z) + \theta_{a'}(f_{a'}^z) && \forall a, a' \in A^z.
\end{aligned}
\tag{4}
$$

The only remaining problem is that we do not know z. However, because there are at most $|A| + 1$ different possibilities (including the case $z = \infty$), we can simply compute a flow f^z for each possible value z and finally return the best flow f that has been encountered. The complete algorithm is summarized in Algorithm 1.

Theorem 2. *Algorithm 1 computes a minimum cost flow f satisfying* (3).

Proof. Let $f = f^z$ be the flow returned by Algorithm 1. Clearly, f is a feasible flow by construction. We argue that f satisfies (3). Consider some $a \in A^+$. Note that $f_{a'} = 0$ for every arc $a' \notin A^z$ and thus $a \in A^z$. Because $(f_a)_{a \in A^z}$ is a feasible solution to (4), we have $\ell_a(f_a) \leq z \leq \ell_{a'}(0) = \ell_{a'}(f_{a'})$ for every $a' \notin A^z$. Moreover, $\ell_a(f_a) \leq \ell_{a'}(f_{a'}) + \theta_{a'}(f_{a'})$ for every $a' \in A^z$. Thus, f satisfies (3).

Let f^* be an optimal flow. We show that $C(f) \leq C(f^*)$. Define z as the minimum value $\ell_a(f_a^*) + \theta_a(f_a^*)$ of a zero-flow arc $a \in A$, i.e., $z = \min\{\ell_a(0) + \theta_a(0) \mid a \in A, f_a^* = 0\}$. Let $z = \infty$ if all arcs have positive flow. Note that $f_a^* = 0$ for every $a \notin A^z$ and thus $C(f^*) = \sum_{a \in A^z} f_a^* \ell_a(f_a^*)$. Observe that $(f_a^*)_{a \in A^z}$ is a feasible solution for the program in (4) with respect to z. Thus $C(f^z) \leq C(f^*)$. Because $C(f) \leq C(f^z)$, this concludes the proof. $\qquad\square$

Finally, observe that the program in (4) is convex if all latency and threshold functions are affine, i.e., of the form $q_1 x + q_0$ with $q_1, q_0 \geq 0$. In particular, the constraints of (4) are linear and the objective function is convex quadratic in this case, so the program can be solved exactly in polynomial time [15].

Corollary 3. *Algorithm 1 computes a minimum cost flow f satisfying* (3) *in polynomial time if all latency and threshold functions are affine.*

Hoefer et al. [12] derived a similar result for the special case that $\theta_a \in \{0, \infty\}$ for every arc $a \in A$.

4   General Efficiency of θ-Restricted Tolls

We provide bounds on the efficiency of θ-restricted tolls for multi-commodity networks with polynomial latency functions of degree p. Our approach is constructive: We show how to compute θ-restricted tolls for a given instance of the restricted network toll problem that guarantee the claimed efficiency bound. The results given in this section hold for dynamic threshold functions.

Let $\mathscr{L}_p$ be defined as the set of all polynomial functions g of the form $g(x) = \sum_{d=0}^{p} q_d x^d$ with non-negative coefficients q_d, $d = 0, \ldots, p$. Moreover, let $\mathscr{M}_d$ refer to the set of all monomial functions of the form $\ell_a(x) = q_d x^d$ with non-negative coefficient q_d. Suppose we are given an arc $a \in A$ with $\ell_a \in \mathscr{L}_p$. We can replace a by a sequence of $p + 1$ arcs with latency functions in $\mathscr{M}_p, \ldots, \mathscr{M}_0$, respectively, in the obvious way. We can therefore assume without loss of generality that all latency functions $(\ell_a)_{a \in A}$ of the given instance are monomials.

The basic idea is very simple. We define toll functions $(\tau_a)_{a \in A}$ as follows:

$$\tau_a(x) := \min\{x \cdot \ell_a'(x), \; \theta_a(x)\}. \tag{5}$$

That is, on each arc $a \in A$, we impose marginal cost tolls $x \cdot \ell_a'(x)$ if this does not exceed the threshold $\theta_a(x)$ and otherwise charge the maximum possible toll $\theta_a(x)$. Clearly, these tolls are θ-restricted. Note that these tolls are dynamic. It is not hard to derive tolls that are static and achieve the same efficiency (details will be given in the full version).

Let $\phi := (\phi_a)_{a \in A}$ be the combined cost, i.e., for every $a \in A$, $\phi_a(x) := \ell_a(x) + \tau_a(x)$ for every $x \geq 0$, and let $f = f^\tau$ be a Nash flow with respect to ϕ. We next derive a bound on the ratio $C(f)/C(f^*)$, where f^* is an optimal flow. We adapt the (λ, μ)-smoothness approach [18] (see also [3,10]). Because f is a Nash flow with respect to ϕ, it satisfies the following variational inequality, i.e., for every feasible flow x, $\sum_{a \in A} \phi_a(f_a) f_a \leq \sum_{a \in A} \phi_a(f_a) x_a$. By the definition of ϕ, we have

$$C(f) \leq \sum_{a \in A} \ell_a(f_a) x_a + \tau_a(f_a)(x_a - f_a) \leq \sum_{a \in A} \omega(\ell_a, \lambda) \ell_a(f_a) f_a + \lambda \ell_a(x_a) x_a, \tag{6}$$

where we define

$$\omega(\ell_a, \lambda) := \sup_{f_a, x_a \geq 0} \frac{(\ell_a(f_a) + \tau_a(f_a) - \lambda \ell_a(x_a)) x_a - \tau_a(f_a) f_a}{\ell_a(f_a) f_a}.$$

We assume by convention that $0/0 = 0$. Finally, let $\omega(\lambda) := \sup_{a \in A} \omega(\ell_a, \lambda)$. With this definition, (6) implies $C(f) \leq \omega(\lambda) C(f) + \lambda C(x)$. Because $\omega(\lambda)$ depends on λ, let Λ refer to the values of λ such that $\omega(\lambda) < 1$. Then for every $\lambda \in \Lambda$, we obtain

$$C(f) \leq \lambda (1 - \omega(\lambda))^{-1} C(x). \tag{7}$$

The goal is to find $\lambda \in \Lambda$ that provides the best upper bound. We omit some proofs in this section due to space restrictions.

Lemma 1. *Let $\ell_a \in \mathscr{M}_d$ and define $\varepsilon_a := \tau_a(f_a)/\ell_a(f_a)$. We have $\omega(\ell_a, \lambda) = \left(\frac{d(1+\varepsilon_a)}{d+1}\right)\left(\frac{1+\varepsilon_a}{(d+1)\lambda}\right)^{1/d} - \varepsilon_a$. Moreover, $\omega(\ell_a, \lambda) < 1$ for $\lambda \geq \left(\frac{1+\varepsilon_a}{d+1}\right)\left(\frac{d}{d+1}\right)^d$.*

We continue to study the values for $\omega(\ell_a, \lambda)$ and λ. Observe that for every arc $a \in A$ with $\ell_a \in \mathcal{M}_d$ there are two possibilities for $\varepsilon_a = \tau_a(f_a)/\ell_a(f_a)$: If $\tau_a(f_a) = f_a \cdot \ell'_a(f_a)$ then $\varepsilon_a = d$; otherwise, $\tau_a(f_a) = \theta_a(f_a) < f_a \cdot \ell'_a(f_a)$ and thus $\varepsilon_a = \theta_a(f_a)/\ell_a(f_a) < d$.

We thus obtain $\left(\frac{1+\varepsilon_a}{d+1}\right)\left(\frac{d}{d+1}\right)^d \leq \left(\frac{1+d}{d+1}\right)\left(\frac{d}{d+1}\right)^d$. Choosing $\lambda = 1$ therefore satisfies the restrictions imposed on λ in the above lemma (and is tight for $d = 0$). Subsequently, we fix $\lambda := 1$. We need to derive an upper bound on $\omega(\ell_a, 1)$: Note that $\omega(\ell_a, 1)$ decreases as ε_a increases. This motivates the following definitions:

$$\bar{\varepsilon}_d = \min\{\varepsilon_a \mid a \in A,\ \ell_a \in \mathcal{M}_d\} \quad \text{and} \quad \omega(d, 1) = d\left(\frac{1+\bar{\varepsilon}_d}{1+d}\right)^{1+1/d} - \bar{\varepsilon}_d. \tag{8}$$

With these definitions, we obtain $\omega(1) = \max_{d=0,\ldots,p} \omega(d, 1)$.

Corollary 4. *Suppose $\bar{\varepsilon}_d = d$. Then $\omega(d, 1) = 0$.*

Observe that if we have $\bar{\varepsilon}_d = d$ for every $d = 0, \ldots, p$ then the above corollary in combination with (7) implies that $C(f) \leq C(x)$ (which actually follows readily from the observation that in this case marginal cost tolls are θ-restricted and induce an optimal flow).

In order to state our results below, it will turn out to be convenient to define

$$\gamma(d, \varepsilon) := \left((1 + \varepsilon)\left(1 - \frac{d}{d+1}\left(\frac{1+\varepsilon}{d+1}\right)^{1/d}\right)\right)^{-1}.$$

Theorem 3. *Given an instance of the restricted network toll problem with latency functions in $\mathcal{L}_p$, the efficiency of the tolls in (5) is no worse than $\max_{d=0,\ldots,p} \gamma(d, \bar{\varepsilon}_d)$.*

Proof. The proof follows from (7) with $\lambda = 1$ ($\lambda \in \Lambda$ as argued above). □

We give some interpretation of the above theorem. Our result suggests that it is more important to impose large tolls on arcs with high degree latency functions than on the ones with low degree functions. As an example, consider the following extreme situation: Suppose the restrictions $(\theta_a)_{a \in A}$ are such that we can impose marginal cost tolls on all arcs $a \in A$ with latency functions of degree larger than t, and no tolls on all other arcs. The above bound then proves that the tolls in (5) achieve an efficiency no worse than the price of anarchy for degree t polynomials, i.e., $\gamma(t, 0)$ (see [19]).

We next show that the bound in Theorem 3 is tight.

Theorem 4. *For every p and every choice of δ with $0 \leq \delta \leq p$ there is a parallel-arc instance of the restricted network toll problem with latency functions in $\mathcal{L}_p$ such that the efficiency of the tolls in (5) is equal to $\gamma(p, \delta)$.*

The next corollary characterizes the efficiency of θ-restricted tolls enforcing that the toll on each arc does not exceed an ε-fraction of the travel time along that arc. This bound matches exactly the *price of stability* of ε-Nash flows shown by Christodoulou, Koutsoupias and Spirakis [5].

Corollary 5. *Given an instance of the restricted network toll problem with latency functions in $\mathcal{L}_p$ and threshold functions of the form $\theta_a(x) = \varepsilon \ell_a(x)$, the efficiency of the tolls in (5) is no worse than 1 if $\varepsilon \geq p$ and no worse than $\gamma(p, \varepsilon)$ otherwise.*

Our approach can also be used to compute θ-restricted tolls that are ρ-approximate, where ρ is the efficiency guarantee stated in Theorem 3 (details will be given in the full version of the paper).

References

1. Awerbuch, B., Azar, Y., Epstein, A.: The price of routing unsplittable flow. In: Proc. 37th ACM Symp. on Theory of Computing, pp. 57–66 (2005)
2. Beckmann, M., McGuire, B., Winsten, C.: Studies in the Economics of Transportation. Yale University Press, New Haven (1956)
3. Bonifaci, V., Harks, T., Schäfer, G.: Stackelberg routing in arbitrary networks. Mathematics of Operations Research 35(2), 1–17 (2010)
4. Christodoulou, G., Koutsoupias, E.: The price of anarchy of finite congestion games. In: Proc. 37th ACM Symp. on Theory of Computing (2005)
5. Christodoulou, G., Koutsoupias, E., Spirakis, P.G.: On the performance of approximate equilibria in congestion games. In: Proc. 17th European Symp. on Algorithms, pp. 251–262 (2009)
6. Cole, R., Dodis, Y., Roughgarden, T.: Pricing network edges for heterogeneous selfish users. In: Proc. 35th Symp. on Theory of Computing, pp. 521–530 (2003)
7. Cole, R., Dodis, Y., Roughgarden, T.: How much can taxes help selfish routing? Journal of Computer and System Sciences 72(3), 444–467 (2006)
8. Fleischer, L.: Linear tolls suffice: New bounds and algorithms for tolls in single source networks. Theoretical Computer Science 348(2-3), 217–225 (2005)
9. Fleischer, L., Jain, K., Mahdian, M.: Tolls for heterogeneous selfish users in multicommodity networks and generalized congestion games. In: Proc. 45th Symp. on Foundations of Computer Science, pp. 277–285 (2004)
10. Harks, T.: Stackelberg strategies and collusion in network games with splittable flow. In: Bampis, E., Skutella, M. (eds.) WAOA 2008. LNCS, vol. 5426, pp. 133–146. Springer, Heidelberg (2009)
11. Harks, T., Schäfer, G., Sieg, M.: Computing flow-inducing network tolls. Technical Report 36-2008, Institut für Mathematik, Technische Universität Berlin, Germany (2008)
12. Hoefer, M., Olbrich, L., Skopalik, A.: Taxing subnetworks. In: Papadimitriou, C., Zhang, S. (eds.) WINE 2008. LNCS, vol. 5385, pp. 286–294. Springer, Heidelberg (2008)
13. Karakostas, G., Kolliopoulos, S.G.: Edge pricing of multicommodity networks for heterogeneous selfish users. In: Proc. 45th Symp. on Foundations of Computer Science, pp. 268–276 (2004)
14. Koutsoupias, E., Papadimitriou, C.H.: Worst-case equilibria. In: Proc. 16th Symp. on Theoretical Aspects of Computer Science, pp. 404–413 (1999)
15. Kozlov, M.K., Tarasov, S.P., Khachiyan, L.G.: The polynomial solvability of convex quadratic programming. USSR Computational Mathematics and Mathematical Physics 20(5), 223–228 (1980)
16. Roughgarden, T.: The price of anarchy is independent of the network topology. J. Comput. Syst. Sci. 67(2), 341–364 (2003)
17. Roughgarden, T.: On the severity of Braess's paradox: Designing networks for selfish users is hard. J. Comput. Syst. Sci. 72(5), 922–953 (2006)
18. Roughgarden, T.: Intrinsic robustness of the price of anarchy. In: Proc. 41st ACM Symp. on Theory of Computing, pp. 513–522 (2009)
19. Roughgarden, T., Tardos, É.: How bad is selfish routing? Journal of the ACM 49(2), 236–259 (2002)
20. Swamy, C.: The effectiveness of Stackelberg strategies and tolls for network congestion games. In: Proc. 18th Symp. on Discrete Algorithms (2007)

Stochastic Selfish Routing

Evdokia Nikolova[1] and Nicolas E. Stier-Moses[2]

[1] MIT CSAIL, Cambridge, MA, USA
nikolova@mit.edu
[2] Columbia Business School, New York, NY, USA
stier@gsb.columbia.edu

Abstract. We embark on an agenda to investigate how stochastic delays and risk aversion transform traditional models of routing games and the corresponding equilibrium concepts. Moving from deterministic to stochastic delays with risk-averse players introduces nonconvexities that make the network game more difficult to analyze even if one assumes that the variability of delays is exogenous. (For example, even computing players' best responses has an unknown complexity [24].) This paper focuses on equilibrium existence and characterization in the different settings of atomic vs. nonatomic players and exogenous vs. endogenous factors causing the variability of edge delays. We also show that succinct representations of equilibria always exist even though the game is non-additive, i.e., the cost along a path is *not* a sum of costs over edges of the path as is typically assumed in selfish routing problems. Finally, we investigate the inefficiencies resulting from the stochastic nature of delays. We prove that under exogenous stochastic delays, the price of anarchy is exactly the same as in the corresponding game with deterministic delays. This implies that the stochastic delays and players' risk aversion do not further degrade a system in the worst-case more than the selfishness of players.

Keywords: Non-additive nonatomic congestion game, stochastic Nash equilibrium, stochastic Wardrop equilibrium, risk aversion.

1 Introduction

Heavy traffic and the uncertainty of traffic conditions exacerbate the daily lives of millions of people across the globe. According to the 2010 Urban Mobility Report [32], "in 2009, congestion caused urban Americans to travel 4.8 billion hours more and to purchase an extra 3.9 billion gallons of fuel for a congestion cost of $115 billion." For a comparison, that congestion cost was $85 billion in 1999. High and variable congestion necessitates drivers to *buffer in extra time* when planning important trips. The recommendation in the report was to consider a buffer of approximately 30% (Los Angeles) to 40% (Chicago) more than the average travel time, and around twice as long as the travel time at night when traffic is light.

A common driver reaction in the face of heavy and uncertain traffic conditions is to look for alternate, sometimes longer but less crowded and less variable routes [15]. With the widespread use of ever-improving technologies for measuring traffic, one might ask: is there a way to game the system? What route should be selected given other

G. Persiano (Ed.): SAGT 2011, LNCS 6982, pp. 314–325, 2011.

drivers' route choices? Considering routing games on networks where delay functions are stochastic, we analyze the resulting equilibria when strategic, risk-averse commuters take into account the *variability* of delays. This approach generalizes the traditional model of Wardrop competition [37] by incorporating uncertainty.

Risk aversion forces players to go beyond considering expected delays. Since it is unlikely that they base their routing decisions on something as complicated as a full distribution of delays along an exponential number of possible paths, it is reasonable that considering expected delays and their standard deviations is a good first-order approximation on route selection. To incorporate the standard deviation of delays into the players' objectives, we consider the traditional *mean-standard deviation* (mean-stdev) objective [14,18] whereby players minimize the cost on a path, defined as the path mean plus a risk-aversion factor times the path standard deviation.[1] By linearity of expectations, the mean of the path equals the sum of the means over all its edges. However, the standard deviation along a path does not decompose as a sum over edges because of the *risk-diversification effect*. Instead, it is given by the square root of the sum of squared standard deviations on the edges of that path. Due to the complicating square root, a single player's subproblem—a shortest path problem with respect to stochastic costs—is a nonconvex optimization problem for which no polynomial running-time algorithms are known. This is in sharp contrast to the subproblem of the Wardrop network game—a shortest path problem, which admits efficient solutions.

A compelling interpretation of this objective in the case of normally-distributed uncertainty is that the mean-stdev of a path equals a percentile of delay along it. This model is also related to typical quantifications of risk, most notably the value-at-risk objective commonly used in finance, whereby one seeks to minimize commute time subject to arriving on time to a destination with at least, say, 95% chance.

Our mean-stdev model works for *arbitrary* distributions with finite first and second moment. To simplify the analysis, throughout this paper we assume that delays of different edges are uncorrelated. Nevertheless, a limited amount of correlation is to be expected in practice; for example, if there is an accident in a location, it causes ripple effects upstream. We remark that local correlations can be addressed with a polynomial graph transformation that encodes correlation explicitly in edges by modifying the standard-deviation functions with correlation coefficients [23]. This results in a graph with independent edge delays where all our results and algorithms carry through.

Related Work. Our model is based on the traditional competitive network game introduced by Wardrop in the 1950's where he postulated that the prevailing traffic conditions can be determined from the assumption that players jointly select shortest routes [37]. The game was formalized in an influential book by Beckmann *et al.* that lays out

---

[1] Another alternative would have been to consider the mean-variance objective. This approach reduces to a deterministic Wardrop network game in which the edge delay functions already incorporate the information on variability. However, the mean and variance are measured in different units so a combination of them is hard to interpret. In addition, under this objective it may happen that players select routes that are stochastically dominated by others. Although this counterintuitive phenomenon may also happen under the mean-stdev objective with some artificially constructed distributions, it is guaranteed *not* to happen under normal distributions.

the mathematical foundations to analyze competitive networks [4]. These models find applications in various domains such as transportation [33] and telecommunication networks [1]. In the last decade, these games received renewed attention with many studies aimed at understanding under what conditions equilibria exist, what uniqueness properties they satisfy, how to compute them efficiently, how expensive they are in relation to a centralized solution, and how to align incentives so the equilibria become optimal. For general references on these topics, we refer the readers to some recent surveys [9,27].

In the majority of models used by theoreticians who study the properties of network games, and by practitioners who compute solutions to real problems, delays have been considered deterministic. Although there are models that incorporate some form of uncertainty [3,5,16,17,19,36], none of these models has become widely accepted in practice, nor have they been extensively studied. Perhaps the only exception is the *stochastic user equilibrium* model, introduced by Dial in the 1970's [11], which has been studied and used in practice (see, e.g., [34,35]). Under it, different players *perceive* each route differently, distributing demand in the network according to a logit model. To reduce route enumeration, the model just takes into account a subset of "efficient routes." In contrast, the objective of the players in the network game we consider is to choose the path that minimizes the mean plus a multiple of the standard deviation of delay. This problem belongs to the class of stochastic shortest path problems (see, for instance, some classic references [2,6] and some newer ones [12,13,25,22]).

In the network games literature, the model most related to our work is that of Ordóñez and Stier-Moses [28]. They introduce a game with uncertainty elements and risk-averse users and study how the solutions provided by it can be approximated numerically by an efficient column-generation method that is based on robust optimization. The main conclusion is that the solutions computed using their approach are good approximations of *percentile equilibria* in practice. Here, a percentile equilibrium is a solution in which percentiles of delays along flow-bearing paths are minimal. The main difference between their approach and ours is that their insights are based on computational experiments whereas the current work focuses on theoretical analysis and also considers the more general settings of endogenously-determined standard deviations and atomic games.

Next, we formally define our model and equilibrium concepts and study the existence of equilibrium under exogenous (Section 3) and endogenous (Section 4) variability of delays. We summarize these results in Table 1. We then prove that equilibria that use polynomially-many paths—referred to as succint—exist (Section 5), and finally we analyze properties of the socially-optimal solution and study the inefficiency of stochastic Wardrop equilibria (Section 6).

2   The Model

We consider a directed graph $G = (V, E)$ with an aggregate demand of d_k units of flow between source-destination pairs (s_k, t_k) for $k \in K$. We let $\mathcal{P}_k$ be the set of all paths between s_k and t_k, and $\mathcal{P} := \cup_{k \in K} \mathcal{P}_k$ be the set of all paths. We encode players decisions as a flow vector $\mathbf{f} = (f_\pi)_{\pi \in \mathcal{P}} \in \mathbb{R}_+^{|\mathcal{P}|}$ over all paths. Such a flow is feasible when demands are satisfied, as given by constraints $\sum_{\pi \in \mathcal{P}_k} f_\pi = d_k$ for all $k \in K$.

Table 1. Equilibria in Stochastic Routing Games

	Exogenous Noise	Endogenous Noise
Nonatomic Users	Equilibrium exists; Solves exponentially-large convex program	Equilibrium exists; Solves variational inequality
Atomic Users	Equilibrium exists; Potential game	No pure strategy equilibrium

For simplicity, when we write the flow on an edge f_e depending on the full flow $\mathbf{f}$, we refer to $\sum_{\pi \ni e} f_\pi$. When we need multiple flow variables, we use the analogous notation $\mathbf{x}, x_\pi, x_e$.

The congestible network is modeled with stochastic delay functions $\ell_e(x_e) + \xi_e(x_e)$ for each edge $e \in E$. Here, $\ell_e(x_e)$ measures the expected delay when the edge has flow x_e, and the random variable $\xi_e(x_e)$ represents the stochastic delay error. The function $\ell_e(\cdot)$ is assumed continuous and non-decreasing. The expectation of $\xi_e(x_e)$ is zero and its standard deviation is $\sigma_e(x_e)$, for a continuous and non-decreasing function $\sigma_e(\ )$. Although the distribution may depend on x_e, we will separately consider the simplified case in which $\sigma_e(x_e) = \sigma_e$ is a constant given exogenously, independent from x_e. We also assume that these random variables are all uncorrelated with each other. Risk-averse players choose paths according to the mean-standard deviation (mean-stdev) objective, which we also refer to as the cost along route π:

$$Q_\pi(\mathbf{f}) := \sum_{e \in \pi} \ell_e(f_e) + \gamma \sqrt{\sum_{e \in \pi} \sigma_e(f_e)^2} , \tag{1}$$

where $\gamma \geq 0$ quantifies the risk aversion of players, assumed homogeneous.

The *nonatomic* version of the game considers the setting where infinite players control an insignificant amount of flow each so the path choice of a player does not unilaterally affect the costs experienced by other players (even though the joint actions of players affect other players). The following definition captures that at equilibrium players route flow along paths with minimum cost $Q_\pi(\cdot)$.

Definition 1. *The* stochastic Wardrop equilibrium *of a nonatomic routing game is a flow* $\mathbf{f}$ *such that for every source-destination pair* $k \in K$ *and for every path* $\pi \in \mathcal{P}_k$ *with positive flow,* $Q_\pi(\mathbf{f}) \leq Q_{\pi'}(\mathbf{f})$ *for every path* $\pi' \in \mathcal{P}_k$.

Instead, the *atomic* version of the game assumes that each player wishes to route one unit of flow. Consequently, the path choice of even one player directly affects the costs experienced by others. There are two versions of the atomic game: in the splittable case players can split their demands along multiple paths, and in the unsplittable case they are forced to choose a single path. In this paper we focus on the *atomic unsplittable* case, which we will sometimes refer to just as *atomic*. The natural extension of Wardrop equilibrium to the atomic case only differs in that players need to anticipate the effect of a player changing to another path. This game always admits a mixed-strategy equilibrium (under the standard expected payoffs with respect to the mixing probabilities) because it is a finite normal-form game [21], so we focus on the existence of pure-strategy equilibria.

Definition 2. *A pure-strategy stochastic Nash equilibrium of the atomic unsplittable routing game is a flow* $\mathbf{f}$ *such that for every source-destination pair* $k \in K$ *and for every path* $\pi \in \mathcal{P}_k$ *with positive flow, we have that* $Q_\pi(\mathbf{f}) \leq Q_{\pi'}(\mathbf{f} + \mathcal{I}_{\pi'} - \mathcal{I}_\pi)$ *for every* $\pi' \in \mathcal{P}_k$ *. Here,* $\mathcal{I}_\pi$ *denotes a vector that contains a one for path* π *and zeros otherwise.*

One of the goals of this work is to evaluate the performance of equilibria. Hence, we define a social cost function that will allow us to compare different flows and determine the inefficiency of solutions. The social cost function is the total cost among players:

$$C(\mathbf{f}) := \sum_{\pi \in \mathcal{P}} f_\pi Q_\pi(\mathbf{f}) . \tag{2}$$

3   Exogenous Standard Deviations

In this section, we consider exogenous noise factors, which result in constant standard deviations $\sigma_e(x_e) = \sigma_e$ that do not depend on the flow on the edge. In this case, the path cost (1) can be written as $Q_\pi(\mathbf{f}) = \sum_{e \in \pi} \ell_e(f_e) + \gamma(\sum_{e \in \pi} \sigma_e^2)^{1/2}$. We investigate the existence of equilibria and provide a characterization. First, we show that an equilibrium always exists, despite the challenge posed by the non-additive cost function. Due to space restrictions, missing proofs can be found in the full version of this paper.

Theorem 1. *A nonatomic routing game with exogenous standard deviations always admits a stochastic Wardrop equilibrium.*

The proof uses a path-based convex programming formulation given by Ordóñez and Stier-Moses [28]:

$$\min \left\{ \sum_{e \in E} \int_0^{x_e} \ell_e(z)dz + \sum_{\pi \in \mathcal{P}} \gamma f_\pi \sqrt{\sum_{e \in \pi} \sigma_e^2} : \text{such that } x_e = \sum_{\pi \in \mathcal{P}:\, e \in \pi} f_\pi \text{ for } e \in E, \right.$$
$$\left. d_k = \sum_{\pi \in \mathcal{P}_k} f_\pi \text{ for } k \in K,\, f_\pi \geq 0 \text{ for } \pi \in \mathcal{P} \right\}. \tag{3}$$

Besides proving existence, the formulation also provides a way to compute this equilibrium using column generation. We remark that this method typically will not use many paths and hence, it is likely to be practical. In addition, the formulation implies that the equilibrium is unique, provided that the objective function (3) is strictly convex:

Corollary 1. *The equilibrium of the stochastic nonatomic routing game with exogenous standard deviations is unique (in terms of edge loads) whenever the expected delay functions are strictly increasing.*

We now return briefly to the question of computation. The convex program (3) contains exponentially-many variables (the flows on all paths) and a polynomial number of constraints. We will see in Section 5 that an equilibrium always has a succinct decomposition that uses at most $|E|$ paths; unfortunately, since we do not know ahead of time which paths these are, we cannot write a succinct version of the convex program. In the case of constant expected delays, the objective (3) coincides with the social cost, and both problems reduce to computing a stochastic shortest path for each source-destination pair. Thus, both the equilibrium and social optimum computation are at least as hard as the stochastic shortest path problem [26,24].

Theorem 2. *When the expected delays and standard deviations are constant for each edge, the equilibrium and social optimum coincide and can be found in time $n^{O(\log n)}$.*

Now, we switch to the atomic unsplittable case and show that the stochastic routing game admits a potential function. We prove this using the characterization given by Monderer and Shapley [20]. The potential game structure implies that an equilibrium always exists.

Theorem 3. *An atomic unsplittable routing game with exogenous standard deviations is potential and, therefore, it always admits a pure-strategy stochastic Nash equilibrium.*

In contrast to the uniqueness of equilibrium in the nonatomic game, the pure-strategy equilibria in the atomic case need not be unique because they are a generalization of those in deterministic games, which admit multiple equilibria. (For example, a game with two players with a unit demand choosing among three parallel edges with $\ell_e(x_e) = x_e$ and $\sigma_e = 0$ admits three equilibria.)

4   Endogenous Standard Deviations

In this section, we consider flow-dependent standard deviations of edge delays. This makes the standard deviations endogenous to the game. We show that equilibria exist in the nonatomic game but they may not exist in the atomic game.

The following example illustrates how an equilibrium changes under endogenous standard deviations. Assume a demand of $d = 1$ and consider a network consisting of two parallel edges with delays $\ell_1(x) = x$ and $\ell_2(x) = 1$, followed by a chain of k edges that players must traverse. This instance admits two paths, each comprising one of the two parallel edges and the chain. We let L denote the expected delay along the chain (a constant since the flow traversing it is fixed) and assume that $\sigma_e(x_e) = sx_e$ for all edges, for some constant $s \geq 0$. Although both the deterministic and the exogenous standard-deviation games are equivalent to Pigou's instance [30,31], the equilibrium with endogenous standard deviations changes significantly. Indeed, it is given by a root of the degree-4 polynomial $(1 - 4s^2)x^4 + 4s^2x^3 + (4s^4 - 2s^2 - 4ks^2)x^2 - 4s^4x + s^4$ such that $x \in [0, 1]$, which in principle might not exist (where x denotes the flow on one of the two paths). An insight with algorithmic implications arising from this example is that an equilibrium in the stochastic game does not decompose to equilibria in subgraphs of the given graph, and in fact it may be quite different from the equilibria in the subgraphs. Hence, it is not immediate how to decompose the problem by partitioning a graph into smaller pieces.

We can show the existence of equilibrium in the general nonatomic setting via a variational inequality. In fact, the following results also hold in the much more general setting where edge-delay functions depend on the full vector of flows, as long as this dependence is continuous.

Theorem 4. *The nonatomic game with endogenous standard deviations admits a stochastic Wardrop equilibrium.*

In contrast to the case of exogenous standard deviations, however, the game with endogenous standard deviations is not *potential* [20] and equilibria cannot be characterized as the solution to a global optimization problem.

Proposition 1. *The stochastic routing game with nonconstant variances does not admit a cardinal potential.*

As in the deterministic case, the stochastic game may have multiple edge-flow equilibria when the delays and variances are not strictly increasing. An open question that remains is whether the equilibrium is unique when the expected delay and/or standard-deviation functions are strictly increasing. The standard approach to establish the uniqueness of a solution to a variational inequality is to show the monotonicity of the path-cost operator (the vector of path cost functions for all paths). However, neither monotonicity, nor a weaker notion of pseudo-monotonicity holds for our problem.

Proposition 2. *The path-cost operator of the nonatomic routing game with endogenous standard deviations is not pseudo-monotone.*

Although we were not able to prove uniqueness in general, we can do so in the extreme cases of players' risk attitudes. We do so by showing that in those cases the stochastic game resembles a deterministic one.

Proposition 3. *In the two extreme settings where players are either risk-neutral or infinitely risk averse, the nonatomic routing game admits a unique stochastic Wardrop equilibrium, for strictly increasing expected edge delays and standard deviations.*

In contrast to the existence of equilibria in nonatomic games, there are atomic games that do not admit a pure-strategy Nash equilibrium. Mixed-strategy equilibria always exist because the game is finite [21].

Proposition 4. *The atomic unsplittable routing game with endogenous standard deviations may not have pure-strategy Nash equilibria, even in the case of a single source-destination pair in a series-parallel network with affine edge mean and standard-deviation functions.*

5   Succinct Representations of Equilibria

We now turn our attention to how one can decompose an equilibrium of the nonatomic game represented as an edge-flow vector to a path-flow vector, and to whether a succinct vector of path flows at equilibrium always exists. The first question is trivial in the deterministic routing game: any path-flow decomposition of an edge flow at equilibrium works since path costs are additive. Instead, path costs of the stochastic game are non-additive and different flow decompositions of the same edge flow may incur different path costs. In particular, for an edge flow at equilibrium, some path-flow decompositions are at equilibrium and others are not. This is captured by the next lemma which illustrates that shortest paths with respect to path costs do not need to satisfy Bellman equations since a subpath of a shortest path need not be shortest.

Lemma 1. *In a nonatomic game, not all path-flow decompositions of an edge flow at equilibrium constitute an equilibrium.*

The previous lemma prompts the question of how one can find a flow decomposition of an equilibrium given as an edge flow. Does a succinct decomposition always exist (namely one that assigns positive flow to only polynomially-many paths)? The next few results provide positive answers to these questions. We show that succinct flow decompositions exist and they can be found in time slightly larger than polynomial, $|V|^{O(\log|V|)}$, which is the best-known running time of an exact algorithm for solving the underlying stochastic shortest path problem [26]. Alternatively, using a fully-polynomial approximation algorithm for the stochastic shortest path problem [24], one can find approximate flow decompositions of equilibria in polynomial time. We first provide a characterization of flow decompositions of equilibria that will enable us to show the existence of succinct decompositions.

Lemma 2. *Consider a flow decomposition $\mathbf{f}^P$ of an edge-flow equilibrium with support $P \subset \mathcal{P}$ (the set of paths with positive flow). Then, every flow decomposition whose support is a subset of P and whose resulting edge flow is the same as that of $\mathbf{f}^P$ is also at equilibrium.*

Using the lemma above, we can prove the existence of succinct equilibrium decompositions.

Theorem 5. *For an equilibrium $(f_e)_{e \in E}$ given as an edge flow, there exists a succinct flow decomposition that uses at most $|E|$ paths. Furthermore, this decomposition can be found in time $|V|^{O(\log|V|)}$, and an ϵ-approximate equilibrium succinct decomposition can be found in polynomial time.*

It remains open whether finding an equilibrium can be done in polynomial time. This is related to the open question of whether the stochastic shortest path subproblem is in P [26].

Corollary 2. *Given an edge flow, we can verify that it is at equilibrium in time $|V|^{O(\log|V|)}$.*

Analogously, we can verify that a given set of player strategies (paths) in the atomic setting forms an equilibrium in time $|V|^{O(\log|V|)}$.

6   Price of Anarchy

In this section we compute bounds for the price of anarchy (POA) for stochastic Wardrop equilibria of the nonatomic game. Recall that the price of anarchy is defined as the supremum over all problem instances of the ratio of the equilibrium cost to the social optimum cost [29]. In the case of exogenous standard deviations, the POA turns out to be the same as in the corresponding deterministic game: it is $4/3$ for linear expected delays and $(1 - \beta)^{-1}$ for general expected delays for an appropriate definition of β as in Correa *et al.* [7] for the corresponding deterministic routing game. The bounds result from a modification of the bounding techniques of Correa *et al.* [7,8].

In the case of endogenous standard deviations, an analysis of the price of anarchy is more elusive and it remains open whether the equilibrium is unique. For this reason, we focus our analysis to the limiting case of extreme risk aversion (the other extreme case, where users are risk neutral, is well-understood). Hence, we assume that path costs are equal to the path standard deviations $Q_\pi(\mathbf{f}) = (\sum_{e \in \pi} \sigma_e(f_e)^2)^{1/2}$. Recall that in this extreme case, Proposition 3 implies that there is a unique equilibrium that can be computed efficiently with a convex program.

We now show that the first order optimality conditions of the optimization problem that defines socially-optimal solutions are satisfied at the equilibrium, when standard-deviation delay functions are monomials of the same degree. Note that in the deterministic case, it is known that the POA is exactly one precisely for that class of delay functions [10].

Theorem 6. *Consider a nonatomic network game with endogenous standard deviations of the form* $\sigma_e(x_e) = a_e x_e^p$ *for some fixed* $p \geq 0$. *An equilibrium of the game is a stationary point of the social-optimum (SO) problem*

$$\min \left\{ \sum_{\pi \in \mathcal{P}} f_\pi Q_\pi(\mathbf{f}) : \sum_{\pi \in \mathcal{P}_k} f_\pi = d_k \ \forall k \in K \text{ and } f_\pi \geq 0 \ \forall \pi \in \mathcal{P} \right\}.$$

As a corollary from the above theorem, whenever the SO problem has a unique stationary point, it would follow that equilibria and social optima coincide and, consequently, the price of anarchy would be 1. Before we identify settings for which convexity of the social cost holds, we show that despite the somewhat misleading square root, the path costs are convex in the edge-flow variables when the standard deviations $\sigma_e(x_e)$ are convex functions.

Proposition 5. *The path costs* $Q_\pi(\mathbf{x}) = (\sum_{e \in \pi} \sigma_e(x_e)^2)^{1/2}$ *are convex whenever the edge standard-deviation functions* $\sigma_e(x_e)$ *are convex.*

As a corollary from Proposition 5 and from the fact that the sum of convex functions is convex, it follows that path costs are also convex in the general case where the path cost is a sum of the mean and standard deviation of the path, as long as the edge mean and standard-deviation functions are convex in the edge flow.

Next, we identify sufficient conditions for the convexity of the social cost, which bear an intriguing resemblance to the sufficient conditions for the uniqueness of equilibrium mentioned earlier.

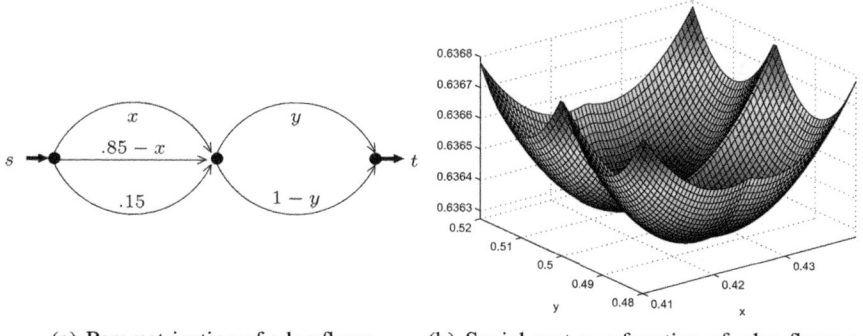

(a) Parametrization of edge flows   (b) Social cost as a function of edge flows x
and y

Fig. 1. Non-convex slice of the social cost function

Proposition 6. *The social cost $C(\mathbf{x}) = \mathbf{x}Q(\mathbf{x}) = \sum_{\pi \in \mathcal{P}} x_\pi Q_\pi(\mathbf{x})$ is convex whenever the path-cost operator Q is monotone and the path costs $Q_\pi(\mathbf{x})$ are convex.*

As established above, the path costs are convex (under convex standard-deviation functions), however the path-cost operator is not necessarily monotone even in the basic case of linear standard deviation functions equal to $\sigma_e(x) = x$, and thus, the social cost may not be convex as shown in Figure 1. Nevertheless, we can still show that the POA is 1 in a network of n pairs of parallel edges connected in series.

Proposition 7. *Consider a nonatomic game on a network of n pairs of parallel edges connected in series with zero mean delays and standard deviation functions equal to $\sigma_e(x) = x$ for all edges. In this case, socially-optimal flows and equilibria coincide.*

Despite the limitation of the hypothesis, the proof requires a careful analysis to bound the social cost, in contrast to our results for exogenous standard deviations under general graphs and costs. For the case of endogenous standard deviations, whether the nonconvexity of the social cost can be circumvented to obtain price of anarchy bounds for more general graphs and delay functions remains open.

7   Conclusions and Open Problems

We have set out to extend the classical theory of Wardrop equilibria and congestion games to the more realistic setting of uncertain delays, focusing on the methodology and questions of algorithmic game theory. The uncertainty of delays calls for models that incorporate players' attitudes towards risk. In this paper, we have focused on the model whereby players seek to minimize a linear combination of the expectation and standard deviation of delays along their chosen route.

The directions pursued in this work have opened a variety of questions which would be interesting to explore in future studies. Some of these questions are:

- What is the complexity of computing an equilibrium when it exists (exogenous standard deviations with atomic or nonatomic players; endogenous standard deviations with nonatomic players)?
- What is the complexity of computing the socially-optimal solution? What is the complexity of computing the socially-optimal flow decomposition if one knows the edge flow that represents a socially-optimal solution?
- Can there be multiple equilibria in the nonatomic game with endogenous standard deviations?
- What is the price of anarchy for stochastic Wardrop equilibria in the setting of nonatomic games with endogenous standard deviations, for general graphs and general classes of cost functions?
- Ordóñez and Stier-Moses considered the case of players with heterogenous attitudes toward risk [28]. Can some of the results in this paper be extended to that setting?

Of course, one could pursue other natural models and player utilities and build on or complement what we have developed here. In particular, our model might be enriched by also considering stochastic demands to make the demand side more realistic.

Acknowledgements. We thank Hari Balakrishnan, José Correa and Costis Daskalakis for helpful discussions. The work of the first author was supported by the National Science Foundation grant number 0931550.

References

1. Altman, E., Boulogne, T., El-Azouzi, R., Jiménez, T., Wynter, L.: A survey on networking games in telecommunications. Computers and Operations Research 33(2), 286–311 (2006)
2. Andreatta, G., Romeo, L.: Stochastic shortest paths with recourse. Networks 18, 193–204 (1988)
3. Ashlagi, I., Monderer, D., Tennenholtz, M.: Resource selection games with unknown number of players. In: Fifth International Joint Conference on Autonomous Agents and Multiagent Systems (AAMAS), Hakodate, Japan, pp. 819–825. ACM Press, New York (2006)
4. Beckmann, M.J., McGuire, C.B., Winsten, C.B.: Studies in the Economics of Transportation. Yale University Press, New Haven (1956)
5. Bell, M.G.H., Cassir, C.: Risk-averse user equilibrium traffic assignment: an application of game theory. Transportation Research 36B(8), 671–681 (2002)
6. Bertsekas, D.P., Tsitsiklis, J.N.: An analysis of stochastic shortest path problems. Mathematics of Operations Research 16(3), 580–595 (1991)
7. Correa, J.R., Schulz, A.S., Stier-Moses, N.E.: Selfish routing in capacitated networks. Mathematics of Operations Research 29(4), 961–976 (2004)
8. Correa, J.R., Schulz, A.S., Stier-Moses, N.E.: A geometric approach to the price of anarchy in nonatomic congestion games. Games and Economic Behavior 64, 457–469 (2008)
9. Correa, J.R., Stier-Moses, N.E.: Wardrop equilibria. In: Cochran, J.J. (ed.) Encyclopedia of Operations Research and Management Science. Wiley, Chichester (2011)
10. Dafermos, S.C., Sparrow, F.T.: The traffic assignment problem for a general network. Journal of Research of the U.S. National Bureau of Standards 73B, 91–118 (1969)
11. Dial, R.B.: A probabilistic multi-path traffic assignment algorithm which obviates path enumeration. Transportation Research 5(2), 83–111 (1971)
12. Fan, Y.Y., Kalaba, R.E., Moore, J.E.: Arriving on time. Journal of Optimization Theory and Applications 127(3), 497–513 (2005)

13. Fan, Y.Y., Kalaba, R.E., Moore, J.E.: Shortest paths in stochastic networks with correlated link costs. Computers and Mathematics with Applications 49, 1549–1564 (2005)
14. Föllmer, H., Schied, A.: Stochastic Finance: An Introduction in Discrete Time. Walter de Gruyter, Berlin (2004)
15. Langer, G.: Traffic in the united states. ABC News, February 13 (2005)
16. Liu, H., Ban, X., Ran, B., Mirchandani, P.: An analytical dynamic traffic assignment model with stochastic network and travelers' perceptions. Transportation Research Record 1783, 125–133 (2002)
17. Lo, H.K., Tung, Y.-K.: Network with degradable links: capacity analysis and design. Transportation Research 37B(4), 345–363 (2003)
18. Markowitz, H.M.: Mean-Variance Analysis in Portfolio Choice and Capital Markets. Basil Blackwell, Cambridge (1987)
19. Mirchandani, P.B., Soroush, H.: Generalized traffic equilibrium with probabilistic travel times and perceptions. Transportation Science 21, 133–152 (1987)
20. Monderer, D., Shapley, L.S.: Potential games. Games and Economic Behavior 14, 124–143 (1996)
21. Nash, J.F.: Noncooperative games. Annals of Mathematics 54(2), 286–295 (1951)
22. Nie, Y., Wu, X.: Shortest path problem considering on-time arrival probability. Transportation Research Part B 43, 597–613 (2009)
23. Nikolova, E.: Strategic algorithms. PhD thesis, Massachusetts Institute of Technology. Dept. of Electrical Engineering and Computer Science (2009)
24. Nikolova, E.: Approximation algorithms for reliable stochastic combinatorial optimization. In: Serna, M., Shaltiel, R., Jansen, K., Rolim, J. (eds.) APPROX 2010, LNCS, vol. 6302, pp. 338–351. Springer, Heidelberg (2010)
25. Nikolova, E., Brand, M., Karger, D.R.: Optimal route planning under uncertainty. In: Long, D., Smith, S.F., Borrajo, D., McCluskey, L. (eds.) Proceedings of the International Conference on Automated Planning & Scheduling (ICAPS), Cumbria, England, pp. 131–141 (2006)
26. Nikolova, E., Kelner, J.A., Brand, M., Mitzenmacher, M.: Stochastic shortest paths via quasi-convex maximization. In: Azar, Y., Erlebach, T. (eds.) ESA 2006. LNCS, vol. 4168, pp. 552–563. Springer, Heidelberg (2006)
27. Nisan, N., Roughgarden, T., Tardos, É., Vazirani, V.V.: Algorithmic Game Theory. Cambridge University Press, Cambridge (2007)
28. Ordóñez, F., Stier-Moses, N.E.: Wardrop equilibria with risk-averse users. Transportation Science 44(1), 63–86 (2010)
29. Papadimitriou, C.H.: Algorithms, games, and the Internet. In: Proceedings of the 33rd Annual ACM Symposium on Theory of Computing (STOC), Hersonissos, Greece, pp. 749–753. ACM Press, New York (2001)
30. Pigou, A.C.: The Economics of Welfare. Macmillan, London (1920)
31. Roughgarden, T., Tardos, É.: How bad is selfish routing? Journal of the ACM 49(2), 236–259 (2002)
32. Schrank, D., Lomax, T., Turner, S.: Annual urban mobility report (2010), Texas Transportation Institute, http://mobility.tamu.edu/ums
33. Sheffi, Y.: Urban Transportation Networks. Prentice-Hall, Englewood Cliffs (1985)
34. Sheffi, Y., Powell, W.: An algorithm for the traffic assignment problem with random link costs. Networks 12, 191–207 (1982)
35. Trahan, M.: Probabilistic assignment: An algorithm. Transportation Science 8(4), 311–320 (1974)
36. Ukkusuri, S.V., Waller, S.T.: Approximate Analytical Expressions for Transportation Network Performance under Demand Uncertainty. Transportation Letters: The International Journal of Transportation Research 2(2), 111–123 (2010)
37. Wardrop, J.G.: Some theoretical aspects of road traffic research. In: Proceedings of the Institute of Civil Engineers, Part II, vol. 1, pp. 325–378 (1952)

Author Index